Y0-DFA-735

The Math Book

NANCY MYERS

Bunker Hill Community College

HAFNER PRESS

A Division of Macmillan Publishing Co., Inc.

NEW YORK

Collier Macmillan Publishers

LONDON

Copyright © 1975, by Hafner Press,
A DIVISION OF MACMILLAN PUBLISHING CO., INC.

All rights reserved. No part of this book may be reproduced or transmitted in any form or by any means, electronic or mechanical, including photocopying, recording or by any information storage and retrieval system, without permission in writing from the Publisher.

HAFNER PRESS
A Division of Macmillan Publishing Co., Inc.
866 Third Avenue, New York, N.Y. 10022
Collier Macmillan Canada Ltd.

Library of Congress Cataloging in Publication Data

Myers, Nancy.
The math book.

1. Mathematics—1961- I. Title.
QA39.2.M88 510 74-14806
ISBN 0-02-849400-x

Illustrations by Deborah Sims
Design by Betty Binns

PRINTED IN THE UNITED STATES OF AMERICA

Contents

5

Arithmetics Without Numbers

6

Calculated Chances

7

Finite Geometries

8

Geometries With A Twist

An invitation

Mathematicians might be one of the most misunderstood minority groups in the world. Some people don't even think they are human. The purpose of this book is to help you explore the world of mathematics and to discover that mathematicians are indeed human and that the work they do is stimulating and exciting.

Mathematics is one of the liberal arts. We sometimes forget this. We forget that—like music, art, literature—mathematics was created by human beings for the enjoyment of human beings. If the word "enjoyment" seems unsuitable to you, even alien to your present idea of mathematics, then it is because you have not really understood mathematics and mathematicians.

How often have you said "I just can't do mathematics!" Probably no more often than you say, "I can't carry a tune," or "I can't draw a straight line," or "I can't put two words together!" But what's so bad about that? It's true that not all people sing very well, and most aren't very good at drawing, and few are talented writers. To perform well in any art takes special talent.

Most people can learn to sing and draw and write adequately—and to do some basic mathematics. But the fact that you can read and understand these words doesn't mean you can recite Hamlet's soliloquy like Lawrence Olivier. Mathematical geniuses, like any other kind, are not made but born, and rarely enough at that. Because you are not a mathematical genius, or can't even do mathematics very well, does not mean you can't learn about it. You can read Shakespeare's Hamlet, even if you can't write like Shakespeare, or recite like Olivier. You can read about mathematics, even if you can't create it, or even if you can't completely understand the creations of mathematical geniuses.

So this book invites you to learn about mathematics; it does not attempt to train you to do mathematics. You will not "do" mathematics in the usual sense, although there are "exercises" for you to work. In this book, the exercises are part of the text, each a kind of guidepost so you won't lose track of what you are reading about and so you can test your understanding. You will read about some of the great concepts of mathematics, and about some of the people whose genius and struggle made them possible. You will see some new concepts which are sources of tremendous excitement and activity in the world of mathematics, but which are hardly known outside of it. This, then, is a book of explorations, not a survey of mathematics; many great mathematicians and great concepts are not mentioned. To explore a relatively limited territory should give you time to reflect, to contemplate, to discuss, and believe it or not, to enjoy your explorations.

Besides reading about mathematics in this book, I hope you will do some exploring on your own. Suggestions for "branching out" are provided at the end of each chapter. You may find among them a path to explore on your own. A number of passages from enjoyable books are quoted in this book. These and other books of interest are listed in a bibliography at the end. The bibliography is purposefully brief, so don't limit yourself to it. A little browsing in your library could open up this new world to you even further.

So I invite you to begin with this book to explore a world of fascinating ideas. Nothing is more exciting than your own discovery—it's what learning is all about. I hope that you will discover some delightful bits of "mathemabelia" and that when you finish your expedition, you will have a better understanding of mathematicians and the world they have created.

Preface

This book is designed for anyone interested in exploring the world of mathematics. Specifically, it is designed for use in the introductory course for liberal arts and elementary education majors. The topics have been selected and the text written in a way designed to appeal to liberal arts students. No attempt is made to train students to "do" mathematics in the usual sense of that term, and no real effort is made to show mathematics as the servant of science, technology, or industry. I have chosen to emphasize mathematics as a realm of ideas. Today everyone has a calculator in his hand and a computer at his elbow, so my aim is to appeal to the student who will not use mathematics to build bridges or to measure metabolism. My goal is to engender an understanding and appreciation of mathematics as structure, concept, an art—a uniquely human creation.

Rather than skim through a long list of standard, currently popular topics, I have chosen a relatively limited number and developed them to considerable depth. Some great mathematical ideas and some great mathematicians are not even mentioned, and someone is likely to find his favorites missing. But, this leaves room for some originality on the part of the instructor—a useful weapon in this course!

Historical vignettes and biographical material complement the presentation, and should help the student to appreciate the ideas and their development. I have tried to approach the material in a relaxed and discursive manner. Indeed, we—the publishers and I—have tried to make all elements of the book inviting, and informal, not textbooky.

About thirty exercises appear in each chapter and an additional ten in the Looking Back sections at the end of each chapter. The starred exercises in the Looking Back sections are not review, but extend the ideas in the chapter. The instructor might use them as a threshold through which to pursue the concepts in a chapter further with an interested class.

Answers to selected exercises appear in the back of the book. Answers to the rest appear in "The Idea Book," the instructor's manual, which is available from the publisher. "The Idea Book," in addition to containing the answers to the exercises which do not appear in the book itself, contains suggestions for using the Branching Out sections, gives answers to all the Looking Back questions, provides sample tests in the spirit of the book, and gives hints and suggestions for teaching the concepts.

The Branching Out sections, also about ten per chapter, provide suggestions for further reading and for library research which students can undertake on their own.

Finally, the publisher has chosen to call this book simply THE MATH BOOK, and not Introduction to Mathematics, or Fundamentals of Mathematics, or Mathematics for the Liberal Arts Student . . . to echo the student's reference to his books as "my psych book," "my history book," and so forth. More importantly, the title is meant to signal the nature of the presentation. I have attempted to write, not just another textbook, but simply an enjoyable book about some significant concepts of mathematics that has no pretentions to being the first word, the last word, or all the word.

I hope that students and instructors who want to use some imagination in exploring one of man's most important disciplines, who want to discuss meaningful, not easy, ideas, who want a non-trivial stimulating course—will enjoy exploring some of these mathematical ideas together.

I would like to thank John Sabol who, as chairman of the mathematics department at the Allegheny Campus of the Community College of Allegheny County, made it possible for me to begin to explore these ideas with my own students. Peggy Porter, research librarian at Allegheny Campus, tracked down all sorts of material. The mathematics department at Bunker Hill Community College embraced the project enthusiastically. Many reviewers, including Philip Nanzetta of Stockton State College, Dorothy Schrader of Southern Connecticut State College, Doris Stockton of the University of Massachusetts, and Robert McGuigan of Westfield State College, provided much helpful advice. Lastly, but definitely first in importance, I want to thank the scores of students at Allegheny Campus and Bunker Hill who helped develop the material, evaluated it, and who were frank in both their praise and their criticism.

NANCY MYERS

To my father,
who inspired me to explore
the world of mathematics

1, 2, 3, 4, 5. . . . . 6?

1

From Counting to Complex Numbers

1

In the Beginning

No one knows, and quite possibly no one will ever know, exactly when and where mathematics began. Like painting on walls of caves or beating out rhythms on drums, primitive forms of mathematics existed long before the beginning of recorded history. From archaeological studies, we know enough about prehistoric societies to know that they must have engaged in many mathematics-like activities. Which of these activities might be called mathematics is more a matter of opinion than of fact.

Perhaps mathematics began when men began to herd instead of hunt animals. Let us imagine a prehistoric shepherd taking his flock out to pasture. It might look like a small flock to us, but to him it appears to be very large, so large he has no name for its size. He can name a single sheep, but two sheep is "many" to him. At the end of the day, how can he tell if all his sheep have come home safely?

A likely solution to the shepherd's problem is to let one object—something that will stay where he puts it—stand for another. He might pick up a pebble as each sheep enters the pasture and make a pile of pebbles. Each pebble represents one sheep and the pile, the entire flock.

Then as each sheep leaves the pasture in the evening, the shepherd removes a pebble from the pile. If a pebble is left over, he knows that he must search for a lost sheep. The shepherd may still have no idea how many of either pebbles or sheep he has, but he does know that he should have the same amount of each.

Soon the shepherd will face a new dilemma. His society will advance another step to trading. Suppose then that he wants to barter his flock. The pile of pebbles tells him each evening that all his sheep are safe, for he chose to use pebbles precisely because they are objects that will stay where he put them. How can he indicate the size of his flock to someone not at the pasture to see the pebbles? His next step must be to associate each pebble with something he can conveniently carry with him, quite likely his fingers.

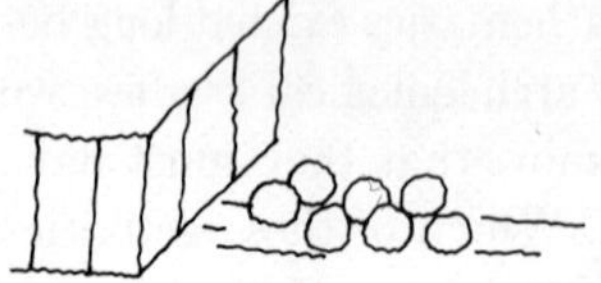

Now he is teetering on the edge of counting . . . and possibly on the edge of recorded history as well.

Man has an irresistible urge to name things. If he develops a new species of rose, he gives it a name. If he discovers a new star, he names it. When he discovers the concept not only of "many," but also of "this amount" as a separate entity, he will eventually give it a name. In this example, we will come to know the name in English as "seven."

The tale of the prehistoric shepherd is strictly hypothetical, but it has introduced you to two ideas. Sophisticated as your knowledge of mathematics may be (and compared with the shepherd, your knowledge of mathematics is sophisticated even if all you know is basic arithmetic), the two ideas are probably new to you unless you have studied so-called modern mathematics.

The first idea is what modern mathematicians call *sets.*

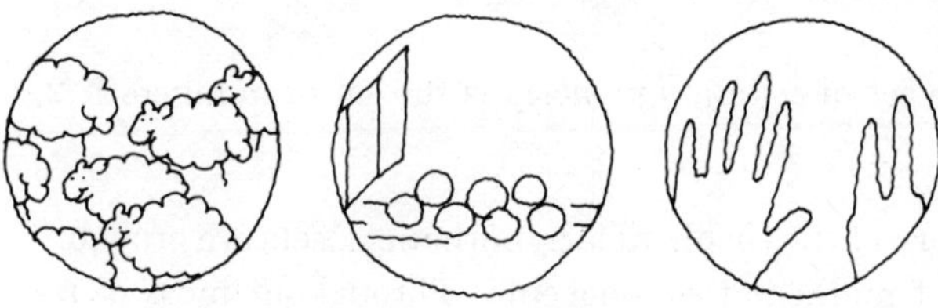

The flock of sheep is a set, the pile of pebbles is another set, and the fingers are also a set. The things in the sets are called *elements* of the sets. Each sheep is an element of the set of sheep, each pebble an element of the set of pebbles, and each finger an element of the set of fingers.

Second, the shepherd has associated each sheep with a pebble and each pebble with a sheep, each pebble with a finger and each finger with a pebble, and also each finger with a sheep and each sheep with a finger. The association of each sheep with a pebble and also each pebble with a sheep is called a *one-to-one correspondence* between the set of pebbles and the set of sheep.

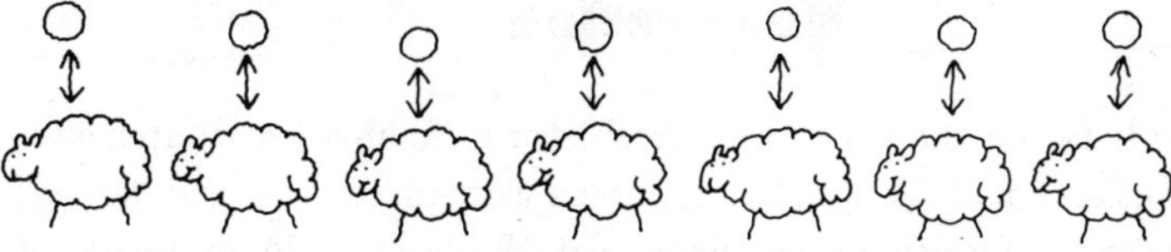

Each pebble corresponds to exactly one sheep, and each sheep to exactly one pebble. The sets of pebbles and fingers are also in one-to-one correspondence, as are the sets of fingers and sheep.

Counting is essentially the act of forming a one-to-one correspondence between a set of objects and a set of number names such as "one," "two," "three," "four," and so forth. What makes counting so much more sophisticated a correspondence than the shepherd's correspondence between sheep and pebbles is that a set of numbers is a set of abstract concepts rather than physical objects. "One" stands for an idea of "oneness," the abstract concept of any single object. "Two" stands for an idea of "twoness," the abstraction of any pair of objects, "three" for any triplet, and so forth. The idea of "sevenness" and the number which we call "seven" can be abstracted from the shepherd's sets of sheep, and pebbles, and fingers, and any other set in one-to-one correspondence with these sets. We call the abstract set of numbers we use for counting the set of *counting*, or *natural*, *numbers*, or *positive integers*.

DEFINITION The set of *counting numbers* is the set of numbers 1, 2, 3, 4, and so forth.

Although the story of the shepherd is hypothetical, sets are probably the most primitive of mathematical concepts. Throughout most of recorded history, numbers have been taken as the beginning concept of mathematics. "God made the whole numbers," said Leopold Kronecker (1823–1891), "all the rest is the work of man." In recent times, however, in fact during Kronecker's lifetime, sets have reappeared as one of the most fundamental of mathematical concepts. We see, incidentally, that some basic ideas of modern mathematics may indeed be very ancient mathematics.

2

Numeration

Somewhere between the time of the prehistoric shepherd and later civilizations that archaeologists study, numbers were discovered and named; and various ways of writing them developed. We call the different ways of writing symbols for numbers *systems of numeration*. The symbols used for numbers in systems of numeration are called *numerals*.

One of the earliest and simplest systems of numeration was part of the ancient Egyptian picture writing called hieroglyphics, developed before 3000 B.C. The Egyptians wrote / for our numeral 1, ∩ for 10, 𓍢 (a scroll) for 100, and 𓆼 (a lotus flower) for 1,000. To form any numeral up to 9,999 in Egyptian numerals, we simply group as many of each Egyptian symbol as are necessary. For example, 2,659 can be written in Egyptian numerals as follows:

𓆼𓆼 𓍢𓍢𓍢𓍢𓍢𓍢 𓎆𓎆𓎆𓎆𓎆 𓏺𓏺𓏺𓏺𓏺𓏺𓏺𓏺𓏺

Because it simply groups symbols, this system of numeration is called a *simple grouping* system. Besides being quite cumbersome to write, an added disadvantage is that a new symbol must be created for 10,000, for 100,000, for 1,000,000, and so on.

On the other hand, it is very easy to add numbers in Egyptian numerals. To do so, we simply group like symbols in each of the numerals together. To add 1,252 and 336, we write

𓆼 𓍢𓍢 𓎆𓎆𓎆𓎆𓎆 𓏺𓏺

𓍢𓍢𓍢 𓎆𓎆𓎆 𓏺𓏺𓏺𓏺𓏺𓏺

𓆼 𓍢𓍢𓍢𓍢𓍢 𓎆𓎆𓎆𓎆𓎆𓎆𓎆𓎆 𓏺𓏺𓏺𓏺𓏺𓏺𓏺𓏺

To do "carrying," if a group of symbols results in more than nine symbols, we can take ten symbols and replace them by one of the next larger symbol. To add 1,858 and 366, we write

𓆼 𓍢𓍢𓍢𓍢𓍢𓍢𓍢𓍢 𓎆𓎆𓎆𓎆𓎆 𓏺𓏺𓏺𓏺𓏺𓏺𓏺𓏺

𓍢𓍢𓍢 𓎆𓎆𓎆𓎆𓎆𓎆 𓏺𓏺𓏺𓏺𓏺𓏺

𓆼 𓍢𓍢𓍢𓍢𓍢𓍢𓍢𓍢𓍢𓍢𓍢 𓎆𓎆𓎆𓎆𓎆𓎆𓎆𓎆𓎆𓎆𓎆 𓏺𓏺𓏺𓏺𓏺𓏺𓏺𓏺𓏺𓏺𓏺𓏺𓏺𓏺

𓆼𓆼 𓍢𓍢 𓎆𓎆 𓏺𓏺𓏺𓏺

The familiar system of Roman numerals is also a simple grouping system. It came into use much later than the Egyptian system, during the last millennium B.C. The Roman system is less cumbersome than the Egyptian because it introduces new symbols for 5, 50, and 500, as well as for 10, 100, and 1,000. In Roman numerals, I is used for our numeral 1, V for 5, X for 10, L for 50, C for 100, D for 500, and M for 1,000. Thus 2,659 in Roman numerals is written MMDCLVIIII. Roman numerals are also somewhat unwieldy to write, however, and they share with Egyptian numerals the disadvantage that new symbols must be created for numbers 5,000 and beyond.

Adding in Roman numerals is again a simple matter of combining groups of symbols. To carry, we replace a group of five symbols, or two symbols for fives, by the next larger symbol; thus IIIII = V, but VV = X. For example, 1,858 plus 366 is written in Roman numerals

```
    MDCCCLVIII
      CCCLXVI
    ─────────────
MDCCCCCCLLXVVIIII
   MDDCCXXIIII
    MMCCXXIIII
```

The so-called subtractive principle of Roman numerals makes writing them even more streamlined. It uses IV = 5 − 1 = 4, whereas VI = 5 + 1 = 6; IX = 10 − 1 = 9, whereas XI = 10 + 1 = 11, and so on. When a smaller numeral precedes a larger, it is subtracted. Thus we can write 2,659 using the subtractive principle as MMDCLIX. The year 1974 can be written MDCCCCLXXIIII without using the subtractive principle, but today it is more commonly written MCMLXXIV. The subtractive principle destroys the simplicity of combining groups of like symbols to add, however. If we write 344 and 36 using the subtractive principle and simply combine groups to add them, we get an incorrect answer:

```
   CCCXLIV
    XXXVI
  ─────────
CCCLXXXXVVII
 CCCLXXXXXII
   CCCLLII
   CCCCII = 402 ≠ 380
```

If we write the longer symbols, not using the subtractive principle, we get the correct result:

CCCXXXXIIII
XXXVI
CCCXXXXXXXVIIIII
CCCLXXVV
CCCLXXX = 380

The reason for the discrepancy is that in the incorrect example, an I and an X are treated as if they were added, when in fact they should be subtracted according to the subtractive principle. We might note that if we do actually subtract the I and X in the first numeral from the like symbols in the second numeral, our answer will be correct:

CCC~~X~~L~~I~~V
XX~~X~~V~~I~~
CCCLXXVV
CCCLXXX

The subtractive principle was used rarely if at all in ancient times. It has been said that the principle was developed much later in order to fit the numerals on clock faces. Later Europeans sometimes did and sometimes did not use the subtractive principle, but they did use Roman numerals for centuries. Amusingly, one of the first parts of Europe to accept our modern notation was Rome, probably around the thirteenth century. There is a story, however, that in the late thirteenth century, the city of Florence prohibited the use of modern numerals because they were thought to be too easy to alter for dishonest purposes. In any event, it is fact that as recently as the sixteenth century, the merchant or banker of northern Europe might do his calculations on an abacus or counting board, but he would record the results in Roman numerals.

An ancient Greek system of numeration, of an entirely different type from the Egyptian and Roman simple grouping systems, also came into use during the first millennium B.C. The ancient Greek system may be called a *cipher* system. The names "simple grouping" and "cipher" for the two types of systems of numeration are from Howard Eves, *An Introduction to the History of Mathematics*.

A cipher system of numeration uses the letters of the alphabet for numerals. Our dictionaries give only twenty-four letters in the Greek alphabet; but fortunately the ancient Greek alphabet had twenty-seven letters. Here is one version of the ancient Greek alphabet, with the numbers associated with each letter:

α	alpha, 1	ι	iota, 10	ρ	rho, 100
β	beta, 2	κ	kappa, 20	σ	sigma, 200
γ	gamma, 3	λ	lambda, 30	τ	tau, 300
δ	delta, 4	μ	mu, 40	υ	upsilon, 400
ϵ	epsilon, 5	ν	nu, 50	ϕ	phi, 500
ς	vau, 6	ξ	xi, 60	χ	chi, 600
ζ	zeta, 7	o	omicron, 70	ψ	psi, 700
η	eta, 8	π	pi, 80	ω	omega, 800
θ	theta, 9	ϙ	koppa, 90	ϡ	sampi, 900

The three "lost" letters are vau, koppa, and sampi. There are some discrepancies concerning these letters. Eves gives the name "digamma" in place of vau and gives no symbol for it. Many books use an F for the symbol as in our alphabet. Morris Kline in *Mathematical Thought from Ancient to Modern Times* gives different symbols for the letters koppa and sampi which are similar to those of the distinguished scholar O. Neugebauer. Another writer on ancient science, B. L. van der Waerden, gives the form above. The discrepancies occur because ancient systems of numeration come to us entirely through archaeological studies which are continuing even now.

To write symbols for the numbers from 1,000 through 9,000, the Greeks used the letters alpha through theta again, with a mark in front to distinguish them from 1 through 9. The mark is $,\alpha$ for 1,000, $,\beta$ for 2,000, and so on. Since letters form both words and numerals, numerals are distinguished by a bar over them or a mark at the end. Thus 2,659 is written $\overline{,\beta\chi\nu\theta}$ in the Greek cipher system. We see that the Greek numeral is more compact than the Egyptian or Roman, but there are many more symbols to remember. Also, addition depends on remembering the meaning of each symbol rather than simply grouping like symbols together. The Greeks used M, called "myriad," for 10,000. Beyond that, new symbols must be created for larger numbers as in the Egyptian and Roman simple grouping systems.

One ancient system of numeration did not require new symbols for increasingly large numbers. The ancient Babylonian civilization, located between the Tigris and Euphrates rivers in what is now Iraq, had developed a system of numeration at about the same time as the Egyptians. Also beginning about 3000 B.C., the Babylonians developed a proficiency in mathematics at least as great as the Egyptians, and perhaps greater. Our concepts of degrees in a circle and minutes in an hour, both involving the special number 60, can be traced to ancient Babylonia.

The Babylonian system of numeration is a simple grouping system, combined with a concept of *place value.* Place value means that a symbol takes part of its meaning from its place in a numeral. The Babylonians used a form of writing called cuneiform, where symbols are made by pressing a wedge-shaped stick into clay. There are just two basic symbols in the system of numeration, for 1 and for 10. These are used to write numerals for the numbers up to 59 by simple grouping. Then, a symbol similar to, and in some books identical to, the symbol for 1 means the special number 60. Sixty is special because it is the *base* of the place-value system of the Babylonians. The first group of symbols means a number between 1 and 59, the next group to the left has its value times 60, the next group its value times $60 \cdot 60 = 3{,}600$, the next group its value times $60 \cdot 60 \cdot 60 = 216{,}000$, and so on. For example,

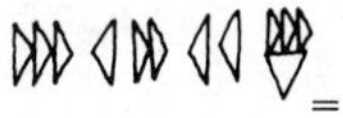

$$= 3 \cdot 3{,}600 + 12 \cdot 60 + 24 = 11{,}544$$

A space left between two groups of symbols means a place value is to be skipped.

$$= 3 \cdot 3{,}600 + 24 = 10{,}824$$

There are two ambiguities in the Babylonian system. If the space left is a bit too large, the numeral above can be taken to mean $3 \cdot 216{,}000 + 24 = 648{,}024$. On the other hand, if the space is a bit too small, the numeral might simply be taken for $3 \cdot 60 + 24 = 204$. Moreover, there is no way to indicate places left empty at the right-hand end of a numeral. Thus  can mean 24, or $24 \cdot 60 = 1{,}440$, or $24 \cdot 3{,}600 = 86{,}400$, and so on. Despite this problem, the Babylonian system of numeration does have the advantage of expressing as large a number as is desired by moving over to higher place values rather than by making up new symbols.

EXERCISE 1 Write each of these numbers in Egyptian numerals. (*a*) 40. (*b*) 44. (*c*) 300. (*d*) 2,340. (*e*) 1,504. (*f*) 2,534. (*g*) 999.

EXERCISE 2 Add these numbers in Egyptian numerals, and check by translating to modern numerals.

(*a*) (*b*) (*c*)

EXERCISE 3 Write each of these numbers in Roman numerals. (*a*) 300. (*b*) 500. (*c*) 2,550. (*d*) 1,812. (*e*) 1,066. Write each of these numbers in Roman numerals in two different ways. (*f*) 1,492. (*g*) 1,942.

EXERCISE 4 Add these numbers in Roman numerals, and check by translating to modern numerals.

(*a*) MMCLXII
MDCXXI

(*b*) MMDCCCLXXXVIII
DCCCXXVIII

(*c*) MCMXLIV
MCCLXVI

EXERCISE 5 Write each of these numbers in Greek numerals. (*a*) 40. (*b*) 44. (*c*) 300. (*d*) 2,340. (*e*) 1,504. (*f*) 2,534. (*g*) 999.

EXERCISE 6 Write each of these Babylonian numerals in modern numerals (assume no place values are left empty at the right-hand end).

(*a*) (*b*) (*c*) (*d*)

(*e*) (*f*) (*g*)

FIGS. 1.11a-g

EXERCISE 7 (*a*) Name three ancient systems of numeration which were simple grouping systems. (*b*) Name an ancient system of numeration which was a cipher system. (*c*) In which type of system of numeration were the numerals cumbersome to write? (*d*) In which type of system of numeration was the symbol for each number difficult to remember? (*e*) Name three ancient systems of numeration in which it is necessary to create new symbols for larger numbers. (*f*) Name an ancient system of numeration which uses the principle of place value. (*g*) What is the advantage of using place value in a system of numeration? (*h*) What is the disadvantage of place value as it was used in one ancient system of numeration?

3
Rationals and Irrationals

Fractions are *ratios*, or divisions, of counting numbers. Although ancient civilizations had at least some symbols for fractions, the really remarkable discovery about them belongs not to ancient numeration but to the ancient Greek geometers. Because of the numerical sophistication of earlier cultures, you might wonder why the Greeks are often considered to be the first real mathematicians. One reason is that the mathematics of Egypt and Babylonia took the form of "recipes" for solving problems. There is no evidence of attempts to "prove" the correctness of formulas or methods, and no development of a general mathematical theory. It was the Greeks who systematized the mathematical facts and formulas, introduced the concepts of deductive logic and proof, and gave a foundation to the formulas and methods.

The Greek development of logical reasoning and deduction, which characterizes modern mathematics, is supposed to have begun with Thales (early sixth century B.C.), a merchant of Miletus. It can certainly be said that he is the first man of mathematics we know by name. Howard Eves, one of the best of the mathematical storytellers as well as historians, says this of Thales in *An Introduction to the History of Mathematics*.[1]

As with other great men, many charming anecdotes are told about Thales, which, if not true, are at least apposite. There was the occasion when he demonstrated how easy it is to get rich; foreseeing a heavy crop of olives coming, he obtained a monopoly on all the oil presses of the region and then later re-

[1] 3d ed. (New York: Holt, Rinehart and Winston, Inc., 1969), pp. 50–51.

alized a fortune by renting them out. And there is the story of the recalcitrant mule which, when transporting salt, found that by rolling over in the stream he could dissolve the contents of his load and thus travel more lightly—Thales broke him of the troublesome habit by loading him with sponges. He answered Solon's query as to why he never married by having a runner appear next day with a fictitious message for Solon stating that Solon's favorite son had been suddenly killed in an accident; Thales then calmed the grief-stricken father, explained everything, and said, "I merely wanted to tell you why I never married." At another time, having fallen into a ditch while observing the stars, he was asked by an old woman how he could hope to see anything in the heavens when he couldn't even see what was at his own feet. Asked how we might lead more upright lives he advised, "By refraining from doing what we blame in others." When once asked what he would take for one of his discoveries he replied, "I will be sufficiently rewarded if, when telling it to others, you will not claim the discovery as your own, but will say it was mine." And when asked what was the strangest thing he had ever seen he answered, "An aged tyrant."

Eves tells these stories in more detail in his book of mathematical stories *In Mathematical Circles*. As he says, these may just be fictional anecdotes, but they probably do illustrate the character of the man. The stories come from the *Eudemian Summary* of Proclus (fifth century A.D.). Thus even though Proclus had access to earlier writings, the actual stories we have were recorded a thousand years after they were supposed to have happened. During that time many fanciful legends could have grown up around an interesting character.

Pythagoras (middle sixth century B.C.) is the Greek geometer whose name is attached to the discoveries about ratios or fractions. Pythagoras and his followers, the Pythagoreans, thought they had systematized not only mathematics, but the universe, and their universe was one of numbers. Although in commerce and everyday life the Greeks used fractions, for theoretical work in mathematics they used the counting numbers only, for the reason that 1 is unity and unity cannot be divided. In place of fractions, the Pythagoreans worked with ratios of the counting numbers. We write such ratios as $a{:}b$ or a/b ("the ratio of a to b"). To Pythagoras and the Pythagoreans, the universe was a harmony of ratios.

Pythagoras is credited with having discovered the theory of harmony in music. Two strings with their lengths in the ratio 2:1 give tones an octave apart. Ratios of 4:3 and 3:2 give other special musical

Many legends have grown up around Pythagoras, mathematician and mystic, but little is really known about him. We do know that he was born on the island of Samos in the latter part of the sixth century B.C. It is thought that he must have studied with Thales, and that his travels brought him to Egypt and Babylon.

After settling in Crotona, a Greek town in southern Italy, Pythagoras founded the school which was the beginning of the secret brotherhood called the Pythagoreans. The school allied itself with the aristocracy and was eventually destroyed for political reasons. Pythagoras fled to Metapontum, where he died around 497 B.C., possibly murdered. The Pythagorean cult continued for at least another hundred years. Since the custom of the brotherhood was to attribute all discoveries to Pythagoras, it is not always known precisely what he did and what was in fact done by other members of the brotherhood.

It is odd that the Pythagorean cult is almost always called a "brotherhood." According to Lynn Osen in Women in Mathematics, there were women among the Pythagoreans; in fact, the order was continued by the wife and daughters of Pythagoras after his death. So what has been called a brotherhood had female members and, in general, women were probably more welcome in the centers of learning in ancient Greece than in any other age from that time until the present.

Photo: David Eugene Smith Collection, Columbia University Library

intervals. The Pythagoreans worked out ratios for the movements of the planets corresponding to their ratios for musical harmonies. You may have heard the phrase "music of the spheres," or "planets," which refers to this relationship.

When a universe is built on ratios, all numbers can be represented by a common measure. In the ratio 2:1, 2 can be represented as twice the measure of 1. In the ratios 4:3 and 3:2, a common measure can be found which if applied four times gives the 4, if applied three times gives the 3, and if applied twice gives the 2. Since 2 was already twice the measure of 1, the common measure is unity. When the numbers are represented by line segments, as they commonly were by the Greeks, unity is represented by a line segment of arbitrary length. Then the line segments representing a ratio are multiples of the basic length, and are called *commensurable.*

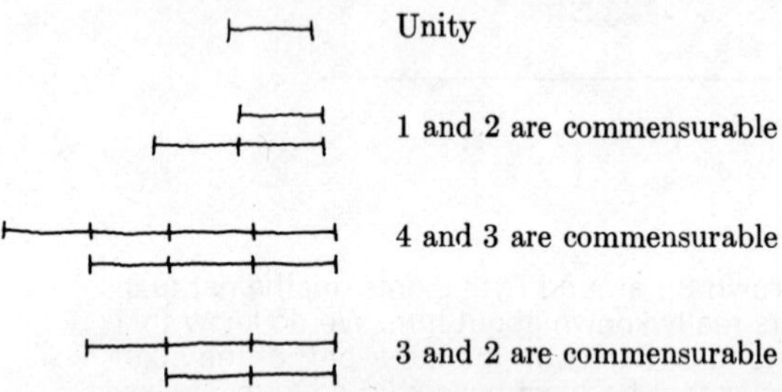

A problem of ancient interest was to find three numbers which are all commensurable, and which also fit a famous formula named for the Pythagoreans. The formula is concerned with *right triangles.* A triangle is a right triangle if two of its sides are perpendicular, forming a *right angle.* The third side of the triangle, opposite the right angle, is called the *hypotenuse.*

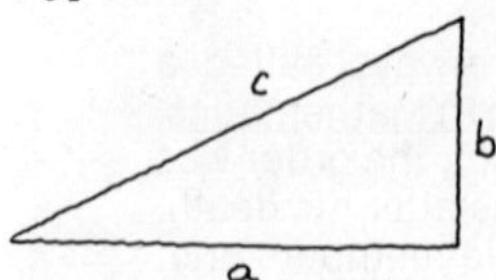

PYTHAGOREAN THEOREM If a and b are sides of a right triangle and c is the hypotenuse, then $a^2 + b^2 = c^2$.
[*Note:* a^2 ("a squared") means $a \cdot a$, and similarly for b^2 and c^2.]

We can find as many triples of numbers as we please which are commensurable and also fit the Pythagorean theorem. The Egyptians knew of a triangle with sides 3, 4, and 5, but there is no evidence that they knew the triangle was a right triangle. There is also no evidence to support the popular story that they used ropes knotted in lengths of 3, 4, and 5 for surveying. The Babylonians, with superior knowledge of algebraic formulas, were familiar with the formula $a^2 + b^2 = c^2$ for the sides of a right triangle. Thus the Pythagorean theorem was known at least a thousand years before Pythagoras.

EXERCISE 8 Check each of these triples in the Pythagorean theorem. (*a*) $a = 3, b = 4, c = 5$. (*b*) $a = 5, b = 12, c = 13$. (*c*) $a = 8, b = 15, c = 17$. (*d*) $a = 12$, $b = 35$, $c = 37$.

The Pythagorean theorem was named for Pythagoras because it was long thought that he discovered it. Even after archaeologists found it was known to the Babylonians, it was thought the Pythagoreans, if not Pythagoras himself, were the first to prove it. Eves says "a proof may well have been given by Pythagoras." Kline believes it could not have been proved even by later Pythagoreans. What is important is not who finally proved the theorem but what the Pythagoreans discovered from it.

From the formula which is named for them, the Pythagoreans discovered *incommensurables*; that is, line segments which are not multiples of some common measure. If we let $a = 1$ and $b = 1$ in the Pythagorean theorem, then $c^2 = a^2 + b^2 = 1^2 + 1^2 = 1 + 1 = 2$. Thus $c^2 = 2$. We write a number c such that $c^2 = 2$ as $c = \sqrt{2}$ ("the square root of 2"). One and $\sqrt{2}$ are incommensurable.

The square root of 2 was not a new number. An easy method for approximating it, as well as other square roots, was known to the Babylonians. If $x = \sqrt{2}$, then x is a number such that $x \cdot x = 2$, and $x = 2/x$. We use any number, say $x = 1$, as a first approximation. If this number is too small, then $2/x$ is too large. If the number is too large, then $2/x$ is too small. In either case, their *average*, $\frac{1}{2}(x + 2/x)$, will give a better approximation. When $x = 1$,

$$\frac{1}{2}\left(x + \frac{2}{x}\right) = \frac{1}{2}\left(1 + \frac{2}{1}\right) = \frac{3}{2}$$

Now we go through the entire process a second time, this time using the new approximation $x = \frac{3}{2}$:

$$\frac{1}{2}\left(x + \frac{2}{x}\right) = \frac{1}{2}\left(\frac{3}{2} + \frac{2}{\frac{3}{2}}\right) = \frac{1}{2}\left(\frac{3}{2} + \frac{4}{3}\right) = \frac{1}{2}\left(\frac{17}{6}\right) = \frac{17}{12}$$

We repeat again, using $x = \frac{17}{12}$:

$$\frac{1}{2}\left(x + \frac{2}{x}\right) = \frac{1}{2}\left(\frac{17}{12} + \frac{2}{\frac{17}{12}}\right) = \frac{1}{2}\left(\frac{17}{12} + \frac{24}{17}\right) = \frac{1}{2}\left(\frac{577}{204}\right) = \frac{577}{408}$$

Using the Babylonian method only three times, we obtain an approximation for $\sqrt{2}$ which is so good it is correct to four decimal places, giving $\sqrt{2} \approx 1.4142\ldots$, where $\approx$ means "approximately equal to," and the three dots, called an ellipsis, mean we can keep finding decimal places indefinitely.

The Babylonians knew how to approximate $\sqrt{2}$, but they did not know and probably did not care that it cannot be written as a ratio of counting numbers. The Pythagoreans may very well have been the first to prove this fact, for the proof depends on properties of *even* and *odd* counting numbers. We must know that the square of any even counting number is even, and that the square of any odd counting number is odd. These facts were known to the Pythagoreans, who were fascinated with special properties of special types of numbers. The proof is still used, in only a very slightly different form, in today's modern mathematics texts.

The proof is *indirect*. Indirect proof means we will show that if $\sqrt{2}$ could be written as a ratio of counting numbers, then some impossible situation would occur. We suppose that $\sqrt{2} = a/b$ for some counting numbers a and b. We also suppose that a/b has been reduced so that a and b have no common factors. Then, writing the ratio a/b as a fraction,

$$\begin{aligned} \sqrt{2} &= \frac{a}{b} \\ 2 &= \frac{a^2}{b^2} \\ 2b^2 &= a^2 \end{aligned}$$

Since a^2 is equal to a number with a factor of 2, a^2 is even and so a is even. Then b must be odd, for a and b have no common factors; and if both were even, there would be a common factor of 2. Furthermore, since a is even, a can be written as twice some other counting number, say c,

$$a = 2c$$
$$a^2 = 4c^2$$
$$2b^2 = a^2 = 4c^2$$
$$2b^2 = 4c^2$$
$$b^2 = 2c^2$$

Thus b^2 is even and so b is even. But then b is both even and odd, which is impossible.

DEFINITION The set of *positive rational numbers* is the set of numbers which can be written as a ratio a/b, where a and b are counting numbers. The set of *positive irrational numbers* is the set of numbers which can be approximated by decimals to any amount of decimal places, but which cannot be written as ratios of counting numbers.

The square root of 2 is an irrational number. You can imagine the Pythagoreans' chagrin at their discovery of numbers which were not ratios of counting numbers. They called the irrational numbers *Alogon*, "the unutterable." The secret of their discovery was not to be revealed. There are many versions of the legend of disaster which befell those who did reveal it. This story is also from Proclus, translated here by Tobias Dantzig.[2]

It is told that those who first brought out the irrationals from concealment into the open perished in shipwreck, to a man. For the unutterable and the formless must needs be concealed. And those who uncovered and touched this image of life were instantly destroyed and shall remain forever exposed to the play of the eternal waves.

Later Greeks were not so concerned about the inviolability of ratios and developed the theory of irrationals. One other irrational number of special interest is the number we call π (pi). This number is the ratio of

[2] *Number, the Language of Science,* 4th ed. (New York: The Macmillan Company, 1954), p. 101.

Yale Babylonian Collection

This little tablet comes from the Old Babylonian period, perhaps around 1700 B.C. The drawing on it is of a square with its diagonals. The marks at the upper left of the tablet and along the horizontal diagonal of the square are Babylonian cunieform numerals. According to Neugebauer (The Exact Sciences in Antiquity, p. 34), these numerals confirm the belief that the ancient Babylonians had a very good approximation for $\sqrt{2}$. This and other tablets from the same period show that the Babylonians could use the Pythagorean theorem for computations, such as finding the length of the diagonal of a square.

The number at the upper left of the tablet is the Babylonian symbol for 30. Neugebauer takes this number, for reasons we shall see, to mean $30 \cdot (1/60) = 1/2$. Departing from Neugebauer for a moment, and using the Pythagorean theorem, we find that if the side of a square is 1/2, then

$$\underline{c}^2 = \left(\frac{1}{2}\right)^2 + \left(\frac{1}{2}\right)^2 = \frac{1}{4} + \frac{1}{4} = \frac{2}{4}$$

and so the diagonal of the square is

$$\underline{c} = \sqrt{\frac{2}{4}} = \frac{\sqrt{2}}{2}.$$

Now, following Neugebauer, the numbers along the diagonal of the square are

$$1 \cdot 216{,}000 + 24 \cdot 3600 + 51 \cdot 60 + 10 = 305{,}470$$

and

$$42 \cdot 3600 + 25 \cdot 60 + 35 = 152{,}735$$

We observe that the second number is exactly one-half the first. Still following Neugebauer, we change the place values, and think of the first number as

$$1 + 24 \cdot \frac{1}{60} + 51 \cdot \frac{1}{3600} + 10 \cdot \frac{1}{216{,}000}.$$

This number is, in decimal approximation,

$$1 + .4 + .014167 + .000046 = 1.414213.$$

The number corresponds to $\sqrt{2} \approx 1.414214\ldots$ correctly through the fifth decimal place, an amazingly accurate approximation. The second number is then, being one-half of the first, $\sqrt{2}/2$ or the length of the diagonal of the square with side 1/2, also correct through the fifth decimal place.

Almost two thousand years later, Ptolemy of Alexandria, who Neugebauer says used the same approximation of $\sqrt{2}$ found on this tablet, also used the Babylonian base of 60 for fractions. (Observe that we still use the base of sixty in computing minutes in an hour or in a degree of a circle.)

the *circumference* of a circle, the distance around the circle, to its *diameter*, the line segment through the center of the circle: $\pi = C/d$:

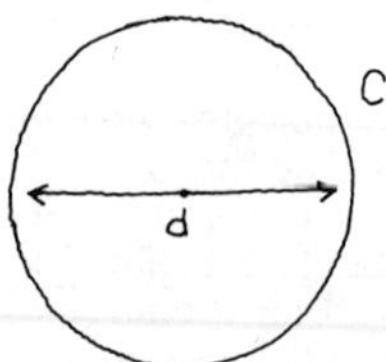

The Greeks approximated π by using inscribed and circumscribed polygons. An *inscribed polygon* is a figure made up of line segments with its *vertices*, the points where the segments meet, on the circle. A *circumscribed polygon* is made up of line segments lying outside the circle and just touching the circle:

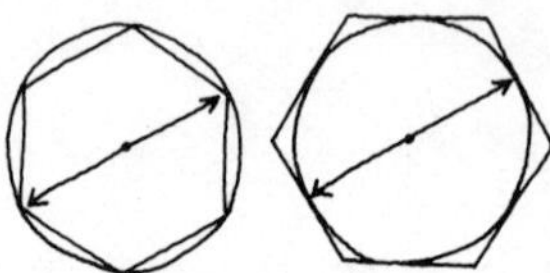

The ratio of the *perimeter*, the sum of its line segments, of the inscribed polygon to d gives a number somewhat less than π. The ratio of the perimeter of the circumscribed polygon to d gives a number somewhat greater than π. When polygons with more and more sides are used, their perimeters become closer and closer to the circumference of the circle, and the ratios closer and closer to π. This is the method of Archimedes (middle third century B.C.), greatest of the Greek mathematicians, physicists, and engineers. Archimedes computed π by this method to be between $\frac{223}{71}$ and $\frac{22}{7}$, or in decimal approximation $\pi \approx 3.14$, an approximation which is still used today. It was not proved that π is actually irrational, however, until the eighteenth century, when Johann Heinrich Lambert (1728–1777) gave the first definitive proof. Using computers, π has now been approximated to hundreds of thousands of decimal places.

EXERCISE 9 (*a*) Suppose $a = 1$ and $c = 2$ in the Pythagorean theorem. Find the value of b. Is b rational? (*b*) Find c in the Pythagorean theorem when a and b are each of these pairs of numbers: (*i*) $a = 1$ and

$b = 2$; (*ii*) $a = 2$ and $b = 3$; (*iii*) $a = 1$ and $b = 4$; (*iv*) $a = 2$ and $b = 4$; (*v*) $a = 3$ and $b = 4$. (*c*) Suppose it has been proved, up to $\sqrt{25}$, which square roots are irrational and which are rational. For each value of c above, is c rational?

EXERCISE 10 Use the Babylonian method for approximating square roots three times to find an approximation for $\sqrt{3}$. For this square root, use $x \cdot x = 3$ and $x = 3/x$. The approximation is correct to three decimal places.

EXERCISE 11 (*a*) What happens if you use the Babylonian method to approximate $\sqrt{4}$ using a first approximation of $x = 1$? (*b*) What happens if you use the Babylonian method to approximate $\sqrt{4}$ using a first approximation of $x = 2$?

EXERCISE 12 Suppose it is known that if $3b^2 = a^2$, then a has a factor of 3. Use this fact, and a method similar to that of the Pythagoreans, to prove that $\sqrt{3}$ is irrational.

EXERCISE 13 (*a*) What happens if you try to use the Pythagorean method of proving that a square root is irrational for $\sqrt{4}$? (*b*) If $4b^2 = a^2$, is it true that a must have a factor of 4?

EXERCISE 14 Suppose a circle has a diameter of 2 inches. Then its radius, half the diameter, is 1 inch. If a regular hexagon is inscribed in the circle, it can be divided into six identical equilateral triangles. (*a*) How many inches is the perimeter of the hexagon? (*b*) What is the ratio of the perimeter of the hexagon to the diameter of the circle?

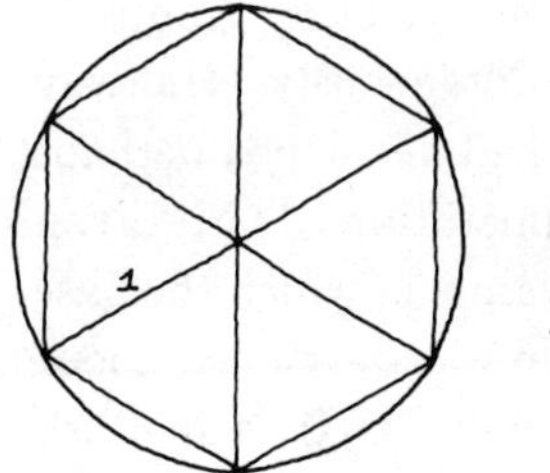

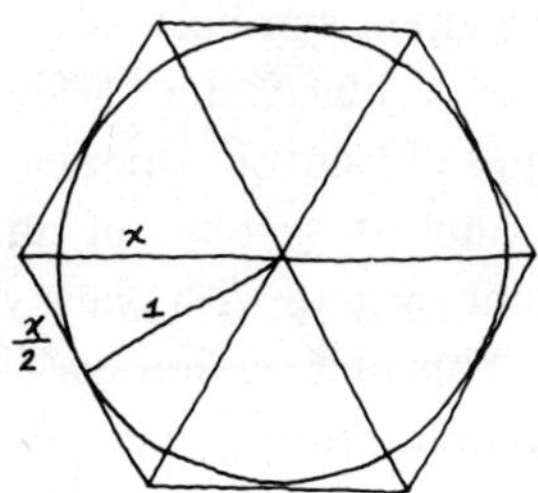

If a regular hexagon is circumscribed about the circle, it can be divided into six identical equilateral triangles. (*c*) Use the Pythagorean theorem to find x, the length of each side of each equilateral triangle. (*d*) How many inches is the perimeter of the hexagon? (*e*) What is the ratio of the perimeter of the hexagon to the diameter of the circle? From the method of Archimedes, observe that a first approximation of π is greater than 3 but less than 3.5 (where $\sqrt{3} \approx 1.732$).

4

The Elusive Zero

Somewhere around 300 B.C., the symbol appeared in the Babylonian cuneiform numerals. We recall that in the ancient Babylonian place-value system of numeration, after the group in the place at the right, the group of symbols in the first place to the left has the place value 60, the group of symbols in the next place to the left has the place value $60 \cdot 60 = 3{,}600$, and so on. There were two ambiguities in the system. The first occurred if a place value between two groups was skipped. The symbol was a placeholder. Thus after 300 B.C., meant $1 \cdot 60 + 10 = 70$, regardless of the amount of space between the symbols, and meant $1 \cdot 3{,}600 + 10 = 3{,}610$. The new symbol solved the problem of the first ambiguity. The second ambiguity, that of indicating empty places at the end of a numeral, was not solved by the new symbol because it was not used to hold places at the end of numerals. Even so, the Babylonians could be said to have discovered a form of zero by inventing a symbol to indicate the absence of a group of symbols between two other groups.

A similar discovery took place almost simultaneously. Halfway around the world, the Mayas of Central America had developed perhaps the most advanced of all ancient systems of numeration. The Mayan system was a simple grouping system with place value, in which the base of the place values was 20. This place-value system was developed about 2000 B.C. In the third or fourth century B.C., a symbol , a mollusk shell, appeared as a placeholder, or zero, perhaps even preceding the similar Babylonian invention.

EXERCISE 15 These numerals represent Babylonian numerals as they were written after about 300 B.C. Write each in modern numerals (assume that no place values are left empty at the right-hand end).

(*a*) (*b*) (*c*)

(*d*) (*e*)

EXERCISE 16 Write each of these numbers in Babylonian numerals, as they would have been written after about 300 B.C. (*a*) 46. (*b*) 64. (*c*) 3,604. (*d*) 216,142. (*e*) 216,240.

The ancient systems of numeration that involved simple grouping probably originated as *tally systems.* In a tally system, a single stroke of some kind, say /, represents 1. Then two strokes of the same kind, //, represent 2, three strokes, ///, represent 3, and so on. You have probably used a tally system yourself to keep track of some kind of count. When you get to 5, you probably write ~~////~~, and then ~~////~~ / for 6. If you grouped your strokes by tens instead of fives, you would have the base of the place-value system of the Hindus of India. The place value of the Hindu system was probably inspired by counting boards, a refinement of the tally system. Instead of a separate stroke for every unit, a separate column of the counting board is kept for units up to 9, for tens, for 10 times tens, or hundreds, and so on. The place value of the counting board, in turn, was probably inspired by counting on ten fingers.

A representation of a counting board from Dantzig, op. cit., p. 31.

The counting board represented by the diagram above has three strokes in the hundreds column, four strokes in the tens column, and two strokes in the ones column. Eventually, the two strokes evolved into a single symbol 2, the three strokes became a single symbol 3, four strokes became 4, and other symbols evolved for the other amounts of strokes up to 9. The number on the counting board is 342. What is important is the *position* of the 3, the 4, and the 2. Their positions determine their meaning. For example, a 2 at the far right means two strokes, but in the next position to the left, a 2 means two tens, or twenty strokes. This is the counting board for 324:

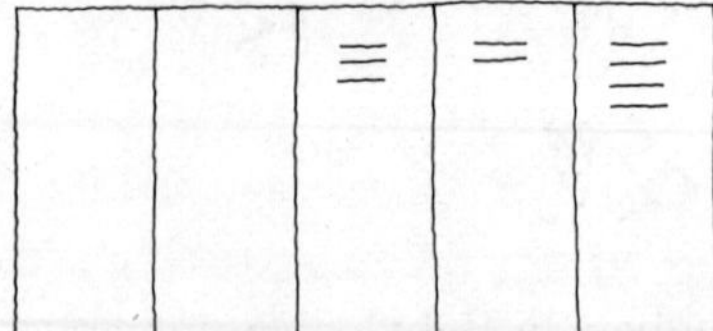

Now, what shall we do when two columns of a counting board have strokes in them with an empty column between?

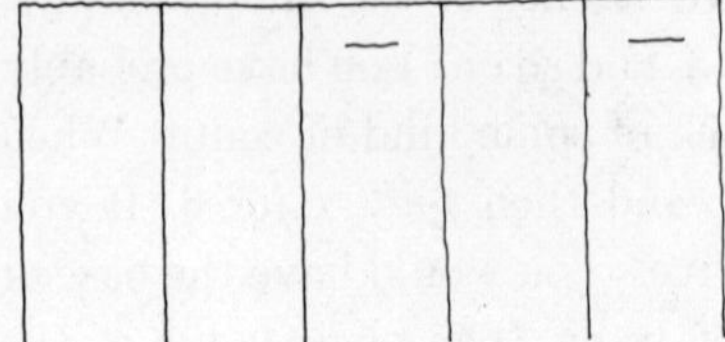

Or when the last column to the right is empty?

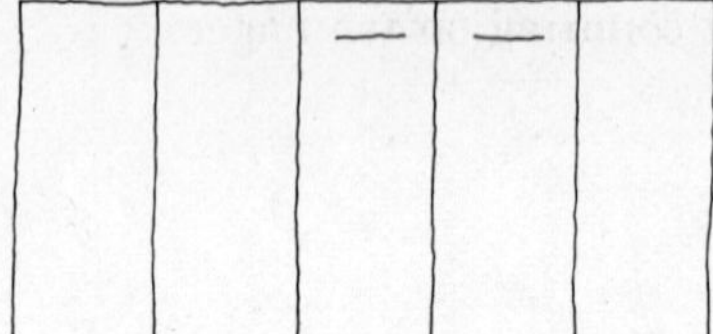

We cannot write 11, for that would mean one 10 and a 1. We must write some symbol for the empty column, say 101 for the first board and 110 for the second.

Since both the Babylonians and the Mayans used their zero symbols only as placeholders and not as numbers in their own right, credit for the discovery of zero as a number is often given to the Hindus sometime before A.D. 825. The date 825 is the date of an arithmetic book by a

Persian, Al-Khowârizmî, which describes the Hindu system. The system was developed by the Hindus but disseminated by the Arabs, so it is called the *Hindu-Arabic* system of numeration. The Hindu-Arabic system is a place-value system in which the base of the place values is 10. It has a separate symbol for each of the numbers 1 through 9. It has a zero both as a placeholder and as a number in the sense that we can compute with it; for example, $5 + 0 = 5$, but $5 \cdot 0 = 0$. The Hindu-Arabic system is our modern system of numeration.

The concepts of place value and the number 0 together enable us to write numerals for numbers as large as we please, without making up new symbols, and without creating ambiguities. Use of place value and zero in the Hindu-Arabic system of numeration, with a base of 10 and separate symbols for each of the numbers 0 through 9, gives us only ten symbols to remember and a compact way to write numerals. The ten symbols, 0, 1, 2, 3, 4, 5, 6, 7, 8, and 9, are called the *digits*. Each digit written alone simply means the number it stands for; thus 5 means the number five. Then, using place value and zero, 50 means five tens plus zero, or $5 \cdot 10 + 0$, 500 means five hundreds plus zero tens plus zero, or $5 \cdot 100 + 0 \cdot 10 + 0$, and so on. The last expression is called the *expanded form* of the numeral. Some other examples of expanded forms of numerals are $55 = 5 \cdot 10 + 5$, $505 = 5 \cdot 100 + 0 \cdot 10 + 5$, $555 = 5 \cdot 100 + 5 \cdot 10 + 5$, and $3{,}025{,}140 = 3 \cdot 1{,}000{,}000 + 0 \cdot 100{,}000 + 2 \cdot 10{,}000 + 5 \cdot 1{,}000 + 1 \cdot 100 + 4 \cdot 10 + 0$.

DEFINITION The set of *whole numbers* is the set of counting numbers with zero; that is, 0, 1, 2, 3, 4, and so forth.

One other invention of a symbol for zero should be mentioned. B. L. van der Waerden writes (*Science Awakening*, p. 39) that the famous Greek astronomer Ptolemy of Alexandria (about A.D. 150) used a symbol for zero, and further (pp. 56–57) that the Hindus might very well have acquired the symbol from the Greek astronomers. Neugebauer shows a very convincing table (*The Exact Sciences in Antiguity*, p. 10) in which Ptolemy uses a symbol similar to the Hindu-Arabic and, moreover, appears to give it meaning as a number 0 preceding the number 1. Some historians give credit for the discovery of a number 0 to the Hindus, perhaps A.D. 500 to 800, some to the Greeks about A.D. 150, and some to the Babylonians or Mayans in the third or fourth century B.C. Whichever date you choose, it may astonish you that such seemingly intricate numbers as the irrationals were well known before even a symbol for zero was invented, and centuries before it was used as a number.

EXERCISE 17 Draw a counting board which represents each of these numbers. (*a*) 234. (*b*) 243. (*c*) 423. (*d*) 432. (*e*) 2,423.

EXERCISE 18 Draw a counting board which represents each of these numbers. (*a*) 20. (*b*) 200. (*c*) 202. (*d*) 220. (*e*) 2,020.

EXERCISE 19 Write the expanded form of each of these Hindu-Arabic numerals. (*a*) 2. (*b*) 20. (*c*) 200. (*d*) 202. (*e*) 234. (*f*) 2,020. (*g*) 2,423. (*h*) 24,230,202.

EXERCISE 20 (*a*) Name three place-value systems of numeration which eventually used a zero symbol as a placeholder. (*b*) Name one cipher system of numeration which may have used zero as a number. (*c*) Name a system of numeration which uses zero both as a placeholder and as a number. (*d*) Name the system of numeration which we use. (*e*) What advantage does the place-value concept of our system of numeration have over cipher systems and simple grouping systems without place value? (*f*) What advantage does the concept of zero in our system of numeration have over simple grouping systems without a zero symbol? (*g*) What advantage does the ten-digit concept of our system of numeration have over simple grouping systems? (*h*) What advantage does the ten-digit concept of our system of numeration have over cipher systems?

5
Negatives and Imaginaries

After the discovery of the number zero, and long after the discovery of the irrationals, the negative numbers were recognized. One reason for this belated recognition takes us back once again to the influence of the Greeks on mathematics. The Greeks used the arithmetic operation of subtraction, and had as an axiom, "if equals be subtracted from equals, the remainders are equal." However, only a smaller quantity could be subtracted from a larger. Since the Greeks represented numbers by line segments, to subtract the larger from the smaller would be absurd because it is geometrically impossible.

To find the source of the negative numbers, we must begin again with the Arab mathematicians, and Al-Khowârizmî. Besides his book on arithmetic, Al-Khowârizmî also wrote a book called *al-jabr w'al-muqâbalah.* Our word "algebra" is derived from the word *al-jabr* in the title. In his book, Al-Khowârizmî collected the algebra known in his time, the operations of addition, subtraction, multiplication, and division, and solutions to linear and quadratic equations (equations in x and x^2). The algebra of the Arabs reached its peak with Omar Khayyâm (about 1100), more famous as the author of the book of poems, the *Rubáiyát.* Khayyâm developed a geometric method for solution of cubic equations (equations in x^3). Although both Al-Khowârizmî and Khayyâm rejected negative solutions to their equations, their contribution to the discovery of negative numbers is in their interpretation of numbers as solutions to equations rather than as line segments.

The Arab interest in equations spread to Italy, along with the Hindu-Arabic system of numeration, by way of trade with Italian merchants. Life Science Library's *Mathematics*[3] tells of one isolated step toward the recognition of negative numbers.

Since it is hard to grasp the idea of negative numbers, a long time elapsed before they were allowed into the parlors of mathematical propriety and common sense. One of the first to give them open-minded consideration was an Italian mathematician, Leonardo da Pisa, otherwise called "Fibonacci," who lived from about 1170 to 1250 A.D. On one occasion, while tackling a financial problem, he saw that it just could not be solved except in terms of a negative number. Instead of shrugging off this number, he looked it squarely in the eye and described it as a financial loss. "This problem," he wrote, "I have shown to be insoluble unless it is conceded that the first man had a debt."

Negative numbers did not receive full recognition until the sixteenth century, when the Italians succeeded in solving cubic and quartic equations (equations in x^3 and x^4) by strictly algebraic methods. During the early sixteenth century it was the custom of the Italian algebraists to hold contests in which one challenged another in solving equations. One

[3] Life Science Library, *Mathematics* by David Bergamini and the Editors of Time-Life Books, © 1963, 1972 Time Inc., p. 67.

Leonardo da Pisa was better known as Fibonacci, "son of Bonaccio." Fibonacci was one of the first Europeans to publish a work on equations, Liber abaci, published in 1202. The book, which urges the Hindu-Arabic system of numeration, covers the arithmetic and algebra brought to Italy by the Arabs.

Fibonacci is probably better known today for the "Fibonacci sequence" of numbers,

1, 1, 2, 3, 5, 8, 13, 21, 34, . . .

Can you see how the Fibonacci sequence is constructed and what the next number should be? (Each number of the Fibonacci sequence is derived by adding the two preceding numbers; the next number is 55.)

Found throughout mathematics, the Fibonacci sequence is also found in surprising places in nature. For example, the diamond shaped sections of a pineapple swirl upward to the left and to the right. According to Life Science Library's Mathematics, there are eight sections one way and thirteen the other—two consecutive numbers of the Fibonacci sequence.

of the most gifted and colorful, and perhaps also unscrupulous, of the equation solvers was Girolamo Cardano [or Jerome Cardan (1501–1576)].

Cardano's great work was the *Ars Magna*, in which he gave solutions to the cubic and quartic equations. There are stories that his solutions to certain types of cubics and quartics may have been stolen from others. In order to keep them from their opponents, the contestants in equation-solving contests often did not publish their methods. Cardano published a method of solution for the cubic which he probably stole from an opponent, Tartaglia, and a solution for a quartic which he had been unable to solve but which was solved by his student Ferrari. In any event, the *Ars Magna*, published in 1545, recognized negative solutions.

DEFINITION The set of *integers* is the set of counting numbers, zero, and the negatives of the counting numbers; that is, 0, 1, -1, 2, -2, 3, -3, 4, -4, and so forth.

We must also name the set which includes the counting numbers, positive rationals, and positive irrationals, along with zero and the negatives of these numbers. We need only note that all these types of numbers can be written as, or approximated by, decimals, if the decimals are allowed to be positive, negative, or zero.

DEFINITION The set of *real numbers* is the set of numbers which can be written as decimals or approximated by decimals to any amount of decimal places.

When negative numbers are accepted, a new problem arises. The real numbers have the properties that the sum of two positives is positive and the sum of two negatives is negative. The product of two positives is positive but the product of two negatives is also positive. For example, the equation $x^2 = 4$ has two solutions, $x = 2$ and $x = -2$, since $2 \cdot 2 = 4$ and also $(-2) \cdot (-2) = 4$. The equation $x^2 = -4$ has no solution among the real numbers, however, since there is no real number x such that $x \cdot x = -4$ or any other negative number. If x is zero, $x \cdot x = 0$; if x is positive, $x \cdot x$ is positive; and if x is negative, then $x \cdot x$ is negative times negative and thus is also positive.

If we wish to have a number which is the solution to equations of the type $x^2 = -4$, we must recognize a new type of number, one which when multiplied by itself gives a negative. The symbol which was invented for the new number is i, where $i^2 = -1$. Then, for example, $(2i) \cdot (2i) = 4i^2 = 4 \cdot (-1) = -4$. We may also say $\sqrt{-4} = 2i$. As another example, $(i\sqrt{2}) \cdot (i\sqrt{2}) = \sqrt{2}\sqrt{2}i^2 = 2 \cdot (-1) = -2$, and we may also say that $\sqrt{-2} = i\sqrt{2}$. Thus numbers of the form bi, where b is a real number, are numbers which when multiplied by themselves give negative real numbers. We may also say that numbers of the form bi are numbers which are the square roots of negative real numbers.

DEFINITION The set of *imaginary numbers* is the set of numbers bi, where b is a real number, which when multiplied by themselves give negative real numbers.

If we add a real number a and an imaginary number bi, we get a number of the form $a + bi$.

DEFINITION The set of *complex numbers* is the set of numbers $a + bi$, which can be written as the sum of a real number and an imaginary number.

If we allow the case of $b = 0$, then the set of complex numbers includes the set of reals, which in turn includes the sets of irrationals, rationals, integers, whole numbers, and counting numbers. If $a = 0$, then the set of complex numbers includes the set of imaginaries. Thus the set of complex numbers includes every kind of number we have defined.

EXERCISE 21 Although the following equations are written in modern symbols, they are types which would have been considered by the Arab algebraists. Which ones have at least one solution the Arabs would have accepted? (*a*) $x - 4 = 0$. (*b*) $x + 4 = 0$. (*c*) $3x - 4 = 0$. (*d*) $3x + 4 = 0$. (*e*) $x^2 - 4 = 0$. (*f*) $x^2 + 4 = 0$.

EXERCISE 22 (*a*) Which ones of these equations have at least one solution that is an integer? (*i*) $2x - 6 = 0$; (*ii*) $2x + 6 = 0$; (*iii*) $2x - 5 = 0$; (*iv*) $2x + 5 = 0$; (*v*) $x^2 - 9 = 0$; (*vi*) $x^2 + 9 = 0$; (*vii*) $x^2 - 10 = 0$; (*vii*) $x^2 + 10 = 0$; (*ix*) $x^2 = 0$. (*b*) If a is a whole number, does the equation $x^2 - a^2 = 0$ have at least one integer solution? (*c*) If a is a whole number, does the equation $x^2 + a^2 = 0$ have at least one integer solution?

EXERCISE 23 Which ones of these equations have at least one solution that is a rational number (not necessarily positive)? (*i*) $2x - 6 = 0$; (*ii*) $2x + 6 = 0$; (*iii*) $2x - 5 = 0$; (*iv*) $2x + 5 = 0$; (*v*) $x^2 - 9 = 0$; (*vi*) $x^2 + 9 = 0$; (*vii*) $x^2 - 10 = 0$; (*viii*) $x^2 + 10 = 0$; (*ix*) $x^2 = 0$.

EXERCISE 24 (*a*) Which ones of these equations have at least one solution that is a real number? (*i*) $2x - 6 = 0$; (*ii*) $2x + 6 = 0$; (*iii*) $2x - 5 = 0$; (*iv*) $2x + 5 = 0$; (*v*) $x^2 - 9 = 0$; (*vi*) $x^2 + 9 = 0$; (*vii*) $x^2 - 10 = 0$; (*viii*) $x^2 + 10 = 0$; (*ix*) $x^2 = 0$. (*b*) If a is a whole number, does the equation $x^2 - a = 0$ have at least one real solution? (*c*) If a is a whole number, does the equation $x^2 + a = 0$ have at least one real solution?

EXERCISE 25 Compute each of these products. (*a*) $3 \cdot 3$. (*b*) $(-3) \cdot (-3)$. (*c*) $\sqrt{3} \cdot \sqrt{3}$. (*d*) $(-\sqrt{3}) \cdot (-\sqrt{3})$. (*e*) $(3i) \cdot (3i)$. (*f*) $(-3i) \cdot (-3i)$. (*g*) $(i\sqrt{3}) \cdot (i\sqrt{3})$. (*h*) $(-i\sqrt{3}) \cdot (-i\sqrt{3})$.

EXERCISE 26 (*a*) Find the solutions, real or imaginary, of each of these equations: (*i*) $x^2 - 9 = 0$; (*ii*) $x^2 + 9 = 0$; (*iii*) $x^2 - 3 = 0$; (*iv*) $x^2 + 3 = 0$. (*b*) Does every equation in Exercises 21 to 24 have at least one solution in the set of complex numbers? [*Note:* The answer to this question is an example of the fundamental theorem of algebra, first proved by the great German mathematician Karl Friedrich Gauss (1777–1855).]

The so-called imaginary and complex numbers have very unfortunate names. They are no more imaginary or complex than are the irrationals which solve equations like $x^2 - 2 = 0$, the negatives which solve equations like $x + 2 = 0$, or for that matter, the number zero itself. Two of the best contemporary writers about mathematics, Edward Kasner and James R. Newman,[4] sum up the situation in their lively style.

Alice was criticizing Humpty Dumpty for the liberties he took with words: "When I use a word," Humpty replied, in a scornful tone, "it means just what I choose it to mean—neither more nor less." "The question is," said Alice, "whether you <u>can</u> make a word mean so many different things." "The question is," said Humpty, "which is to be master, that's all."

[4] *Mathematics and the Imagination.* Copyright © 1940 by Edward Kasner and James R. Newman. Reprinted by permission of Simon and Schuster, Inc.

Those who are troubled (and there are many) by the word "imaginary" as it is used in mathematics, should hearken unto the words of H. Dumpty. At most, of course, it is a small matter. In mathematics familiar words are repeatedly given technical meanings. But as Whitehead has so aptly said, this is confusing only to minor intellects. When a word is precisely defined, and signifies only one thing, there is no more reason to criticize its use than to criticize the use of a proper name. Our Christian names may not suit us, may not suit our friends, but they occasion little misunderstanding. Confusion arises only when the same word packs several meanings and is what Humpty D. calls a "portmanteau."

Semantics, a rather fashionable science nowadays, is devoted to the study of the proper use of words. Yet there is much more need for semantics in other branches of knowledge than in mathematics. Indeed, the larger part of the world's troubles today arise from the fact that some of its more voluble magnificoes are definitely anti-semantic.

An imaginary number is a precise mathematical idea. It forced itself into algebra much in the same way as did the negative numbers. We shall see more clearly how imaginary numbers came into use if we consider the development of their progenitors—the negatives.

Negative numbers appeared as roots of equations as soon as there were equations, or rather, as soon as mathematicians busied themselves with algebra. Every equation of the form $\underline{ax} + \underline{b} = 0$, where $\underline{a}$ and $\underline{b}$ are greater than zero, has a negative root.

The Greeks, for whom geometry was a joy and algebra a necessary evil, rejected negative numbers. Unable to fit them into their geometry, unable to represent them by pictures, the Greeks considered negative numbers no numbers at all. But algebra needed them if it were to grow up. Wiser than the Greeks, wiser than Omar Khayyám, the Chinese and the Hindus recognized negative numbers even before the Christian era. Not as learned in geometry, they had no qualms about numbers of which they could draw no pictures. There is a repetition of that indifference to the desire for concrete representation of abstract ideas in the contemporary theories of mathematical physics, (relativity, the mechanics of quanta, etc.) which, although understandable as symbols on paper, defy diagrams, pictures, or adequate metaphors to explain them in terms of common experience.

Cardan, eminent mathematician of the sixteenth century, gambler, and occasional scoundrel, to whom algebra is vastly indebted, first recognized the true importance of negative roots. But his scientific conscience twitted him to the point of calling them "fictitious." Raphael Bombelli of Bologna carried on from where Cardan left off. Cardan had talked about the square roots of negative numbers, but he failed to understand the concept of imaginaries. In a work published in 1572, Bombelli pointed out that imaginary quantities were essential to the solution of many algebraic equations. He saw that equations of the form $\underline{x}^2 + \underline{a} = 0$, where $\underline{a}$ is any number greater than 0, could not be solved except with the aid of imaginaries. In trying to solve a simple equation $\underline{x}^2 + 1 = 0$, there are two alternatives. Either the equation is meaningless, which is absurd, or $\underline{x}$ is the square root of -1, which is equally absurd. But mathematics thrives on absurdities, and Bombelli helped it along by accepting the second alternative.

Looking Back

1 Write each of these numbers in Egyptian numerals, Roman numerals (two ways when possible), Greek numerals, and Babylonian numerals (after about 300 B.C.). (*a*) 40. (*b*) 100. (*c*) 149. (*d*) 1,360. (*e*) 3,644.

2 Consider the terms simple grouping, cipher, place value, and zero symbol. Which ones apply to each of these systems of numeration. (*a*) Egyptian. (*b*) Roman. (*c*) Early Greek. (*d*) Early Babylonian. (*e*) Early Mayan. (*f*) Babylonian after about 300 B.C. (*g*) Mayan after about the third century B.C. (*h*) Greek after about A.D. 150. (*i*) Hindu-Arabic after A.D. 825.

3 (*a*) What were two disadvantages of the Egyptian system of numeration? (*b*) What were two disadvantages of the Roman system of numeration? (*c*) What were two disadvantages of the Greek system of numeration? (*d*) What were two disadvantages of the early Babylonian system of numeration? (*e*) What are the two major concepts which enable the Hindu-Arabic system of numeration to overcome the disadvantages of the systems above?

***4** Suppose you are a member of a society which writes / for 1, // for 2, and so on up to 9. Your society uses a place-value system in which the base of the place values is 10, and a symbol · for 0. The symbol for 10 is /·, and the symbol for 11 is / / (not to be confused with // for 2). Write these numbers as your society would write them. (*a*) 20. (*b*) 22. (*c*) 200. (*d*) 202. (*e*) 220. (*f*) 222. (*g*) 2,000. (*h*) 2,020.

5 (*a*) State the Pythagorean theorem. (*b*) A Pythagorean, Theodorus of Cyrene (about 425 B.C.), showed that besides $\sqrt{2}$, also $\sqrt{3}$, $\sqrt{5}$, $\sqrt{6}$, $\sqrt{7}$, $\sqrt{8}$, $\sqrt{10}$, $\sqrt{11}$, $\sqrt{12}$, $\sqrt{13}$, $\sqrt{14}$, $\sqrt{15}$, and $\sqrt{17}$ are irrational. B. L. van der Waerden says it has been suggested that Theodorus might have used a figure like this; with the Pythagorean theorem, to derive the square roots (*Science Awakening*, p. 142):

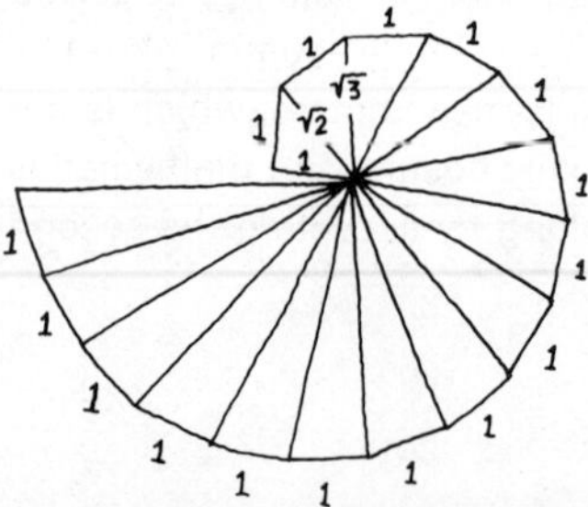

In the first triangle,

$$c^2 = a^2 + b^2 = 1^2 + 1^2 = 1 + 1 = 2 \qquad \text{and} \qquad c = \sqrt{2}$$

In the second triangle, side c of the first triangle becomes side a:

$$c^2 = a^2 + b^2 = 2 + 1 = 3 \qquad \text{and} \qquad c = \sqrt{3}$$

Find the length of each of the remaining "spokes." Observe that this figure does not explain why there is anything special about $\sqrt{4}$, $\sqrt{9}$, and $\sqrt{16}$.

6 (*a*) Use the Babylonian method for approximating square roots to find a decimal approximation for $\sqrt{5}$. Four applications will give $\sqrt{5} \approx 2.236\ldots$, which is correct to three decimal places. (*b*) Use the Babylonian method to approximate $\sqrt{5}$ starting with $x = 2$ as a first approximation. (Since $\sqrt{4} = 2$ and $\sqrt{9} = 3$, $\sqrt{5}$ is between 2 and 3 and closer to 2.) Three applications will give an approximation correct to three decimal places. (*c*) Use the method of the Pythagoreans to show that $\sqrt{5}$ is irrational. You may assume that if $5a^2 = b^2$, then a has a factor of 5.

***7** The formula $(a + b)^2 = a^2 + 2ab + b^2$ was known to the Babylonians. It could have been proved either by them or by the Pythagoreans using the figure to the left below. Only two formulas are needed: the area of a rectangle is the product of its length and width; and the area of a square is the square of its side. Thus the area of the outside square is $(a + b)^2$. (*a*) What is the area of the smaller inside square? (*b*) What is the area of the larger inside square? (*c*) What is the area of each of the inside rectangles? (*d*) What is the sum of the areas of the inside squares and rectangles? Since the sum in part d is the same as the area of the outside square, observe that $(a + b)^2 = a^2 + 2ab = b^2$.

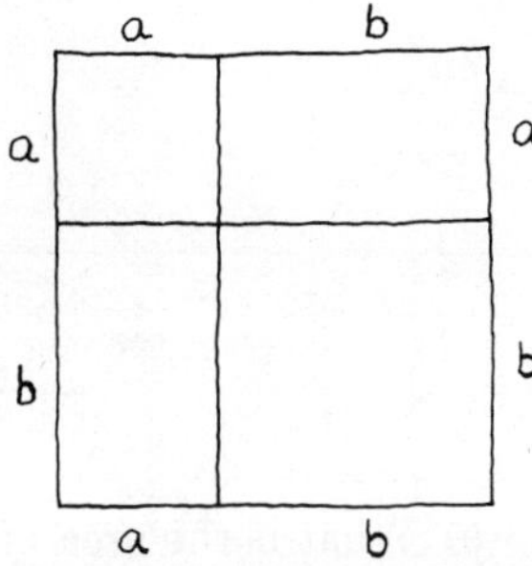

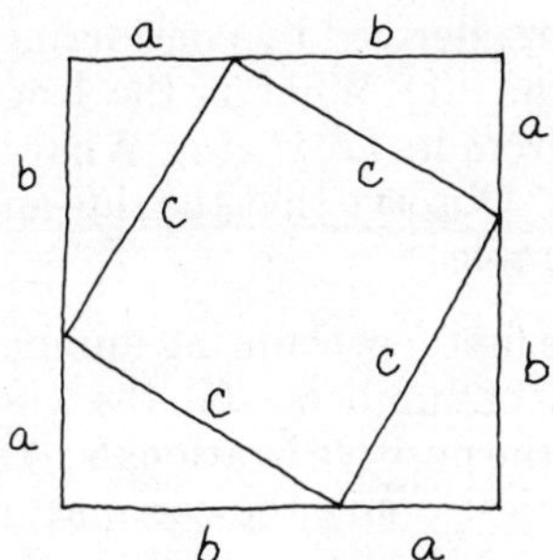

Howard Eves suggests that the Pythagoreans could have proved the Pythagorean theorem similarly, using the figure to the right (*History of Mathematics*, p. 57). (*e*) What is the area of the outside square? (*f*) What is the area of the inside square? (*g*) What is the area of each triangle, if it is known that the area of a right triangle is one-half the product of its two sides (not the hypotenuse)? (*h*) What is the sum of the areas of the inside square and triangles? Set your first and last answers equal to derive the Pythagorean theorem.

*__8__ Sir Thomas Heath suggests that the Pythagoreans could have discovered incommensurable line segments by comparing the side of a square with its diagonal, without using the Pythagorean theorem (*A Manual of Greek Mathematics*, p. 98, by permission of The Clarendon Press, Oxford). Consider this figure, representing one square placed inside another with its corners exactly in the middle of the sides of the outside square. The diagonals of the inside square, the dotted lines, are equal in length to the sides of the outside square. The length of the side of the inside square is unity.

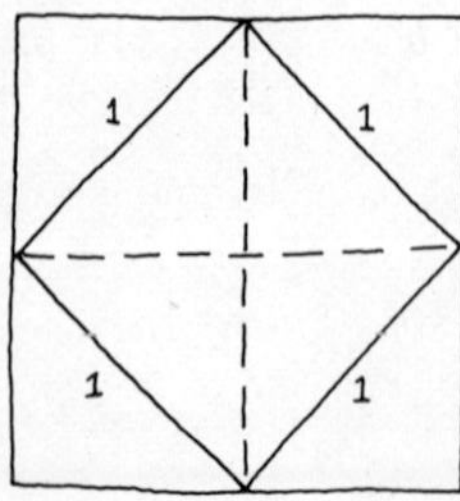

(*a*) What is the area of the inside square? (*b*) What is the area of the outside square, derived by considering how much of it is covered by the inside square? (*c*) What is the length of the side of the outside square, derived from its area? (*d*) What is the length of the diagonal of the inside square? Observe that the side and diagonal of the inside square are incommensurable.

9 (*a*) Write the first few counting numbers. (*b*) Write the definition of each of these sets of numbers: (*i*) the whole numbers; (*ii*) the positive rationals; (*iii*) the positive irrationals; (*iv*) the integers; (*v*) the reals; (*vi*) the imaginaries; (*vii*) the complex numbers.

10 Say whether each of these statements is true or false.
(*a*) Any counting number is also a whole number.
(*b*) Any whole number is also a rational.
(*c*) Any whole number is also an integer.
(*d*) Any integer is also a rational (not necessarily positive).
(*e*) Any rational is also an irrational.
(*f*) Any rational is also a real number.
(*g*) Any irrational is also a real number.
(*h*) Any real number is also an imaginary.
(*i*) Any real number is also a complex number.
(*j*) Any imaginary number is also a complex number.

11 Consider these equations.

$5x - 10 = 0 \qquad 5x + 10 = 0$
$5x - 12 = 0 \qquad 5x + 12 = 0$
$4x^2 - 9 = 0 \qquad 4x^2 + 9 = 0$
$2x^2 - 3 = 0 \qquad 2x^2 + 3 = 0$

(*a*) Which ones have a solution that is a counting number? (*b*) Which ones have a solution that is an integer? (*c*) Which ones have a solution that is a rational number? (*d*) Which ones have a solution that is a real number? (*e*) Which ones have a solution that is a complex number?

***12** A formula for solving quadratic equations, which can be derived from methods that go back to the Babylonians, is the *quadratic formula*: If

$$ax^2 + bx + c = 0$$

then

$$x = \frac{-b \pm \sqrt{b^2 - 4ac}}{2a}$$

The part under the square root, $b^2 - 4ac$, is called the *discriminant*. If a, b, and c are real numbers, with a not zero, and the discriminant is positive or zero, the solutions are also real. If the discriminant is negative, the solutions are either imaginary numbers or complex numbers which are neither real nor imaginary. Say whether the solutions of each of these quadratic equations are real numbers, imaginary numbers, or complex numbers which are neither real nor imaginary. Find the solutions to each equation. (*a*) $x^2 - 1 = 0$. (*b*) $x^2 + 1 = 0$. (*c*) $x^2 + 2x + 1 = 0$. (*d*) $x^2 + 2x - 1 = 0$. (*e*) $x^2 + 2x + 2 = 0$.

Branching Out

1 We have looked at the systems of numeration of the Egyptians and Babylonians but not at the mathematics they developed. The Babylonians were very advanced in algebra, but less so in geometry. Despite their accurate approximation of $\sqrt{2}$, they had a bad approximation for π. The Egyptians had a better approximation for π, and must have had a practical knowledge of geometry to have built the famous pyramids. (The pyramids of Egypt alone offer a fruitful field for study.) More general summaries of Egyptian and Babylonian mathematics are given in histories of mathematics such as Eves' *An Introduction to the History of Mathematics* and Kline's *Mathematical Thought from Ancient to Modern Times.* There are very detailed but far more technical discussions in van der Waerden's *Science Awakening* and Neugebauer's *The Exact Sciences in Antiquity.* The first is a bit more detailed and lighter in style, whereas Neugebauer's book is recognized as a scholarly account of ancient mathematics.

2 The Mayan civilization of Central America was highly developed when the Spanish explorers arrived. The Mayans had two systems of numeration, one used by the priests and one by the common people. Both systems had zeros. The system of the people was a tally-type system; that of the priests was a beautifully artistic set of pictures. The Mayans are also celebrated for their calendar. You will find little about the Mayans in histories of mathematics, for their culture did not influence our mathematics. However, there are many fine books about the Mayan civilization which include discussions of their mathematics.

3 The development of our system of numeration, the Hindu-Arabic system, has two interesting aspects. The evolution of the symbols for the digits 1 to 9, and 0, from their earliest forms to those we use today, is traced in Newman's *The World of Mathematics*, van der Waerden, and other references. There are also entertaining stories about Europe's reluctance from the thirteenth to fifteenth centuries to accept the new notation. Again, van der Waerden is an excellent reference, or you might start with Life Science Library's *Mathematics.*

4 The story of zero is, of course, one of the most interesting in mathematics. Cultures such as the Babylonian came close to it. The later Greeks may have come close or actually used it. The Hindus definitely developed a zero. The etymology of the word "zero" is itself interesting, as is that of the word "cipher." Dantzig's *Number, the Language of Science* has a section which will get you started on the words "zero" and "cipher."

5 The number π is another number with an interesting history. The Egyptians had a good approximation, the Babylonians had a bad one, and the Greeks developed one of the best. If you persevere, you can trace π (although not under that name) all the way back to the Bible. Kasner and Newman have a nontechnical section on π in *Mathematics and the Imagination.*

6 The Pythagoreans, famous for the Pythagorean theorem and $\sqrt{2}$, developed a great deal of geometry and number theory. They were particularly interested in "mystical" properties of numbers and "magic" figures. You might wish to concentrate on their mathematics, their mysticism, or the interplay of the two. You can use standard histories of mathematics, such as Eves' *An Introduction to the History of Mathematics*, to start with.

7 One of the most colorful of the Greek mathematicians, and one of the three greatest mathematicians of all time, was Archimedes. Check various books to find out the background of the remark, "Give me a place to stand on and I will move the earth," the famous "Eureka!" story, Archimedes' consideration of very large numbers in "The Sand Reckoner" (see also Chapter 2, Branching Out), and the story of Archimedes' death in the Roman conquest of Syracuse in 212 B.C.

8 The three greatest mathematicians of all time are considered to be Archimedes, Sir Isaac Newton, and Karl Friedrich Gauss. The first and last have been mentioned, and we shall see Gauss again. We will not study Newton's work in this book, for an understanding of it requires a reasonably strong background in algebra and physics. If you have such a background, you will find that there are books in which you can read about Newton's work on your own. In any event, you can read about his life, and the legends about him. The article in E. T. Bell's *Men of Mathematics* is a good place to start.

"Aleph null miles to...

2

Counting Beyond the Counting Numbers

1
Sets and Subsets

In its most general sense, a *set* is any collection of objects. Of course, the objects may be abstract concepts such as numbers, or they may be physical objects. Whether or not the objects are abstract is immaterial to the concept of a set. What is material is that the objects in the set are described unambiguously. For any object, the description of the objects in the set should tell us whether or not that particular object is in the set. When this is the case, the set is *well-defined.* The objects in the set are the *elements* of the set.

If we wanted to talk about the set of all animals, that seems simple enough. If you have a dog or a cat, it is an element of the set. What if you have a goldfish? For the set to be well-defined, we must decide whether the description means animals in the general sense, as opposed to vegetables or minerals. In the rest of this chapter, we will assume that animals means animals in the general sense. Now your goldfish is in the set.

Would a dinosaur be an element of the set? If we want to include it, we must say that the set includes all animals that ever lived. For the rest of this chapter, we will assume that unless it is given otherwise, the time meant is the moment you have read the description of the set. The dinosaur is not in the set.

Finally, suppose we wish to exclude any animal life that may exist elsewhere—on Mars, for example. Then we must say that the set is the set of animals of the earth. Now the set is well-defined. A good rule of thumb is to ask yourself "what," "when," and "where." If you cannot answer these three questions, then the set is probably not well-defined.

In considering "what" when describing a set, we must watch out for words which might be controversial. For example, suppose we said the set of all *large* animals of the earth. We would probably agree that a blue whale is an element of the set, but not the goldfish. What about a horse? Relative to the blue whale, a horse is pretty small, so perhaps it is not in the set. Relative to the goldfish, however, the horse is large, so perhaps it is in the set. The use of the controversial word "large" makes it impossible to determine exactly which potential elements are in the set and which are not, so the set is not well-defined. If we want to describe a set of a particular type, we must use a word or phrase which will mean the same thing to anyone who uses it. In this instance, we might say the set of animals of the earth measuring 5 feet or more in length, or the set of animals of the earth weighing 1 ton or more.

In discussing the set of all animals of the earth, we mentioned dogs, cats, goldfish, whales, and horses. Each of these kinds of animals itself forms a set in which the elements are also elements of the original set. For example, every element of the set of dogs is also an element of the set of animals of the earth. We say that one set is a *subset* of another if every element of the first set is also an element of the second. Of course, since a subset is itself a set, it too must be well-defined. The set of all large animals of the earth is not a subset of the set of animals of the earth because it is not well-defined. The set of all human beings that ever lived on the earth is a subset of the set of animals that ever lived on the earth. It is potentially well-defined, although anthropologists have not yet established the exact dividing line in the evolution of advanced nonhuman primates to the first human beings. The set of all human beings alive right now is a subset of the set of animals of the earth. Each human being alive now is an animal of the earth, and it is possible to determine, biologically at least if not always by their behavior, just which animals are human beings.

We use capital letters such as A or B to stand for sets. We use lowercase letters such as a or b or x to stand for elements of sets.

DEFINITION Set A is a *subset* of set B if whenever x is an element of A, then x is an element of B.

The great advantage of a definition in symbols such as the one above is that instead of writing out specific sets, we can now talk about a set A where A is *any* set. For example, *for any set A, A is a subset of itself.* In this case, the set B in the definition is also the set A. The definition becomes, "set A is a subset of set A if whenever x is an element of A, then x is an element of A." The condition for the claim to be true is that the part following the word "if" be true; that is, "whenever x is an element of A, then x is an element of A." You cannot fail to agree with this totally redundant statement!

Besides itself, any set A has another subset called the *empty*, or *null*, set. The empty set has no elements at all. An example might be the set of all dinosaurs alive on the earth. It is well-defined, for you will recall that well-defined means we can tell whether any potential element is in or is not in the set. The empty set is certainly well-defined, since for any potential element whatsoever we know it is not in the set. A special symbol $\emptyset$ is used to stand for the empty set. (This symbol for the empty set is the same as a Scandinavian letter and is not, as is often supposed, the Greek letter phi.)

It was claimed that *the empty set is a subset of any set A.* To prove this claim, the definition of a subset becomes, "$\emptyset$ is a subset of A if whenever x is an element of $\emptyset$, then x is an element of A." To deny the if part, "whenever x is an element of $\emptyset$, then x is an element of A," we would have to find an element in $\emptyset$ which is not in A. But to do that, we must find an element in the empty set!

EXERCISE 1 Which ones of these sets are well-defined? For each that is not, say why not. (*a*) The set of all animals native to warm countries. (*b*) The set of all warm-blooded animals of the earth. (*c*) The set of all big states of the United States. (*d*) The set of all states of the United States with total area (land and water) more than 75,000 square miles. (*e*) The set of all counting numbers between 1 and 10. (*f*) The set of all counting numbers between 1 and 10 inclusive; that is, including 1 and 10.

EXERCISE 2 Which ones of these sets are subsets of the set of all animals that ever lived on the earth? For each that is not, say why not. (*a*) The set of all amoebae. (An amoeba is animal as opposed to vegetable or mineral.) (*b*) The set of all earth amoebae. (*c*) The set of all extinct animals of the earth. (*d*) The set of all animals of the United States. (*e*) The set of all living things of the United States. (*f*) The set of all living dinosaurs in the United States.

EXERCISE 3 Which ones of these sets are examples of the empty set? For those which are not, say why not. Are there any for which you cannot tell, even though the set is well-defined? Are there any which may be empty at one time and not at another? (*a*) The set of all counting numbers between 1 and 2 inclusive. (*b*) The set of all counting numbers between 1 and 2 exclusive; that is, not including 1 and 2. (*c*) The set of living things known to exist outside the solar system. (*d*) The set of all living things outside the solar system, whether known or not. (*e*) The set of all earth animals ever to have been outside the limits of the earth's atmosphere. (*f*) The set of all earth animals currently outside the limits of the earth's atmosphere.

2
Counting Subsets

The empty set has exactly one subset, itself. Any nonempty set has at least two subsets, itself and the empty set.

When a set has just one element a, we can use the symbol $\{a\}$ for the set. This symbol means the set containing only the element a. The set $\{a\}$ has exactly two subsets, itself and the empty set, $\{a\}$ and $\emptyset$.

The symbol $\{a, b\}$ means the set containing the two elements a and b. This set has itself and $\emptyset$ as subsets, and also the sets $\{a\}$ and $\{b\}$. It has four subsets, $\{a, b\}$, $\{a\}$, $\{b\}$, and $\emptyset$.

Next we list the subsets of $\{a, b, c\}$. It is best to use some kind of pattern to avoid missing any. In keeping with the pattern we have been using, we will first list the set itself, then all the subsets with two elements, then all the subsets with one element, and finally the empty set. There are eight subsets in all: $\{a, b, c\}$, $\{a, b\}$, $\{a, c\}$, $\{b, c\}$, $\{a\}$, $\{b\}$, $\{c\}$, and $\emptyset$.

Our results so far can be listed as a chart.

Number of elements	0	1	2	3	4	5
Number of subsets	1	2	4	8	?	?

The sequence of numbers 2, 4, 8, 16, 32, and so forth, may be familiar to you. They are the *powers* of 2.

$2 = 2^1$	Two to the first power
$4 = 2^2$	Two to the second power, or two *squared*
$8 = 2^3$	Two to the third power, or two *cubed*
$16 = 2^4$	Two to the fourth power
$32 = 2^5$	Two to the fifth power

The numbers 1, 2, 3, 4, 5, and so forth, which indicate the power of 2, are called *exponents*. The total number of subsets of a set appears to be a power of 2, where the exponent is the number of elements in the set.

For each element of a set, we can put the element in a subset or not put the element in a subset. Thus, in constructing a subset, we have two choices for each element: in or not in. By counting all the possible choices for all the elements of a set, we can count the total number of subsets. For the set $\{a\}$ there are just two choices: to put the element a in the subset or not to put a in the subset. The first choice gives the set itself, $\{a\}$, and the second choice gives $\emptyset$.

a	***Subsets***
In	$\{a\}$
Not in	$\emptyset$

For the set $\{a, b\}$, there are two choices for the first element a: in or not in. For each of these choices, there are also two choices for b. For a in, we can have b either in or not in. For a not in, we can have b either in or not in.

a	*b*	***Subsets***
In	In	$\{a, b\}$
	Not in	$\{a\}$
Not in	In	$\{b\}$
	Not in	$\emptyset$

This diagram is called a *tree diagram* because each column branches out from the preceding one like the branches of a tree. The tree diagram provides another proof that for any set A, the set itself and the empty set are always subsets. The top line of the tree diagram will always have every element in, which gives the set itself. The bottom line of the tree diagram will always have every element not in, which gives the empty set.

a	*b*	*c*	***Subsets***
		In	$\{a, b, c\}$
	In	Not in	$\{a, b\}$
In			
		In	$\{a, c\}$
	Not in	Not in	$\{a\}$
		In	$\{b, c\}$
	In	Not in	$\{b\}$
Not in			
		In	$\{c\}$
	Not in	Not in	$\emptyset$

This is the tree diagram for the subsets of $\{a, b, c\}$. Each tree diagram has exactly as many columns as there are elements in the set. Also, the last column of each tree diagram has exactly as many branches as there are subsets of the set. The set with one element has exactly two subsets. Each time we put an additional element in the set, we double the branches of the tree. This doubling is the same as multiplying the number of branches by 2, which is why we get the powers of 2. Therefore, our chart can be written in powers of 2.

Number of elements	0	1	2	3	4	5
Number of subsets	1	2^1	2^2	2^3	2^4	2^5

If we define $2^0 = 1$ because the empty set has just one subset, then we will have a rule which gives the number of subsets of any set.

RULE FOR SUBSETS If n is the number of elements in any set A, then 2^n is the number of subsets of A.

We do not need to have many elements in a set to have quite large numbers of subsets. You can easily verify that a set with 10 elements has 1,024 subsets, since $2^{10} = 1{,}024$.

EXERCISE 4 (*a*) How many subsets does $\{a, b, c, d\}$ have? (*b*) List all the subsets of $\{a, b, c, d\}$. (*c*) How many subsets does $\{a, b, c, d, e\}$ have? (*d*) List all the subsets of $\{a, b, c, d, e\}$.

EXERCISE 5 Make a copy of the tree diagram for $\{a, b, c\}$ in the text. Leave several spaces between each branch in the last column. (*a*) Extend the tree diagram to $\{a, b, c, d\}$. List the subset represented by each branch. (*b*) Extend the tree diagram to $\{a, b, c, d, e\}$. List the subset represented by each branch.

One hundred is not an exceptionally large number of elements to have in a set. There are probably at least 100 students in your school. The number of ways that these students could be distributed in courses if a course could have all the students in it, or none, or just one, or any other subset of the set of students, is the number of subsets of the set of students. If there are 100 students, then the number of subsets is 2^{100}. How large a number do you think 2^{100} is? about ten thousand? one million? a trillion? a hundred trillion?

We can get a very rough estimate of the size of 2^{100} by observing that $2^{10} = 1{,}024$ and $10^3 = 1{,}000$. Allowing that 1,024 and 1,000 are reasonably close, we write

$2^{10} \approx 10^3$ (where $\approx$ means "approximately equal to")
$(2^{10})^{10} \approx (10^3)^{10}$

If you have learned the rules for exponents in algebra, you know that $(2^{10})^{10} = 2^{100}$ and $(10^3)^{10} = 10^{30}$, so

$2^{100} \approx 10^{30}$

The symbol 10^{30} means 10 to the thirtieth power, which is thirty tens all multiplied together. This number is represented by 1 followed by thirty zeros:

1,000,000,000,000,000,000,000,000,000,000

For comparison, this is

Ten thousand	10,000
One million	1,000,000
A trillion	1,000,000,000,000
A hundred trillion	100,000,000,000,000

We see that 10^{30} is a very large number indeed. It is called a *nonillion.*

If you guessed that 2^{100} was a much smaller number than approximately a nonillion, you are not alone. There are many tales of people who were fooled by very large numbers. George Gamow,[1] physicist and writer of both fanciful and serious books about science, tells one of the most famous.

One victim of overwhelming numbers was King Shirham of India, who, according to an old legend, wanted to reward his grand vizier Sissa Ben Dahir for inventing and presenting to him the game of chess. The desires of the clever vizier seemed very modest. "Majesty," he said kneeling in front of the king, "give me a grain of wheat to put on the first square of this chessboard, and two grains to put on the second square, and four grains to put on the third, and eight grains to put on the fourth. And so, oh King, doubling the number for each succeeding square, give me enough grains to cover all 64 squares of the board."

"You do not ask for much, oh my faithful servant," exclaimed the king, silently enjoying the thought that his liberal proposal of a gift to the inventor of the miraculous game would not cost him much of his treasure. "Your wish

[1] From *One Two Three . . . Infinity.* Copyright 1947, © 1961 by George Gamow. Reprinted by permission of The Viking Press, Inc. and of Macmillan, London and Basingstoke .

will certainly be granted." And he ordered a bag of wheat to be brought to the throne.

But when the counting began, with 1 grain for the first square, 2 for the second, 4 for the third and so forth, the bag was emptied before the twentieth square was accounted for. More bags of wheat were brought before the king, but the number of grains needed for each succeeding square increased so rapidly that it soon became clear that with all the crop of India the king could not fulfill his promise to Sissa Ben. To do so would have required 18,446,744,-073,709,551,615 grains!

That's not so large a number as the total number of atoms in the universe, but it is pretty big anyway. Assuming that a bushel of wheat contains about 5,000,000 grains, one would need some 4000 billion bushels to satisfy the demand of Sissa Ben. Since the world production of wheat averages about 2,000,000,000 bushels a year, the amount requested by the grand vizier was that of the world's wheat production for the period of some two thousand years!

Thus King Shirham found himself deep in debt to his vizier and had either to face the incessant flow of the latter's demands, or to cut his head off. We suspect that he chose the latter alternative.

[Gamow adds in a footnote] The number of wheat grains that the clever vizier had demanded may be represented as follows:
$1 + 2 + 2^2 + 2^3 + 2^4 + \ldots + 2^{62} + 2^{63}$

A very large number which has been given a name is called a *googol.* The googol is 10^{100}, which is 1 followed by one hundred zeros. It looks like this:

10,000,000,000,000,000,000,000,000,000,000,000,000,000,000,000,000,000,-
000,000,000,000,000,000,000,000,000,000,000,000,000,000,000,000,000

The googol is so big that there is no set of physical objects known that amounts to as much as a googol. Gamow says that if the entire universe visible by telescope were filled with grains of sand packed tightly together, the number of grains of sand would be more than a googol. However, the universe is not actually packed tightly with anything; and, in fact, it is estimated that on the average there is just about one atom per cubic meter of space. The total number of atoms in the universe was once estimated at $3(10^{74})$.

On the other hand, an abstract concept like the set of subsets of a set can easily amount to a googol or more. We could think of a set which has so many elements that it has a googol of subsets. How many elements do you think such a set should have?

The calculation is similar to the one we did before. The rule for subsets tells us that if the number of elements in the set is n, then the number of subsets is 2^n. We must determine n so that $2^n \approx 10^{100}$.

A little maneuvering with the exponents gives

$$2^{3n/100} \approx 10^3$$

Recalling that $2^{10} \approx 10^3$, we have

$$2^{3n/100} \approx 2^{10}$$

Since both sides are now powers of 2, the exponents must be approximately equal. Thus,

$$\begin{aligned} 3n/100 &\approx 10 \\ 3n &\approx 1{,}000 \\ n &\approx 333 \end{aligned}$$

If you are looking for $\frac{1}{3}$ in the last step, a fraction of an element would not make sense; and besides, this is only a very rough estimate. A set needs only about 333 elements to have approximately a googol of subsets!

The googol was named by Edward Kasner's nine-year-old nephew. At the same time, he named a much larger number, the *googolplex*. The googolplex is 10^{googol}, or 1 followed by a googol of zeros. Kasner and Newman, in *Mathematics and the Imagination*, p. 23, say there would not be room to write all the zeros in a googolplex, a googol of zeros after the 1, even between here and the furthest star.

EXERCISE 6 (*a*) How many subsets does a set with eleven elements have in terms of a power of 2? (*b*) Compute the exact value of the number in part *a*. (*c*) How many subsets does a set with twenty elements have in terms of a power of 2? (*d*) Find an approximate value for the number in part *c*. (*e*) How many subsets does a set with thirty elements have in terms of a power of 2? (*f*) Find an approximate value for the number in part *e*.

EXERCISE 7 (*a*) Find an approximate value for 2^{200} as a power of 10. (*b*) Write out the power of 10 which is in part *a*.

3
Equivalent Sets

A *one-to-one correspondence* is a useful relationship between two sets. If we have a set of numbers, such as social security numbers, and a set of people such that each person corresponds to exactly one number and each number to exactly one person, then the two sets are in one-to-one correspondence. If we have a set such as birth dates, however, and a set of people where two or more people have the same birth date, then the two sets are not in one-to-one correspondence. One of the most common of one-to-one correspondences is formed by the ordinary act of counting, where the one-to-one correspondence is between the elements of a set and the set of number names "one," "two," "three," and so forth.

Our society has a compulsion to count. The proclivity toward counting is so strong we often think counting must be the most fundamental type of record-keeping available. This notion is far from true. In Chapter 1, we saw how a member of a primitive society might have kept track of things without actually counting. Furthermore, societies which were far from primitive have found it convenient, and sometimes necessary, to use record-keeping methods other than counting.

The following passage from the script of a motion picture, *Ivan the Terrible,*[2] provides an example.

Shot 284. Long shot from above of the army stretched out like three serpents on the plain below.
CHOIR off: 'Misery, sorrow,
The tartar Steppes . . .
Shot 285. A closer shot of the army, from above: two serpentine lines of foot soldiers with halberds, marching slowly forwards, and a third line of cavalry. The CHOIR goes on singing.
Shot 286. Medium shot of a bearded SOLDIER in the foreground, holding a large metal dish filled with small coins. As the armed soldiers march past, each throws in a coin. The sound of the coins landing in the dish mingles with the CHOIR'S song.

[2] A film by Sergei Eisenstein (New York: Simon and Schuster, London: Lorrimer Publishing Limited, 1970), p. 58.

Shot 287. Medium close-up of IVAN and RASMUSSEN contemplating the men. The song fades slightly for a few seconds.
IVAN: The unclaimed coins at the end of the battle will indicate the number of our losses.

The set of soldiers starting out and the set of coins in the bearded soldier's dish are in one-to-one correspondence. Once those returning have reclaimed their coins, the set of those lost and the set of coins remaining in the dish are in one-to-one correspondence. Here it is convenient to use a method other than counting to keep track of the entire army, and actually to count only those lost.

In Africa, one-to-one correspondence has been a highly developed mathematical tool. It should be understood that modern Africa uses the system of numeration developed by the Hindus just as we do, along with most of the rest of the modern world. In African tradition, however, correspondence of sets has been more important than has appeared to be the case in other cultures. One reason might be that African societies have had different priorities from ours of emphasis on written records and thus have not had as strong a need for number names and symbols. A second reason is that differences in languages and methods of recording numbers among African states has made it easier to communicate by other means, including comparison to a common measure such as a corresponding set. Claudia Zaslavsky's *Africa Counts*[3] is possibly the first comprehensive study of African mathematics accessible to the general reader. In the first of the two following passages, she gives another interesting reason for the use of one-to-one correspondence; and in the second, she furnishes an example of its use, which is remarkably similar to the one above.

There is a widespread fear among Africans that the counting of human beings, domestic animals and valuable possessions will lead to their destruction. To circumvent the taboo, counting is done indirectly, by setting up a one-to-one correspondence with some type of counting device.

Taboos on counting are found throughout the world and throughout the ages. The Jews require that ten men, a minyan, be present in the temple for religious services. The word minyan itself means "count." To ascertain that the quota has been filled, a ten-word sentence is recited, establishing a one-

[3] (Boston: Prindle, Weber & Schmidt, Inc., 1973), pp. 52–53, 96.

to-one correspondence between the worshippers in the synagogue and the words of the sentence.

One can recall, too, the Biblical story of King David, who acted against the advice of Joab and the other army captains in ordering a census of his people. After the count had been brought to him, he realized his error, but it was too late. Jehovah visited a terrible pestilence on the people, and many died. Only the offering of a sacrifice succeeded at last in staying his hand.

The superstition about counting is treated at length by A. Seidenberg in his article, "The Ritual Origin of Counting."

Yet a centralized state has to collect taxes, and therefore it needs an accurate count of the number of people living in its domains. This was accomplished in precolonial African states by various forms of indirect census. Herskovits describes the procedures in the highly organized kingdom of Dahomey, where, by the way, the women of the royal court had control over all transactions. On certain market days, a crier would be sent by the chief priest of the powerful spirit of the sacred river, to announce that the spirit threatened disaster to crops and livestock unless the people did as he bade. Every man and woman was to bring to the palace a cowrie shell for each animal he owned, and to deposit the shells in separate piles for sheep, goats, and cattle. First he must touch the animal with the cowrie, to transfer the danger from the animal to the shell. The king contributed an equal number of cowries, and retained a pebble for each shell he contributed. Thus the royal bureaucracy secured an accurate count of livestock, by kind and by village, to be used as a basis for fiscal computations.

Hunters in Dahomey were counted indirectly during the rituals for the gods of the hunt. The heads of all slain animals had to be sent to the palace, thus controlling the amount of game killed. Peasants, ironmongers, weavers, traders—all were assessed a certain part of their products or their sales, thus furnishing a count of the donors and an estimate of the gross national product.

The Feast of the New Yam is the most significant Igbo festival, marking the harvest and the beginning of the New Year. This is the occasion for the annual census. Every grown man brings to the sacred shrine one yam for each member of his household. The number of yams contributed by the village is counted and announced to the assembled populace by the priest. There is great joy if the population has increased.

The colonial invaders had little understanding of and no sympathy with this fear on the part of the African people. No doubt many open struggles were precipitated by the administrators' insistence that they be permitted to count people, houses, or livestock.

. . . .

Counting Beyond the Counting Numbers

In an earlier section we described the indirect census of the population based on the collection of yams or cowrie shells. This system applied also to conscription of the army. During the nineteenth century the kingdom of Dahomey was engaged in frequent warfare with its Yoruba neighbors. Both sides were supplied with European guns. In 1840 Dahomey had a standing army of 12,000, including 5000 of the famous women warriors. The government also had at its command a number of reserve units. Each draftee was represented by a pebble, and every province sent to the capital a box containing the number of pebbles equivalent to the number of soldiers on call.

Equivalence is simply a way of saying that two sets have the same amount of elements without counting the elements.

DEFINITION Two sets A and B are *equivalent* if a one-to-one correspondence exists between A and B.

We often use equivalent sets in the form of a *tally* when we write / for 1, // for 2, up to ~~////~~ for 5, and so on. We might then count the marks, but we often just compare two sets of marks for one-to-one correspondence. For example, suppose that we have two people's records of the votes received by their union against another in a bargaining election:

~~////~~ ~~////~~ ~~////~~ ~~////~~ ~~////~~ ~~////~~ /

~~////~~ ~~////~~ ~~////~~ ~~////~~ ~~////~~ ~~////~~ ///

Since the two sets are clearly not equivalent, if the election was close but their group lost, they should call for a recount without bothering to count the tallies.

Two sets can be shown to be equivalent by drawing double-headed arrows between corresponding elements of a one-to-one correspondence. The set of letters in the English alphabet and the set of counting numbers from 1 to 26 inclusive are equivalent:

a	b	c	d	e	f	g	h	i	j	k	l	m
↕	↕	↕	↕	↕	↕	↕	↕	↕	↕	↕	↕	↕
1	2	3	4	5	6	7	8	9	10	11	12	13

n	o	p	q	r	s	t	u	v	w	x	y	z
↕	↕	↕	↕	↕	↕	↕	↕	↕	↕	↕	↕	↕
14	15	16	17	18	19	20	21	22	23	24	25	26

However, the sets of letters in the word "alphabet" (we list the *a* only once) and the word "number" are not equivalent:

a	l	p	h	b	e	t
↕	↕	↕	↕	↕	↕	
n	u	m	b	e	r	

If a set is too large to list all its elements, three dots, called an *ellipsis*, may be used; but we must be careful that the elements replaced by the ellipsis really do correspond. The set of odd counting numbers from 1 through 99 inclusive and the set of even counting numbers from 2 through 100 inclusive are equivalent since it is clear that each odd number can be paired with the even number which follows it, and each even number can be paired with the odd number which precedes it:

1	3	5	7	...	99
↕	↕	↕	↕		↕
2	4	6	8	...	100

EXERCISE 8 Which ones of the following pairs of sets are equivalent? For each pair, compare tallies of the sets rather than counting the elements. (*a*) The set of colors in the United States flag and the set of letters in the word "set." (*b*) The set of colors in the United States flag and the set of colors in the rainbow. (*c*) The set of colors in the rainbow and the set of days of the week. (*d*) The set of days of the week and the set of months with the letter *r* in their names. (*e*) The set of months with no *r* in their names and the set of months with 30 days. (*f*) The set of colors in the rainbow and the set of states in New England.

EXERCISE 9 Which ones of the following pairs of sets are equivalent? For each pair, list the elements of each set and decide if there is a one-to-one correspondence rather than counting them. (*a*) The set of counting numbers from 1 through 10 inclusive, and the set of even counting numbers from 2 through 10 inclusive. (*b*) The set of even counting numbers from 2 through 10 inclusive and the set of odd counting numbers from 1 through 9 inclusive. (*c*) The set of even counting numbers from 2 through 10 inclusive and the set of counting numbers from 1 through 5 inclusive.

For the next sets, use the ellipsis. (*d*) The set of counting numbers from 1 through 1,000,000 inclusive and the set of even counting numbers from 2 through 1,000,000 inclusive. (*e*) The set of even counting numbers from 2 through 1,000,000 inclusive and the set of odd counting numbers from 1 through 999,999 inclusive. (*f*) The set of even counting numbers from 2 through 1,000,000 inclusive and the set of counting numbers from 1 through 500,000 inclusive.

EXERCISE 10 The sets in each of these pairs of sets are equivalent. Demonstrate a one-to-one correspondence between each pair by drawing double-headed arrows between corresponding elements. (*a*) The set of colors in the United States flag and the set of letters in the word "set." (*b*) The set of letters in the word "four" and the set of letters in the word "five." (*c*) The set of even counting numbers from 2 through 10 inclusive and the set of odd counting numbers from 1 through 9 inclusive.

EXERCISE 11 Can more than one one-to-one correspondence exist between a pair of equivalent sets? Demonstrate a second one-to-one correspondence for each pair of sets in Exercise 10.

4

Cardinal Numbers

For many purposes it is sufficient to know that two sets have the same amount of elements, that is, that they are equivalent. For other purposes, we may need to know what that amount is.

DEFINITION The *cardinal number* of a set is the amount of elements in the set.

The cardinal number of ∅ is 0, because ∅ has no elements. If the amount of elements in a nonempty set is a counting number n, then that cardinal number can be demonstrated by showing a one-to-one correspondence between the set and a set $\{1, 2, 3, \ldots, n\}$. For example, we write

$$\begin{array}{ccccc} a & b & c & \ldots & z \\ \updownarrow & \updownarrow & \updownarrow & & \updownarrow \\ 1 & 2 & 3 & \ldots & 26 \end{array}$$

to demonstrate that the cardinal number of the set of letters of the English alphabet is 26.

We observe that cardinal numbers of sets, if they are any type of number familiar to us, must be whole numbers, that is, zero or a counting number. Further, we observe that for sets with cardinal numbers which are whole numbers, equivalent sets have the same cardinal number.

RULE I FOR CARDINAL NUMBERS Suppose two sets A and B have cardinal numbers which are whole numbers. If A is equivalent to B, then the cardinal number of A is the same as the cardinal number of B; and also, if the cardinal number of A is the same as the cardinal number of B, then A is equivalent to B.

Now suppose a set has a whole number as its cardinal number. Then any subset of that set also has a whole number as its cardinal number and, moreover, the subset will have a cardinal number which is the same whole number or else a smaller whole number. For example, $\{a, b, c\}$ has cardinal number 3. Its subsets are itself, which again has cardinal number 3; $\{a, b\}$, $\{a, c\}$, and $\{b, c\}$, which have cardinal number 2; $\{a\}$, $\{b\}$, and $\{c\}$, which have cardinal number 1; and $\emptyset$, which has cardinal number 0.

RULE II FOR CARDINAL NUMBERS Suppose set B has a cardinal number which is a whole number. If set A is a subset of B, then the cardinal number of A is the same or a smaller whole number than the cardinal number of B.

Finally, there appear to be sets that have cardinal numbers which are not whole numbers. Consider the set of all counting numbers. We may write this set as $\{1, 2, 3, 4, \ldots\}$, where the ellipsis with no number following it means that the numbers keep going forever; that is, there is no last counting number. Since the set is not the empty set, its cardinal number is not zero. All other whole numbers are themselves counting numbers. If one of them, say a counting number N, were the cardinal number of $\{1, 2, 3, 4, \ldots\}$, then this set would be equivalent to $\{1, 2, 3, 4, \ldots N\}$. But then N would be the last counting number. Since we have assumed that there is no last counting number, there can be no such N. In later sections we will prove that the cardinal number of the set of counting numbers is not itself a counting number.

EXERCISE 12 What is the cardinal number of each of these sets? For each set, demonstrate a one-to-one correspondence between the set and a set $\{1, 2, 3, \ldots, n\}$, where n is the cardinal number of the set. (*a*) The set of colors in the United States flag. (*b*) The set of letters in the word "four." (*c*) The set of even counting numbers from 2 through 10 inclusive. (*d*) The set of even counting numbers from 2 through 1,000,000 inclusive.

EXERCISE 13 What is the cardinal number of each set in these pairs of sets? For each pair, determine whether the pair is an example of rule I or rule II for cardinal numbers. (*a*) The set of letters in the word "set" and the set of colors in the United States flag. (*b*) The set of letters in the word "set" and the set of letters in the word "letters." (*c*) The set of letters in the word "number" and the set of states in New England. (*d*) The set of states in New England and the set of states in the United States. (*e*) The set of letters in the word "letters" and the set of letters in the word "trestle."

EXERCISE 14 The *converse* of an "if . . ., then" statement is formed by interchanging the "if" part and the "then" part. The part of rule I for cardinal numbers which follows the phrase "and also" is the converse of the part which precedes it. The converse of rule II for cardinal numbers would say: If the cardinal number of A is the same or a smaller whole number than the cardinal number of B, then A is a subset of B. (*a*) What is the cardinal number of the set of letters in the word "number"? (*b*) What is the cardinal number of the set of letters in the word "numeral"? (*c*) Explain why the converse of rule II for cardinal numbers is not true.

5
Infinite Sets

Although they are very large, the nonillion and googol and googolplex are *finite numbers*. They are cardinal numbers of *finite sets*. It has appeared that the cardinal number of the set of all counting numbers is not itself a counting number because there is no last counting number. We think of sets such as the set of counting numbers as *infinite sets*. Can you think of an example of a set which is infinite but which does, in a sense, have an end?

We must distinguish between "infinite in extent" and "infinite in amount." Since we are talking about infinite sets, we mean the *amount* of elements in the set, or *infinite in amount*.

This line segment is 1 inch long. It is finite in extent. The set of points it contains is, however, infinite in amount. We will be able to prove this fact in this section.

First we return to the set {1, 2, 3, 4, . . .} of counting numbers and separate it into two of its subsets, the set of *evens*, {2, 4, 6, 8, . . .}, and the set of *odds*, {1, 3, 5, 7, . . .}. You will surely agree that the set of counting numbers has been equally divided; that is, the sets of evens and odds are equal in amount. We cannot actually count these sets to show that they have the same amount of elements. Instead, we show that they are equivalent.

Evens	2	4	6	8	. . .	$2n$	. . .
	↕	↕	↕	↕		↕	
Odds	1	3	5	7	. . .	$2n - 1$	. . .

Using $2n$ to represent any even guarantees an even number whenever n is a counting number. Each even $2n$ is paired with the odd $2n - 1$ which precedes it. Each odd $2n - 1$ is paired with the even $2n$ which follows it. In this way a one-to-one correspondence between the two sets is established and the sets are equivalent.

Now let us compare the set of evens and the set of counting numbers. Surely there should be more counting numbers, for we have included only half the counting numbers in the set of evens.

Evens	2	4	6	8	. . .	$2n$	. . .
	↕	↕	↕	↕		↕	
Counting numbers	1	2	3	4	. . .	n	. . .

The sets of evens and of counting numbers are equivalent! Each even $2n$ is paired with the counting number n which is half of it and each counting number n is paired with the even $2n$ which is twice it, and thus a one-to-one correspondence is established. The set of odds is also equivalent to the set of counting numbers.

The set of *whole numbers* is {0, 1, 2, 3, 4, . . .}, the set of counting numbers along with zero. It appears as though the set of whole numbers has one more element than the set of counting numbers. These sets are, in fact, also equivalent.

In each of these cases, the evens and the counting numbers, the odds and the counting numbers, and the counting numbers and the whole numbers, a set is equivalent to another set which is only part of it. Since a *subset* can include the whole set, we need a new definition for a set which is strictly part of another set.

DEFINITION A is a *proper subset* of B if A is a subset of B and also there is at least one element in B which is not an element of A.

The requirement that A be a subset of B guarantees that A will be part of B. The additional requirement that B have at least one element not in A guarantees that A will not be all of B.

EXERCISE 15 Consider these sets: the evens, the odds, the counting numbers, and the whole numbers. (*a*) Which ones, if any, are proper subsets of the set of odds? (*b*) Which ones, if any, are proper subsets of the set of evens? (*c*) Which ones, if any, are proper subsets of the set of counting numbers? (*d*) Which ones, if any, are proper subsets of the set of whole numbers?

EXERCISE 16 (*a*) Show that the set of odds $\{1, 3, 5, 7, \ldots\}$ and the set of counting numbers $\{1, 2, 3, 4, \ldots\}$ are equivalent. (*b*) Explain how the elements of the sets correspond in the one-to-one correspondence that you use to demonstrate that the sets are equivalent.

EXERCISE 17 (*a*) Show that the set of counting numbers $\{1, 2, 3, 4, \ldots\}$ and the set of whole numbers $\{0, 1, 2, 3, 4, \ldots\}$ are equivalent. (*b*) Explain how the elements of the sets correspond in the one-to-one correspondence that you use to demonstrate that the sets are equivalent.

DEFINITION A set is *infinite* if it has a proper subset to which it is equivalent.

The set of counting numbers is infinite because it has a proper subset, for example, the evens, to which it is equivalent.

The definition of an infinite set also applies to the set of points in a line segment. Suppose we draw a 1-inch line segment. Any proper subset of points which is a shorter segment, say the first half-inch, may be used. We redraw the first half-inch above the 1-inch segment. It must be assumed that the sets of points in the original first half-inch and the new half-inch segment are equivalent.

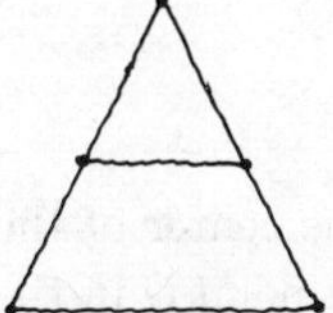

The ends of the 1-inch segment and the new half-inch segment are joined to form a triangle.

We must show that the set of points in the 1-inch segment, which is the base of the triangle, is equivalent to the set of points in the half-inch segment, which lies across the triangle. For any point P in the 1-inch segment, we draw a line connecting P to the top of the triangle. The line will meet the half-inch segment at some point P'. In this way, any point in the 1-inch segment can be paired with a point in the half-inch segment.

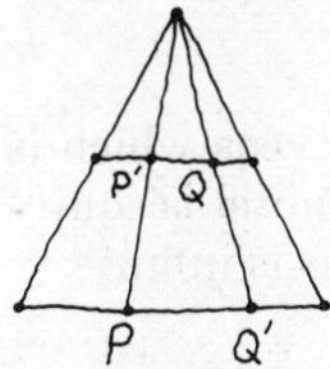

Now start with a point Q in the half-inch segment. Connect it with the top of the triangle and extend the line down to the base. It will meet the base at some point Q'. In this way, any point in the half-inch segment can be paired with a point in the 1-inch segment. Thus we have established a one-to-one correspondence between the sets of points in the two segments, and the sets of points are equivalent.

By varying the size and shape of the triangle, we can show that the sets of points in any two line segments of different lengths are equivalent. Since the set of points in the shorter segment is a proper subset of the set of points in the longer segment, we have also proved that the set of points in any line segment is infinite.

Moreover, the set of points in any line segment is equivalent to the set of points in a whole line! To show this, we take any segment and bend it into a semicircle. We must assume that the amount of points in the segment is not changed by bending it into a semicircle. We delete the end points of the semicircle and draw a line below it.

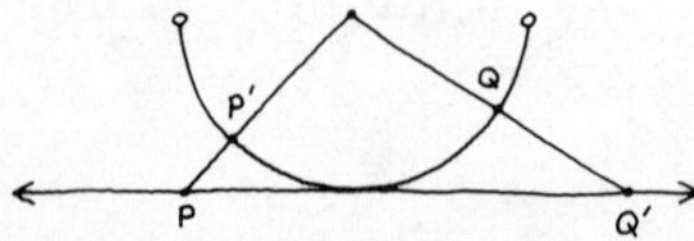

For any point P in the line, a line connecting P with the center of the semicircle will meet the semicircle at a point P'. For any point Q in the semicircle, a line connecting Q with the center of the semicircle and extended to the line below will meet the line at a point Q'. Thus a one-to-one correspondence is established between the set of points in the semicircle and the set of points in the line.

Of course, the set of points in the line segment from which the semicircle was formed may be considered to be a proper subset of the set of points in the line, and therefore the set of points in the line is infinite.

EXERCISE 18 (*a*) Name three proper subsets of the set of whole numbers which are equivalent to the set of whole numbers. (*b*) Explain why the set of whole numbers is infinite.

EXERCISE 19 (*a*) Find a proper subset of the set of evens which is equivalent to the set of evens. (*b*) Demonstrate that the subset is equivalent to the set of evens. (*c*) Explain why the set of evens is infinite.

EXERCISE 20 (*a*) Show that the set of points in a 2-inch line segment and the set of points in a 1-inch line segment are equivalent. (*b*) Show that the set of points in a 3-inch line segment and the set of points in a 1-inch line segment are equivalent.

EXERCISE 21 Find a way to show that the set of points in this square and the set of points in the circumscribed circle are equivalent.

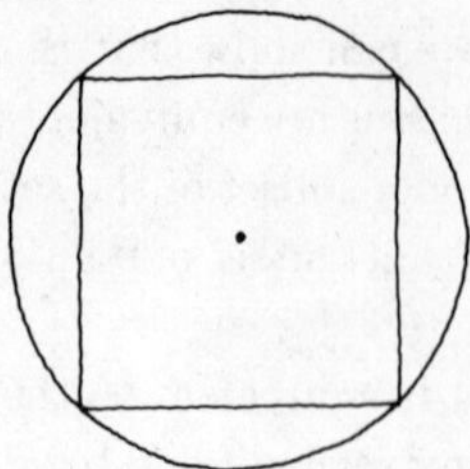

(*Hint:* Draw lines from the common center of the square and circle across the square to the circle.) Since the square is made up of line segments, observe that there is no change made in the amount of points in a line segment by bending it into a semicircle or circle.

6

Countable Sets

We have seen that the sets of evens and odds are each equivalent to the set of counting numbers. Also, the set of whole numbers is equivalent to the set of counting numbers.

DEFINITION An infinite set is *countable* if it is equivalent to the set of counting numbers.

The sets of evens and odds, the set of counting numbers itself, and the set of whole numbers are all countable sets.

EXERCISE 22 (*a*) Demonstrate a one-to-one correspondence which shows that the set of numbers of the form $a/1$, where a is a counting number, is countable. (*b*) Demonstrate a one-to-one correspondence which shows that the set of numbers of the form $1/a$, where a is a counting number, is countable.

The set of *integers* contains the counting numbers, their negatives, and zero. Do you think the set of integers is countable?

If we arrange the set of integers so that we start with zero and then write each counting number followed by its negative, we can show a one-to-one correspondence between the set of integers and the set of counting numbers.

$$\begin{array}{cccccccccccccc} 0 & 1 & -1 & 2 & -2 & 3 & -3 & 4 & -4 & 5 & -5 & \dots & n & -n & \dots \\ \updownarrow & \updownarrow & \updownarrow & \updownarrow & \updownarrow & \updownarrow & \updownarrow & \updownarrow & \updownarrow & \updownarrow & \updownarrow & & \updownarrow & \updownarrow & \\ 1 & 2 & 3 & 4 & 5 & 6 & 7 & 8 & 9 & 10 & 11 & \dots & 2n & 2n+1 & \dots \end{array}$$

The integer 0 is paired with the counting number 1. Each positive integer n is paired with an even counting number $2n$. Each negative integer $-n$ is paired with an odd counting number $2n + 1$. This time we use $2n + 1$, not $2n - 1$, because the counting number 1 is already paired with 0. In this way, we have shown that the set of integers is equivalent to the set of counting numbers and thus is countable.

The set of *rationals* is the set of all numbers which can be written in the form a/b, where a and b are integers and b is not 0. All integers are rational numbers because any integer a can be written in the form $a/1$. There are also a great many rationals which are not integers.

To show that a set is countable, we must show a way to list its elements so that there is a specific first element, a specific second element, a specific third, fourth, and so on. The first element is paired with the counting number 1, the second with 2, the third with 3, and so forth, so that a one-to-one correspondence with the counting numbers is established. Try to find some pattern in which you can list the rationals with a specific first, second, third, fourth, and so on. Do you think the set of rationals is countable?

We begin with only the positive rationals and form a two-dimensional array:

1/1	2/1	3/1	4/1	5/1	6/1	7/1	8/1	...
1/2	2/2	3/2	4/2	5/2	6/2	7/2	8/2	...
1/3	2/3	3/3	4/3	5/3	6/3	7/3	8/3	...
1/4	2/4	3/4	4/4	5/4	6/4	7/4	8/4	...
1/5	2/5	3/5	4/5	5/5	6/5	7/5	8/5	...
1/6	2/6	3/6	4/6	5/6	6/6	7/6	8/6	...
1/7	2/7	3/7	4/7	5/7	6/7	7/7	8/7	...
1/8	2/8	3/8	4/8	5/8	6/8	7/8	8/8	...
⋮	⋮	⋮	⋮	⋮	⋮	⋮	⋮	

The array continues indefinitely to the right and down. It is possible to locate any positive rational in the array if you go across and down far enough. For example, $\frac{21}{101}$ is in the twenty-first column and the one-hundred-and-first row.

Our problem now is to find some way to write this array as a list. We cannot start out merely listing the numbers across the first row, for the first row never ends, and we would never get to the second row. We would certainly never get to the one-hundred-and-first row, so $\frac{21}{101}$ would not be in the list. Similarly, we cannot start down the first column, for we would never get to the second. However, the diagonals going up from left to right do end. We can list all the positive rationals by going along these diagonals, following the directions of the arrows.

1/1	2/1	3/1	4/1	5/1	6/1	7/1	8/1	...
1/2	2/2	3/2	4/2	5/2	6/2	7/2	8/2	...
1/3	2/3	3/3	4/3	5/3	6/3	7/3	8/3	...
1/4	2/4	3/4	4/4	5/4	6/4	7/4	8/4	...
1/5	2/5	3/5	4/5	5/5	6/5	7/5	8/5	...
1/6	2/6	3/6	4/6	5/6	6/6	7/6	8/6	...
1/7	2/7	3/7	4/7	5/7	6/7	7/7	8/7	...
1/8	2/8	3/8	4/8	5/8	6/8	7/8	8/8	...
⋮	⋮	⋮	⋮	⋮	⋮	⋮	⋮	

We skip numbers which are equal to a preceding number, such as $\frac{2}{2} = \frac{1}{1}$ and $\frac{2}{4} = \frac{1}{2}$. The resulting list is $\frac{1}{1}$, $\frac{1}{2}$, $\frac{2}{1}$, $\frac{1}{3}$, $\frac{3}{1}$, $\frac{1}{4}$, $\frac{2}{3}$, $\frac{3}{2}$, $\frac{4}{1}$, $\frac{1}{5}$, $\frac{5}{1}$, $\frac{1}{6}$, $\frac{2}{5}$, $\frac{3}{4}$, $\frac{4}{3}$, $\frac{5}{2}$, $\frac{6}{1}$, $\frac{1}{7}$, ... This list shows that the set of positive rationals is countable.

EXERCISE 23 Show how to write the entire set of rationals, positive, negative, and zero, as a list. (*Hint:* Start with the list of the positive rationals above. Find a way to include zero, written in the form $\frac{0}{1}$. Then find a way to include the negative of each positive in the list.)

7

Uncountable Sets

Every rational number can be written as a terminating or repeating decimal. An example of a terminating decimal is $\frac{3}{4} = 0.75$. We can express $\frac{3}{4}$ as a nonterminating decimal by writing $\frac{3}{4} = 0.75000\ldots$ or $\frac{3}{4} = 0.74999\ldots$. By a repeating decimal, we mean a nonterminating decimal where the digits repeat in some pattern, such as $\frac{1}{3} = 0.333\ldots$, or $\frac{200}{111} = 1.801801\ldots$. A decimal such as $\frac{3}{4} = 0.75000\ldots$ is considered to be repeating, since it repeats after the first two digits. Thus any rational number can be written as a repeating decimal.

There are also irrational numbers, such as $\sqrt{2}$, which cannot be written in the form a/b. Written as decimals, irrational numbers are nonterminating, but they do not repeat in a pattern. Although irrational numbers may be approximated to as many decimal places as we wish, they cannot be written exactly as decimals since they are nonterminating and also nonrepeating. For example, $\sqrt{2} = 1.41421318054\ldots$ and $\pi = 3.14159265358\ldots$ to eleven decimal places.

The set of *real numbers* is the set of all rationals and all irrationals. We can think of the set of real numbers as the set of all nonterminating decimals, whether repeating or not. Do you think the set of real numbers is countable?

By now you may be guessing that every infinite set is countable. This is far from the truth, as we shall see. In fact, we will prove that the set of real numbers just between 0.000 . . . and 0.999 . . . is not countable. The proof is a famous one called Cantor's diagonal proof. You should take care not to confuse it with the proof in the preceding section, which also uses diagonals of an array, and was also invented by Georg Cantor (1845–1918). Otherwise, Cantor's diagonal proof is very different. It uses the *main diagonal* of the array, which goes downward from left to right starting at the upper left corner.

The proof is an indirect proof. We begin by supposing that the set of real numbers between 0.000 . . . and 0.999 . . . is indeed countable. If that were the case, we could list them as we did the rationals. Since each real number is represented by a nonterminating decimal which goes across the page, we will make the list of decimals going down the page.

$$
\begin{array}{l}
0.a_{11}a_{12}a_{13}a_{14}a_{15}a_{16}a_{17}a_{18}\ldots \\
0.a_{21}a_{22}a_{23}a_{24}a_{25}a_{26}a_{27}a_{28}\ldots \\
0.a_{31}a_{32}a_{33}a_{34}a_{35}a_{36}a_{37}a_{38}\ldots \\
0.a_{41}a_{42}a_{43}a_{44}a_{45}a_{46}a_{47}a_{48}\ldots \\
0.a_{51}a_{52}a_{53}a_{54}a_{55}a_{56}a_{57}a_{58}\ldots \\
0.a_{61}a_{62}a_{63}a_{64}a_{65}a_{66}a_{67}a_{68}\ldots \\
0.a_{71}a_{72}a_{73}a_{74}a_{75}a_{76}a_{77}a_{78}\ldots \\
0.a_{81}a_{82}a_{83}a_{84}a_{85}a_{86}a_{87}a_{88}\ldots \\
\quad\vdots\;\;\vdots\;\;\vdots\;\;\vdots\;\;\vdots\;\;\vdots\;\;\vdots\;\;\vdots
\end{array}
$$

The letters a stand for the digits in each decimal. The numbers to the lower right of each a are called *subscripts.* The first number in each subscript tells us which decimal of the list the digit is in. The second number in each subscript tells us the decimal place of the digit. For example, a_{23} is in the second decimal on the list, and in the third place of that decimal. The digits in the main diagonal are a_{11}, a_{22}, a_{33}, and so forth.

If there were any possible way to replace the a's on our list with actual digits without missing any decimal between 0.000 . . . and 0.999 . . ., then the set would be countable. We must show that there will always be a decimal missing from the list.

Suppose that a_{11} is any number except 2. Then let $\bar{a}_{11} = 2$. If $a_{11} = 2$, then let $\bar{a}_{11} = 3$. Then, in any case, $a_{11} \neq \bar{a}_{11}$. Simlarly, let $\bar{a}_{22} = 3$ if $a_{22} = 2$, and $\bar{a}_{22} = 2$ otherwise; let $\bar{a}_{33} = 3$ if $a_{33} = 2$, and $\bar{a}_{33} = 2$ otherwise; and so forth, for every digit a_{nn} on the main diagonal of the array. Then the decimal $0.\bar{a}_{11}\bar{a}_{22}\bar{a}_{33}\bar{a}_{44}\bar{a}_{55}\bar{a}_{66}\bar{a}_{77}\bar{a}_{88} \ldots$ consists entirely of 2s and 3s. It is in the set of decimals from 0.000 . . . to 0.999 . . . . However, it is not on our list.

The decimal $0.\bar{a}_{11}\bar{a}_{22}\bar{a}_{33}\bar{a}_{44}\bar{a}_{55}\bar{a}_{66}\bar{a}_{77}\bar{a}_{88} \ldots$ could not be the first decimal on the list, for its first digit, $\bar{a}_{11}$, is different from a_{11}. It could not be the second, because its second digit, $\bar{a}_{22}$, is different from a_{22}. It could not be the third because $\bar{a}_{33}$ is different from a_{33}, and so forth. No actual list is possible because $0.\bar{a}_{11}\bar{a}_{22}\bar{a}_{33}\bar{a}_{44}\bar{a}_{55}\bar{a}_{66}\bar{a}_{77}\bar{a}_{88} \ldots$ will always be missing, and so the set of decimals between 0.000 . . . and 0.999 . . . is uncountable.

EXERCISE 24 (*a*) Describe a one-to-one correspondence between the set of reals from 0.000 . . . through 0.999 . . . and the set of reals from 1.000 . . . through 1.999 . . . . (*b*) Since the set of reals from 0.000 . . . through 0.999 . . . is uncountable, what do you conclude about the set of reals from 1.000 . . . through 1.999 . . . ? (*c*) What do you conclude about the set of reals from 0.000 . . . through 1.999 . . . ? (*d*) What do you conclude about the set of all reals?

EXERCISE 25 A *real-number line* is formed by exhibiting a one-to-one correspondence between the set of real numbers and the set of points of a line. We choose a point to correspond to 0, and a point any distance to its right, say 1 inch, for 1. Then the point 2 inches to the right corresponds to 2, $\frac{1}{2}$ inch to the right to $\frac{1}{2}$, 1 inch to the left to -1, and so forth. We place irrationals such as $\sqrt{2}$ and π by using decimal approximations.

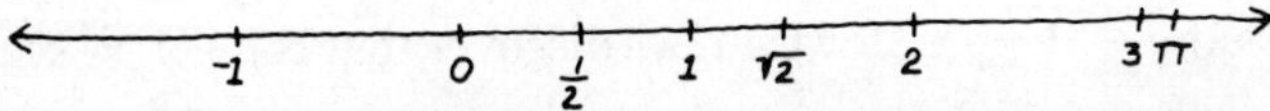

(*a*) Explain why the set of points in a 1-inch line segment is uncountable. (*b*) Explain why the set of points in a 2-inch line segment is uncountable. (*c*) Explain why the set of points in the entire line is uncountable.

8

Transfinite Numbers

The cardinal numbers of finite sets are counting numbers. Rule I for cardinal numbers says that equivalent sets have the same cardinal number. When we discover a class of equivalent finite sets, we discover a counting number. For example, all sets equivalent to the set of fingers on our two hands have the cardinal number 10.

The countable sets are a class of infinite sets equivalent to the set of counting numbers. Despite appearances, the number of elements in any countable set must be the same as the number of elements in the set of counting numbers; for example, the amounts of evens, odds, counting numbers, whole numbers, integers, and rationals are the same. We have discovered a new cardinal number which is the number of elements in the set of counting numbers, and also in any other countable set. Since each counting number is the cardinal number of a finite set, this new cardinal number is not a counting number.

DEFINITION A *transfinite number* is the cardinal number of an infinite set.

Georg Cantor named the cardinal number of the countable sets $\aleph_0$ (aleph-null). *Aleph* is the first letter of the Hebrew alphabet. Cantor proved that there is no infinite set smaller than the countable sets, so $\aleph_0$ is the smallest transfinite number. Cantor meant the subscript 0 (null) to indicate that $\aleph_0$ is the smallest transfinite number.

For Cantor's new number, we discover a new arithmetic, with rules very different from ordinary arithmetic. There is no finite number a for which it is true that $a + 1 = a$. However, we discover that $\aleph_0 + 1 = \aleph_0$. To justify this fact, consider the sets of counting numbers and whole numbers. The set of whole numbers is just the set of counting numbers with one more element, zero. Since the sets of counting numbers and whole numbers are equivalent, they have the same cardinal number. That cardinal number is the transfinite number $\aleph_0$.

Further, we find that $\aleph_0 + \aleph_0 = \aleph_0$. Consider the sets of evens and odds, each of which is equivalent to the set of counting numbers. Together they make the set of counting numbers. All three sets have the same cardinal number, the transfinite number $\aleph_0$.

Finally, $\aleph_0 \cdot \aleph_0 = \aleph_0$. We go back to the pattern we used to show that the set of positive rationals is equivalent to the set of counting numbers. The set of numbers $\frac{1}{1}, \frac{2}{1}, \frac{3}{1}, \ldots$ across the top is equivalent to the set of counting numbers, and so is the set of numbers $\frac{1}{1}, \frac{1}{2}, \frac{1}{3}, \ldots$ down the edge. Using the familiar idea of width times height, the product of the element across and down is the area, or all the elements in the array. However, this set is the set of positive rationals. All these sets are equivalent to the set of counting numbers and have cardinal number $\aleph_0$.

By now you may be thinking that no matter what we do we will end up with $\aleph_0$. Just as it turned out that there exist sets which are uncountable, there exist cardinal numbers other than $\aleph_0$. The set of reals is not countable, so it must have a different cardinal number. The transfinite number which is the cardinal number of the set of reals is called c, the *continuum*. Its name comes from the fact that the set of reals is equivalent to the set of points in a line, so it is "continuous," with no gaps. The cardinal number of the set of points in a line is c. Also, the set of points in any line segment is equivalent to the set of points in a line, so the cardinal number of the set of points in any line segment is c.

Cantor proved that $2^{\aleph_0} = c$. Although the proof is too complicated to give here, it is not hard at least to believe that $2^{\aleph_0}$ is greater than $\aleph_0$. We know that for any finite set, if the set has n elements, then it has 2^n subsets. We also know that as n gets large, 2^n is a great deal larger than n. If n is 100, we recall that 2^n is approximately 10^{30}. If n is only about 333, 2^n is approximately a googol. It would seem, then, that for transfinite numbers 2^n would be so very much larger than n, that $2^{\aleph_0}$ would be a larger transfinite number than $\aleph_0$.

EXERCISE 26 (*a*) Given $\aleph_0 + 1 = \aleph_0$, show that $\aleph_0 + 2 = \aleph_0$. (*b*) Given $\aleph_0 + 2 = \aleph_0$, show that $\aleph_0 + 3 = \aleph_0$. (*c*) What do you conclude about $\aleph_0 + n$ for any counting number n?

David Eugene Smith Collection, Columbia University Library

GEORG CANTOR

Georg Cantor is often credited with the creation of a whole new field of mathematics. Now popularly called "foundations of mathematics," it embraces Cantor's theories of sets and transfinite numbers, the study of paradoxes of sets and logic, and two major schools of thought of mathematical logic. E.T. Bell says, Cantor's " 'positive theory of the infinite' precipitated, in our own generation, the fiercest frog-mouse battle (as Einstein once called it) in history over the validity of traditional mathematical reasoning."[1]

[1] E. T. Bell, *Men of Mathematics*, Simon and Schuster, New York, 1937, p. 556.

David Eugene Smith Collection, Columbia University Library

LEOPOLD KRONECKER

The other half of the frog-mouse battle was Leopold Kronecker. Kronecker allowed only the use of constructive methods, in which a mathematical object must be constructed in a finite number of steps. Indirect proofs were not allowed in Kronecker's system. Much of Cantor's work in the theories of sets and transfinite numbers, on the other hand, is based on nonconstructive methods. Kronecker opposed Cantor's theories, considering them unsound and mathematically dangerous. But Cantor's work withstood these vigorous attacks and now permeates mathematics. It was the sensitive Cantor who suffered from them.

Kronecker, a first-rate mathematician in his own right, was a professor at the University of Berlin. Cantor, more than twenty years younger, had studied at Berlin. Yet he taught at the University of Halle, a less distinguished post, and he longed for a professorship at Berlin. Whether or not it was due to Kronecker's influence no one knows, but the professorship never came. In 1884 Cantor began to suffer from a series of breakdowns. Although he continued to write brilliant papers between illnesses, Cantor died in a mental hospital, still in Halle, in 1918.

EXERCISE 27 (*a*) Use the sets of whole numbers and counting numbers to justify that $\aleph_0 - 1 = \aleph_0$. (*b*) What do you conclude about $\aleph_0 - 2$? (*c*) What do you conclude about $\aleph_0 - n$ for any counting number n? (*d*) Find sets which justify that $\aleph_0 - \aleph_0 = 0$. (*e*) Find sets which justify that $\aleph_0 - \aleph_0 = \aleph_0$. Observe that we cannot define subtraction of transfinite numbers uniquely simply by considering examples of countable sets.

EXERCISE 28 If c is the continuum, and recalling the arithmetic of $\aleph_0$, guess: (*a*) What is $c + 1$? (*b*) What is $c + \aleph_0$? (*c*) What is $c + c$? (*d*) What is $c \cdot c$? (*e*) Find sets which justify that $c + c$ is the result you claim. (*f*) What can you guess about 2^c?

Cantor had intended to name the second transfinite number $\aleph_1$, the third $\aleph_2$, and so on, but he was unable to prove that c is the next transfinite number after $\aleph_0$. In much of his work, however, he assumed this was the case. This assumption is called the *continuum hypothesis*. For almost a century no one could prove the continuum hypothesis true, but neither could anyone prove that it was false.

A mathematical system is *consistent* if there is no statement in the system which can be both proved and disproved. For many years Cantor's followers tried to prove his system of transfinite numbers consistent, in order to justify the logic on which the system is based. In 1931, Kurt Gödel (1906–), an Austrian mathematician and logician, showed that it is not possible to establish the internal consistency of a system of the size of even ordinary arithmetic. We can only say one such system is consistent if another is.

In 1963, an American mathematician, Paul Cohen (1934–), showed that if the theory of sets is itself consistent, then consistent systems can be derived from both the continuum hypothesis and its denial. That is, we can assume that the continuum hypothesis is true, that c is indeed the next transfinite number after $\aleph_0$, and get a consistent system. Alternatively, we can assume the continuum hypothesis is false; that there are other transfinite numbers between $\aleph_0$ and c (although we do not know what they are) and get a different but also consistent system.

EXERCISE 29 Show that the cardinal number of the set of irrational numbers must be greater than $\aleph_0$ by this indirect proof. Suppose the cardinal number of the set of irrational numbers is $\aleph_0$. (*a*) What is the cardinal number of the set of rationals? (*b*) What set of numbers consists of the rationals and irrationals together? (*c*) What would the cardinal number of the set in part *b* be under the supposition above? (*d*) Why is the conclusion in part *c* impossible? (*e*) Explain why the cardinal number of the set of irrationals must be *c* if the continuum hypothesis is assumed.

9

Russell's Paradox

A paradox is something which appears to contradict itself. When the theories of sets and transfinite numbers were new, and not much was known about them, questions were asked which could not be answered within these systems. Such questions led to paradoxes.

So far we have seen two transfinite numbers: $\aleph_0$, the cardinal number of countable sets; and *c*, the cardinal number of the continuum. But there must be a larger transfinite number than *c*, because the set of all subsets of a set with cardinal number *c* would have a larger cardinal number. Furthermore, the set of all subsets of a set with that larger cardinal number would have an even larger cardinal number, and so on without end. Thus the amount of transfinite numbers is certainly greater than two, and indeed is not finite.

We might ask, as Cantor did in 1899, what is the cardinal number of the set of all transfinite numbers? The question leads to one of the first paradoxes of set theory, called *Cantor*'s *paradox*. Another way of stating this paradox is to ask what is the cardinal number of the set of all sets? This set has as its elements all possible sets, including itself. It cannot be finite, for even the set of subsets of the set of counting numbers is not finite. If we call it infinite and call its cardinal number a transfinite number, there can be no larger set than the set of all sets and its cardinal number will be the largest transfinite number. But there is no largest transfinite number, and there is the paradox!

Meanwhile, the eminent British philosopher and mathematician Bertrand Russell (1872–1970) found a similar paradox of set theory called *Russell's paradox*. Russell was very enthusiastic about Cantor's work. He realized, however, that limitations on sets and the underlying logic of set theory would be necessary. Paradoxes, and the limiting principles to avoid them, are an important part of the *Principia Mathematica*, a monumental attempt to systemize all mathematics by recreating the foundations of mathematics from a few fundamental principles. Russell and Alfred North Whitehead (1861–1947) published the *Principia* in three volumes from 1910 to 1913. Russell's paradox has been given in many versions. The form which follows is a somewhat different type of paradox from the original. It was given by Russell in 1918:

In a certain village there is a barber who makes it a rule that he shaves all those and only those people who do not shave themselves. Who shaves the barber? If he shaves himself, he breaks his rule by shaving someone who shaves himself. If he does not shave himself, then there is someone in the village whom he does not shave but who does not shave himself.

Russell's, and also Cantor's, paradox results from a situation which Russell and Whitehead call the *vicious-circle principle*. The vicious-circle principle says that a set must not contain any element which can only be described by using the description of the set itself. In Russell's paradox, the barber is such an element. His role can only be described by using the description of the set itself, namely, all the people who are shaved.

This is another version of the same form of Russell's paradox:

A library has a collection of bibliographies. Among the bibliographies are some which list bibliographies. And among these is one which lists all those bibliographies which do not list themselves. Does this bibliography list itself?

The third version, which follows, is the original Russell's paradox, stated in his earlier book of 1903. It is worthy of the stronger name *antinomy*, an actual as opposed to a seeming contradiction. Like Cantor's paradox, which is also an antinomy, it involves a set which is an element of itself.

Two classes of sets are called M and N. Class M contains all those sets which have themselves as an element. Class N contains all those sets which are not elements of themselves. Is N an element of itself?

We can make another example of the vicious-circle principle using the set of all sets. This set is a set, and it contains all sets as elements, so

Bertrand Russell envisioned all mathematics as an extension of logic. This statement does not say simply that mathematics is logical or uses logic, as we know it is and does, but that mathematics is logic. From this idea grew the logistic school in mathematics.

The goal of logicism was to start from the principles of symbolic logic (Chapter 3), and from them to derive all of mathematics. Russell first presented these ideas in 1903 in the Principles of Mathematics. This book was followed a few years later by the three volume Principia Mathematica, which he wrote with Alfred North Whitehead. The Principia, an extraordinarily detailed attempt to carry out the logistic concept of mathematics, made a profound impression on the scientific world.

In spite of the fact that many scholars consider logicism no longer to be significant as a school of mathematical thought, and even though the work of developing all mathematics from logic was never completed, the Principia probably did more than any other work since Cantor's to advance the branch of mathematics called foundations.

it must contain itself as an element. But then it has an element, itself, which can only be described by using the description of the set itself.

Finally, we return to the cardinal number of the set of all transfinite numbers. The cardinal number of the set of counting numbers is not itself a counting number, but the transfinite number $\aleph_0$. Similarly, the cardinal number of the set of transfinite numbers must not itself be a transfinite number, for then we would also have the cardinal number of the set of all sets, and the set of all sets falls into the vicious-circle principle. The number we seek must be some further type of number, one which lies even beyond the transfinite numbers.

EXERCISE 30 Consider the bibliography version of Russell's paradox. (*a*) Explain what happens if the bibliography described in this version does list itself. (*b*) Explain what happens if the bibliography described in this version does not list itself. Consider the version of Russell's paradox concerning the class of sets M and N. (*c*) Explain what happens if the class N described in this version is an element of itself. (*d*) Explain what happens if the class N described in this version is not an element of itself.

EXERCISE 31 Suppose that A is the set of all cardinal numbers of finite sets (including $\emptyset$) and B is the set of all cardinal numbers which are not cardinal numbers of finite sets. (*a*) What set of numbers is described by set A? (*b*) Explain why the cardinal number of A is an element of B. (*c*) Explain why the cardinal number of B is an element of B. (*d*) Explain why B is not the same set as the set of all cardinal numbers of infinite sets. (*e*) Do "not finite" and "infinite" mean the same thing?

Looking Back

1 Two sets are equal if they have exactly the same elements. A repeated element counts only once. For example, $\{a, a, b, b, c, c\} = \{a, b, c\}$. Consider the sets:

$\{e, i, t, h, e, r\}$, $\{t, h, e, i, r\}$, $\{t, h, e, r, e\}$, $\{e, a, r, t, h\}$, $\{t, r, e, e\}$

(*a*) Which ones are equal? (*b*) Which ones are equivalent? (*c*) Which ones are a subset of another? (*d*) Which ones are a proper subset of another?

2 (*a*) Write the definition of equivalent sets. (*b*) Show that each of the following sets is equivalent to the set of counting numbers: (*i*) the set of evens; (*ii*) the set of odds; (*iii*) the set of whole numbers; (*iv*) the set of integers; (*v*) the set of positive rationals.

3 The square of a counting number is $n^2 = n \cdot n$. Show that the set of all squares of counting numbers is equivalent to the set of counting numbers.

4 Show that the sets of points in two line segments of different lengths are equivalent.

5 Show that the sets of points in two circles with the same centers but different radii are equivalent. The "circle" consists only of the points on the circumference.

6 (*a*) Write the definition of an infinite set. (*b*) Explain why each of the following sets is infinite: (*i*) the set of counting numbers; (*ii*) the set of whole numbers; (*iii*) the set of integers; (*iv*) the set of rationals; (*v*) the set of real numbers; (*vi*) the set of points in a line segment; (*vii*) the set of points in a line.

7 (*a*) Write the definition of countable for an infinite set. (*b*) Which ones of the sets in Exercise 6 are countable? (*c*) What is the cardinal number of each of the sets in Exercise 6?

8 (*a*) Write the definition of a transfinite number. (*b*) Name two transfinite numbers. (*c*) Are there more than two transfinite numbers?

9 (*a*) State the continuum hypothesis. (*b*) If the continuum hypothesis is assumed, what is the cardinal number of the set of irrationals? (*c*) What do you know about the cardinal number of the set of irrationals otherwise?

10 According to Russell and Whitehead (*Principia Mathematica*, p. 60), the oldest paradox of the type of Russell's paradox is from the *Epimenides*. The Greek philosopher Epimenides (sixth century B.C.) is supposed to have said, "All Cretans are liars." But Epimenides was himself a Cretan. (*a*) Explain what happens if Epimenides is telling the truth. (*b*) Explain what happens if Epimenides is lying.

Branching Out

1 The ancient Greeks had symbols using the letters of their alphabet for the numbers from 1 to 9,999. For 10,000 they used M, meaning "myriad," a large indefinite number. Archimedes was one of the few ancients, perhaps the only, to work with the concept of very large finite numbers. In "The Sand Reckoner" he deals with large numbers. Find out, or review in Chapter 1, Section 2, how the Greeks wrote numbers. Find out how Archimedes wrote large numbers. "The Sand Reckoner" is available in Newman's *The World of Mathematics.*

2 There are many large-number stories besides the one about King Shirham and his vizier Sissa Ben. There are also estimates of large amounts including amounts of atoms, snow crystals, and Archimedes' grains of sand, which involve large numbers. You might want to work with large numbers in general in fact and fiction or with some specific types. Gamow's *One Two Three . . . Infinity* will get you started.

3 Although Cantor invented the transfinite numbers, he was not the first to realize that an infinite set can be equivalent to a proper subset. Galileo, more famous for the telescope and astronomy, is the first person known to have written down such an equivalence, in "Dialogs Concerning the New Sciences" in 1636. In *Mathematical Thought from Ancient to Modern Times,* Kline says he also worked with a triangular diagram, such as we have used, to study line segments of different lengths. Find out about the origins of Cantor's theories. There is a good article in Dantzig's *Number, the Language of Science,* which includes Galileo.

4 Zeno of Elea, an ancient Greek, raised questions about infinity in four famous paradoxes. You will find Zeno and his paradoxes in many books. There are also many other paradoxes of the infinite. Kasner and Newman have a chapter on paradoxes in *Mathematics and the Imagination;* but be careful: some are not paradoxes of the infinite. E. T. Bell has Zeno's four paradoxes in *Men of Mathematics,* but again two of them are not paradoxes of the infinite.

5 Many people find paradoxes of various types a fascinating subject. As we mentioned above, Kasner and Newman have a chapter which includes paradoxes of many different types. Eves and Newsom discuss paradoxes similar to those we have seen in this chapter in *An Introduction to the Foundations and Fundamental Concepts of Mathematics,* but their book is a more difficult one than Kasner and Newman's. You will find a few different paradoxes in Kline's *Mathematical Thought from Ancient to Modern Times,* these also related to Cantor's and Russell's paradoxes. There is an article about the different types of paradoxes and antinomies,

"Paradox" by W. V. Quine, in *Mathematics in the Modern World*, a collection of articles from *Scientific American* edited by Kline. This article is also quite difficult reading.

6 The schism in mathematics begun by Cantor's work and Kronecker's objections exists today in two schools of mathematical logic. The first, the formalists, led by the late David Hilbert, follows Cantor. The other, the intuitionists, led by Brouwer, follows Kronecker. There was also a third, begun by the *Principia Mathematica* of Russell and Whitehead, called the logistic school. You will find all three discussed in Eves and Newsom. There is a more complete discussion of the formalists and intuitionists in Meschkowski, *Evolution of Mathematical Thought. Beware:* This is a difficult subject. To tackle it, you should have a strong background in mathematics, and possibly some philosophy as well.

7 Bertrand Russell was philosopher, mathematician, and general man of letters. His interests ranged from foundations of mathematics to, more recently, participation in antiwar activities. You will find books and articles about Russell interesting. In addition, Russell was a commentator on mathematics as well as a mathematician. You will find an example in his article "Mathematics and the Metaphysicians" in Newman's *The World of Mathematics*.

"Contrariwise,...
if it was so, it might be;
and if it were so, it would be;
but
as it isn't so, it ain't.
That's logic."

3

Calculating with Statements

1

Logic as a Mathematical Object

The objects you are used to thinking of as the subject matter of mathematics are such things as sets, numbers, equations, points, and lines. Logic is a tool you use to study these objects. Such was the case until about the middle of the nineteenth century. Logic was a tool in the study of mathematics but was not itself considered to be an object of mathematics. During the second half of the nineteenth century there was a great change in the role of logic in mathematics. The extent of this change can be seen in the following two quotations.

Augustus De Morgan (1806–1871), himself a leading figure in the development of mathematical logic, said, probably somewhat facetiously,

We know that mathematicians care no more for logic than logicians for mathematics. The two eyes of exact science are mathematics and logic: the mathematical sect puts out the logical eye, the logical sect puts out the mathematical eye; each believing that it can see better with one eye than with two.[1]

Bertrand Russell wrote, just about fifty years later,

Pure mathematics was discovered by Boole in a work which he called The Laws of Thought. . . . His work was concerned with formal logic, and this is the same thing as mathematics.[2]

Perhaps Russell's comment, like De Morgan's, is a bit of an exaggeration. However, the British mathematician George Boole (1815–1864) did make logic an object as well as a tool of mathematics.

The basic concept of logic as a mathematical object is a type of sentence called a *statement*. A statement is a sentence which is either true or false, that is, one or the other but not both. We do not have to know which it actually is, just so long as we know it must be one or the other. "Are you here?" and "Go home!" are not statements. There is no true-false criterion to a question or a command. "Politicians are people" is a statement. So are "Abraham Lincoln was a president of the United States" and "Abraham Lincoln was the first president of the United States," although the second happens to be false.

The most basic type of statement is the *simple statement*. A simple statement is of the form "*X* is *Y*," where *X* and *Y* are objects, physical or abstract, or sets of objects. For example, consider the simple statement "Abraham Lincoln was a president of the United States." *X* is Abraham Lincoln and *Y* is the set of all presidents of the United States. The statement actually says that Abraham Lincoln *is an element of* the set of all presidents of the United States. The simple statement "Abraham Lincoln was the first president of the United States" not only is false but also is a different type of statement. Here *X* is still Abraham Lincoln but *Y* is a single element, the person who was the first president of the United States, and the statement says that the two elements are the same. This type of simple statement actually says that Abraham Lincoln *is identical*

[1] Howard W. Eves, *In Mathematical Circles* (Boston: Prindle, Weber & Schmidt, Inc., 1969), 308°.
[2] Ibid., 289°.

Historical Pictures Service, Chicago

George Boole's brief work, The Mathematical Analysis of Logic, published in 1847, signaled the new place of logic in mathematics. Seven years later Boole published a much longer work with a much longer title: An Investigation of the Laws of Thought, on which are founded the Mathematical Theories of Logic and Probabilities. In these works he developed what he called "the calculus of logic," in which he used the symbols and operations of algebra to express the fundamental principles of logic and to apply them to the process of logical reasoning.

The logic used by mathematicians was that written down by the Greek philosopher Aristotle (fourth century B.C.), which had come down to us virtually unchanged. Mathematicians used it as a tool, but Aristotelian logic belongs to philosophy. Boole's work made logic a part of mathematics. The culmination of his work came in the Principia Mathematica of Russell and Whitehead (Chapter 2, Section 9) when symbolic logic was developed and became a branch of mathematics.

to the first president of the United States. Finally, we might say "Presidents of the United States (so far) are men." Now X is the set of presidents of the United States and Y is the set of all men. This simple statement says that the set of presidents of the United States *is a subset of* the set of men. The example "Politicians are people" is of the same type. X is the set of politicians and Y is the set of people, and the simple statement says that the set of politicians is a subset of the set of people.

Often in a simple statement the type of "is," or indeed the "is" itself, can be disguised. For example, instead of "Politicians are people" we might say "A politician is a person." Since no specific politician is indicated, and no specific person, the simple statement still is of the type, the set of politicians is a subset of the set of people. If, however, we said "Politician P is a person," then we have specified a particular element of the set of politicians, and so the statement is now of the type, politician P is an element of the set of people. Finally, if we said "Politician P is the mayor of city C," we have also specified a particular element of the set of people, and so the statement is now of the type, politician P is identical to the mayor of city C.

Further, the "is" might not actually appear at all. Consider the statements "Politician P has a limousine" and "Politicians argue with one another." Both are simple statements. The first actually means that politician P is a person who has a limousine. It is of the type, politician P is an element of the set of people who have limousines. The second means that politicians are people who argue with one another and is of the type, the set of politicians is a subset of the set of people who argue with one another.

These examples should convince you that there are actually three distinct uses of the word "is" and that every simple statement can be identified with one of the three uses. Related to the three types of "is" are diagrams called *Euler* (pronounced "oiler") *diagrams,* named for the Swiss mathematician Leonhard Euler (1707–1783). They are also sometimes called *Venn diagrams* for the British logician John Venn (1834–1923); or, if one wishes to compromise, *Venn-Euler diagrams.* In a Venn-Euler diagram, if X is a set, it is usually represented by a circle; and if it is a single element, it is represented by an x. "Is an element of" means that an object x is in the set X and thus has the following Venn-Euler diagram.

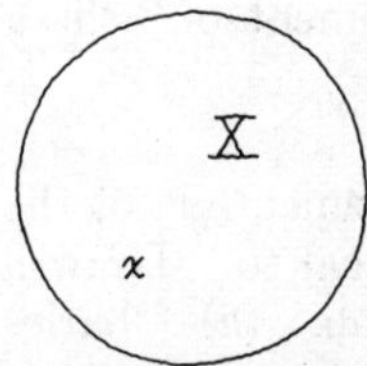

"Is a subset of" means that one set X is included in another set Y and therefore has this Venn-Euler diagram.

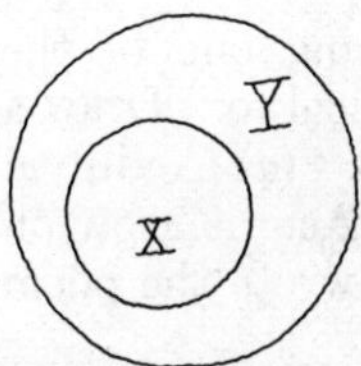

The following are the Venn-Euler diagrams for "Abraham Lincoln was a president of the United States" and "Presidents of the United States (so far) are men."

There is no Venn-Euler diagram for the third type of is, "is identical to."

EXERCISE 1 Which of these sentences are statements? For each that is not, say why not. (*a*) Amoebae are animals. (*b*) What is an amoeba? (*c*) Answer the question. (*d*) An amoeba is a microscopic, single-celled life form. (*e*) If the cell under this microscope is an amoeba, then the cell is a protozoan and protozoa are primitive animal organisms.

EXERCISE 2 Consider this statement: If the cell under this microscope is an amoeba, then the cell is a protozoan and protozoa are primitive animal organisms. The statement is not a simple statement; however, it contains three simple statements. Write each of the three simple statements. Say whether each is of the type "is an element of," "is a subset of," or "is identical to."

EXERCISE 3 Say whether each of these simple statements is of the type "is an element of," "is a subset of," or "is identical to." Draw a Venn-Euler diagram for each statement which has one. (*a*) Charles Darwin was the author of *On the Origin of Species*. (*b*) *On the Origin of Species* is a book about evolution. (*c*) *On the Origin of Species* is the most famous book ever written about evolution. (*d*) Books about evolution are about animals. (*e*) Dinosaurs are extinct animals.

EXERCISE 4 Rewrite each of these statements using one of the phrases "is an element of," "is a subset of," or "is identical to." Draw a Venn-Euler diagram for each statement which has one. (*a*) Audubon painted pictures of birds. (*b*) Birds live in trees. (*c*) Acorns grow on trees. (*d*) An acorn can grow into a tree. (*e*) Kilmer wrote the poem called "Trees."

2

Truth Values

Every simple statement is assigned a *truth value*. There are just two truth values, T for true and F for false, since every statement is either true or false. We may not actually know the truth value for a specific statement, such as, "It is raining in Afghanistan right now"; but since the statement must be either true or false, there must be a truth value, either T or F.

Compound statements are combinations of simple statements. Boole's calculus of logic constructs formulas which will calculate the truth values of compound statements from those of simple statements. In modern symbolic logic, the results of these calculations are often given in *truth tables*. Truth tables are similar to addition or multiplication tables, except that there are only two numbers, T and F. You may think this analogy somewhat far-fetched at first, but the similarity is actually so close that numbers are indeed sometimes used. Boole used 0 for false and 1 for true. The system of formulas derived is called a Boolean algebra.

The easiest variation of a simple statement "*X* is *Y*" is "It is false that *X* is *Y*," or "*X* is not *Y*." This statement is called the *negation* of the original statement. We use the single letter p for "X is Y." Then the negation of p is written $\sim p$ ("negation of p," or simply "not p").

It is clear that if p is true, then $\sim p$ must be false, and if p is false, then $\sim p$ must be true. For example, "You are a student" is probably true, so its negation, "You are not a student," is false. If, on the other hand, "You are a student" is false, then "You are not a student" is true. We display these results in *the truth table for* $\sim p$:

p	$\sim p$
T	F
F	T

The truth table for $\sim p$ gives the truth values of $\sim p$ for any statement p.

Other compound statements combine two simple statements p and q. We might say "You are a student and you enjoy the fine arts." Then p is the simple statement "You are a student" and q is the simple statement "You enjoy the fine arts." The compound statement is called the *conjunction of* p *and* q, written $p \wedge q$ ("p and q").

For the compound statement $p \wedge q$ to be true, both the statement p and the statement q must be true. If you are a student but do not enjoy the fine arts, then the statement "You are a student and you enjoy the fine arts" is false. Similarly, if you are not in fact a student, the statement is false regardless of your feelings about the fine arts. Of course, if you are neither a student nor one who enjoys the fine arts, then the statement "You are a student and enjoy the fine arts" is obviously false.

To construct the truth table for $p \wedge q$ we must record all the possible combinations of truth values for two simple statements p and q. Following the sequence of the preceding example, p and q can both be true, p can be true with q false, p can be false with q true, and p and q can both be false. We have seen that the conjunction of p and q is true only in the case that both p and q are true; so *the truth table for* $p \wedge q$ *is*

p	q	$p \wedge q$
T	T	T
T	F	F
F	T	F
F	F	F

The compound statement which connects p with q using "or" is called the *disjunction of p and q*, written $p \vee q$ ("p or q"). Skipping for a moment the case where p and q are both true, it is clear that either p true or q true is sufficient to have $p \vee q$ true. For example, if you are a student, then the statement "You are a student or you enjoy the fine arts" is true even though you may not in fact enjoy the fine arts. If you do enjoy the fine arts, then even if you are not a student the statement is also true. Of course, if you are neither a student nor a person who enjoys the fine arts, then the statement is false. So far the truth table is

p	q	$p \vee q$
T	T	?
T	F	T
F	T	T
F	F	F

Special consideration must be given the case where both p and q are true. There are two distinct meanings of the word "or." The statement "To understand this book you must be good at math or a good reader" means that one skill or the other is sufficient; but if you have both, so much the better. This use is the *inclusive-or*, meaning "p or q or both." On the other hand, you might say to a child, "You may have the chocolate bar or the lollipop," meaning that the child may choose one or the other but not both. This use is the *exclusive-or*, meaning "p or q but not both." The Romans had two different words, one for each type of or. In Latin, *vel* means "at least one," which is our inclusive-or, whereas *aut* means "only one," our exclusive-or. In symbolic logic, or always means the inclusive-or unless the other use is clearly indicated by some phrase such as "only one" or "but not both." You can remember the distinction by thinking of the symbol in $p \vee q$ as a v for *vel.*

Since we use the inclusive-or, $p \vee q$ means "p or q or both." If you are a student and also enjoy the fine arts, then the statement "You are a student or you enjoy the fine arts" is true. The disjunction of p and q is true in every case except when both p and q are false, so *the truth table for*

$p \vee q$ is as follows

p	q	$p \vee q$
T	T	T
T	F	T
F	T	T
F	F	F

The truth tables for $\sim p$, $p \wedge q$, and $p \vee q$ are building blocks from which truth tables for more complicated compound statements are constructed.

EXERCISE 5 Construct the truth table for the exclusive-or, "p or q, but not both," written $p \underline{\vee} q$.

EXERCISE 6 Use "You are good at math" for p, and "You are good at English" for q. Write the symbolic form for each of these compound statements. Construct the truth table for each statement. (*a*) It is false that you are good at math. (*b*) You are good at math or English. (*c*) You are good at only one of math or English. (*d*) You are good at math and English.

3
Logical Equivalence

Suppose we said, "It is false that you are good at math or English." Since this compound statement is the negation of the compound statement $p \vee q$, it is written $\sim(p \vee q)$. To construct its truth table, we first construct the truth table for $p \vee q$. Then we negate every truth value by changing each T to F and each F to T.

p	q	$p \vee q$
T	T	T
T	F	T
F	T	T
F	F	F

p	q	$\sim(p \vee q)$	
T	T	F	T
T	F	F	T
F	T	F	T
F	F	T	F

The final results, the truth values in the box, are the truth values for $\sim(p \vee q)$.

You may feel that "It is false that you are good at math or English" could be worded some other way. Perhaps we could say "You are not good at math or not good at English" or "You are not good at math and not good at English." Which of these alternatives, if either, is an accurate rewording of the original statement?

To find out if one of the rewordings is really the same as the original, we write the statements in symbolic form. Since p is the statement "You are good at math," the statement "You are not good at math" is $\sim p$. Similarly, "You are not good at English" is $\sim q$. Then the two proposed compound statements are $(\sim p) \vee (\sim q)$ and $(\sim p) \wedge (\sim q)$. To construct truth tables for these compound statements, we must first negate each truth value of p and also each truth value of q by changing each T to F and each F to T.

p	q	$(\sim p)$	$(\sim q)$
T	T	F	F
T	F	F	T
F	T	T	F
F	F	T	T

Now we combine $(\sim p)$ with $(\sim q)$, remembering that "and" is true only if it combines two T's, but "or" is true if at least one of the truth values is a T. The truth table for $(\sim p) \wedge (\sim q)$ is

p	q	$(\sim p)$	$\wedge$	$(\sim q)$
T	T	F	F	F
T	F	F	F	T
F	T	T	F	F
F	F	T	T	T

Again, the truth values in the box are the truth values for the compound statement. Observe that the truth values for $(\sim p) \wedge (\sim q)$ are exactly the same as the truth values for $\sim(p \vee q)$, which means that the compound statement "You are not good at math and not good at English" is an accurate rewording of the compound statement "It is false that you are good at math or English."

DEFINITION Two statements are *logically equivalent* if they have exactly the same truth values.

The statements $(\sim p) \wedge (\sim q)$ and $\sim(p \vee q)$ are logically equivalent.

EXERCISE 7 Use "You are good at math" for p, and "You are good at English" for q. Write each of these statements in symbolic form. (*a*) It is false that you are good at math and good at English. (*b*) You are not good at math or you are not good at English. (*c*) Construct the truth table for each statement. (*d*) Explain why the statements are logically equivalent.

EXERCISE 8 Write the negation of each of these statements without using the phrase "it is false that." (*a*) George Boole was a philosopher or a mathematician. (*b*) Aristotle was a philosopher and a logician.

EXERCISE 9 Construct truth tables for each statement in these pairs of statements. Which pairs are logically equivalent? (*a*) $(\sim p) \vee q$ and $p \vee (\sim q)$. (*b*) $(\sim p) \wedge q$ and $p \wedge (\sim q)$. (*c*) $(\sim p) \vee q$ and $\sim[p \wedge (\sim q)]$. (*d*) $(\sim p) \wedge q$ and $\sim[p \vee (\sim q)]$.

When two statements are logically equivalent we connect them by a double-headed arrow. Thus we may write $[(\sim p) \wedge (\sim q)] \leftrightarrow [\sim(p \vee q)]$ and $[(\sim p) \vee (\sim q)] \leftrightarrow [\sim(p \wedge q)]$. These two logical equivalences are called *De Morgan's Laws* after Augustus De Morgan.

Suppose that p and q are two simple statements. Then $p \leftrightarrow q$ ("p if and only if q") is itself a compound statement. Since p and q are logically equivalent when they have the same truth values, $p \leftrightarrow q$ is true when p and q are both true or both false, and false when p and q have different truth values. Thus *the truth table for* $p \leftrightarrow q$ is

p	q	$p \leftrightarrow q$
T	T	T
T	F	F
F	T	F
F	F	T

Since logically equivalent statements have all the same truth values, when two logically equivalent statements are combined with the logical equivalence symbol, the result will be all T's. For example, we used truth tables to show that $\sim(p \vee q)$ and $(\sim p) \wedge (\sim q)$ are logically equivalent. If we combine these truth tables with the logical equivalence symbol, the resulting truth table is

p	q	$[\sim$	$(p \vee q)]$	$\leftrightarrow$	$[(\sim p)$	$\wedge$	$(\sim q)]$
T	T	F	T	T	F	F	F
T	F	F	T	T	F	F	T
F	T	F	T	T	T	F	F
F	F	T	F	T	T	T	T

The truth values in the dotted boxes are the truth values of the two separate compound statements. The truth values in the solid box are the truth values for the entire statement combined with the logical equivalence symbol. Since each pair of truth values of the separate statements is the same, the values in the box are all T's.

DEFINITION A compound statement is a *tautology* if its truth values are all T's.

The truth table above shows that the compound statement $[\sim(p \vee q)] \leftrightarrow [(\sim p) \wedge (\sim q)]$ is a tautology.

An example of a very simple tautology is the statement $p \vee (\sim p)$. In words this statement is "p or the negation of p." Since the negation of p is true means that p is false, the statement means that either p is true or p is false. This was assumed to be a basic property of any statement, so it should always be true. The truth table is

p	p	$\vee$	$(\sim p)$
T	T	T	F
F	F	T	T

Notice that the truth table has just two lines because only one statement p is involved. The truth values in the box are both T's, so the statement is a tautology.

The statement $p \wedge (\sim p)$ is always false. It means p and the negation of p, or p is true and p is false. But it was assumed for any statement p, that p is either true or false, but not both.

p	p	$\wedge$	$(\sim p)$
T	T	F	F
F	F	F	T

DEFINITION A compound statement is a *contradiction* if all its truth values are F's.

The statement $p \wedge (\sim p)$ is a contradiction, since the two values in the box in its truth table are F's.

EXERCISE 10 Use truth tables to show that p if and only if q, $p \leftrightarrow q$, is logically equivalent to the negation of the exclusive-or, $\sim(p \veebar q)$, of Exercise 5.

EXERCISE 11 Recall that $\sim(p \wedge q)$ and $(\sim p) \vee (\sim q)$ are logically equivalent. Use a truth table to show that the statement $[\sim(p \wedge q)] \leftrightarrow [(\sim p) \vee (\sim q)]$ is a tautology.

EXERCISE 12 Construct a truth table for each of these statements. Say whether or not each statement is a tautology. (*a*) $[(\sim p) \vee q] \leftrightarrow [p \vee (\sim q)]$. (*b*) $[(\sim p) \wedge q] \leftrightarrow [p \wedge (\sim q)]$. (*c*) $[(\sim p) \vee q] \leftrightarrow \sim[p \wedge (\sim q)]$. (*d*) $[(\sim p) \wedge q] \leftrightarrow \sim[p \vee (\sim q)]$.

EXERCISE 13 Construct a truth table for each of these statements. Say whether or not each statement is a contradiction. (*a*) $[(\sim p) \vee q] \leftrightarrow [p \vee (\sim q)]$. (*b*) $[(\sim p) \wedge q] \leftrightarrow [p \wedge (\sim q)]$. (*c*) $[(\sim p) \vee q] \leftrightarrow [p \wedge (\sim q)]$. (*d*) $[(\sim p) \wedge q] \leftrightarrow [p \vee (\sim q)]$.

4
The Conditional

Many mathematical statements can be reduced to some form of compound statement called the *conditional*, written $p \rightarrow q$ ("if p then q"). There are many different ways of saying if p then q. Some of them are "p implies q," "p hence q," "p therefore q," "p is sufficient for q"; and also "q follows from p," "q is a consequence of p," and "q is necessary for p." In all these variations, to have $p \rightarrow q$ true the concern is that *if* one statement p is true, another statement q must surely also be true. Whether the statement p is in fact true is immaterial. It matters only that if it were, then the statement q must be true as well.

In constructing the truth table for $p \rightarrow q$, we must be especially careful to consider the statement as a whole and not fall into the error of considering whether p alone, or q alone, is true. Suppose we entered into a contract that said, "If I make a million dollars or more, then I will give you half a million." If I make a million or more and do give you $500,000, I have fulfilled the contract. If I make the million and do not give you at least $500,000, you would probably sue me, for I have broken the contract. Using p to stand for the simple statement "I make a million or more," and q for "I give you $500,000," we see that when p is true and q is true, the conditional $p \rightarrow q$ is true; but when p is true and q is false, $p \rightarrow q$ is false. The first two lines of the truth table must be

p	q	$p \rightarrow q$
T	T	T
T	F	F
F	T	?
F	F	?

Now suppose I do not make the million. The contract you hold is worthless. I could give you $500,000 anyway, if I had it, or half of whatever I make, or nothing at all. In any case you could not sue because, when I do not make a million or more, I do not break the contract. (My brother-in-law, who is a lawyer, tells me that in law this example is called a "condition precedent," meaning that if a certain condition happens you gain certain rights. You do not have those rights unless the specified condition is fulfilled.) Since the contract is not broken, $p \rightarrow q$ is true. Whenever p is false, $p \rightarrow q$ is true! With this result recorded, *the truth table for* $p \rightarrow q$ is

p	q	$p \rightarrow q$
T	T	T
T	F	F
F	T	T
F	F	T

EXERCISE 14 Suppose we have a contract which says, "I will give you half of whatever I make unless I make a million dollars or more." (This is a "condition subsequent," meaning that you have certain rights unless a certain condition happens, in which case you lose those rights.) Use "I make less than a million dollars" for p, and "I give you half" for q. (*a*) What is the wording of $p \rightarrow q$? (*b*) Under what condition is the contract broken? (*c*) Under what condition is the contract worthless?

The statements p and q play very different roles in the conditional $p \rightarrow q$. The statement p is called the *premise* of the conditional, and q is called the *conclusion*. If we reverse the premise and the conclusion, the resulting conditional may have a different meaning. One way to show this result is to construct the truth table for $q \rightarrow p$, remembering that the conditional is true unless it goes from true to false.

p	q	q	$\rightarrow$	p
T	T	T	T	T
T	F	F	T	T
F	T	T	F	F
F	F	F	T	F

The truth values in the box are not the same as the truth values for $p \rightarrow q$, so $p \rightarrow q$ and $q \rightarrow p$ are not logically equivalent. The statement $q \rightarrow p$ is called the *converse* of the conditional.

It is not hard to think of examples where $p \rightarrow q$ and its converse $q \rightarrow p$ are obviously different. When you cross the border from Canada into the United States, the official often asks, "Where were you born?" Suppose you answer that you were born in New York City. The official would reason that if you were born in New York City, then you are a United States citizen, and from one question would learn both your place of birth and your citizenship. Were he to ask, "Are you a United States citizen?" he would get just one piece of information, for the converse of the official's conditional is "If you are a United States citizen, then you were born in New York City," which is not true.

A third arrangement of the conditional is the statement $(\sim p) \rightarrow (\sim q)$, which we will call the *obverse* of the conditional. (Many people use the word "inverse," but we will be using that word in a different situation in Chapter 4.) To construct the obverse of a conditional, we negate the premise and the conclusion and then make the conditional of those negations. The truth table for the obverse of the conditional is

p	q	$(\sim p)$	$\rightarrow$	$(\sim q)$
T	T	F	T	F
T	F	F	T	T
F	T	T	F	F
F	F	T	T	T

The truth values of the obverse and the converse are the same, so the obverse and the converse of a conditional are always logically equivalent. The obverse of our example is "If you were not born in New York City, then you are not a United States citizen," which is again false.

There is one more arrangement of the conditional, $(\sim q) \rightarrow (\sim p)$, called the *contrapositive* of the conditional. The contrapositive of the conditional is very important because it is logically equivalent to the original conditional. To show this result, we construct the truth table for the contrapositive:

p	q	$(\sim q)$	$\rightarrow$	$(\sim p)$
T	T	F	T	F
T	F	T	F	F
F	T	F	T	T
F	F	T	T	T

The contrapositive of "If you were born in New York City, then you are a United States citizen," is "If you are not a United States citizen, then you were not born in New York City." Allowing for changes of national allegiance, both are true.

EXERCISE 15 Use "Today is Monday" for p, and "You go to your math class" for q. (*a*) Write the conditional, converse, obverse, and contrapositive. (*b*) Which statements are true and which are false? (*c*) Which pairs of statements are logically equivalent?

EXERCISE 16 Write the converse, obverse, and contrapositive of each of these conditionals. (*a*) If you are good at math, then you will pass this course. (*b*) If you like English, then you will like this course. (*c*) If you are not good at English, then you will fail this course.

5
The Biconditional

The converse of the conditional plays a special role in mathematical statements. Given a conditional and its converse, $p \rightarrow q$ and $q \rightarrow p$, we see that together they could be thought of as making the logical equivalence symbol $p \leftrightarrow q$. We have made a poor choice of symbols unless $p \leftrightarrow q$ is logically equivalent to the statement consisting of the conditional and its converse, $(p \rightarrow q) \wedge (q \rightarrow p)$.

p	q	$p \leftrightarrow q$
T	T	T
T	F	F
F	T	F
F	F	T

p	q	$(p \rightarrow q)$	$\wedge$	$(q \rightarrow p)$
T	T	T	T	T
T	F	F	F	T
F	T	T	F	F
F	F	T	T	T

The conditional, $p \rightarrow q$, may be read "p only if q." The converse, $q \rightarrow p$, may be read "p if q." Thus we read the symbol $p \leftrightarrow q$, "p if and only if q." Since $p \leftrightarrow q$ is logically equivalent to the conjunction of two conditionals, it is called the *biconditional.*

EXERCISE 17 Use "Today is Monday" for p, and "Tomorrow is Tuesday" for q. (*a*) Write the conditional, converse, obverse, contrapositive, and biconditional. (*b*) Which statements are true and which are false? (*c*) Are statements p and q logically equivalent?

EXERCISE 18 Use truth tables to show that $p \leftrightarrow q$ and $[(\sim p) \rightarrow (\sim q)] \wedge [(\sim q) \rightarrow (\sim p)]$ are logically equivalent.

It is possible to construct a symbolic logic that avoids the conditional and biconditional altogether. In fact, we can use only the two symbols $\sim$ and $\vee$. For example, the conditional $p \rightarrow q$ is logically equivalent to $(\sim p) \vee q$.

p	q	$p \rightarrow q$
T	T	T
T	F	F
F	T	T
F	F	T

p	q	$(\sim p)$	$\vee$	q
T	T	F	T	T
T	F	F	F	F
F	T	T	T	T
F	F	T	T	F

Thus we can write $(\sim p) \vee q$ in place of $p \rightarrow q$. Similarly, $q \rightarrow p$ is logically equivalent to $p \vee (\sim q)$, and $p \wedge q$ is logically equivalent to $\sim[(\sim p) \vee (\sim q)]$. Thus we can write $p \vee (\sim q)$ in place of $q \rightarrow p$, and $\sim[(\sim p) \vee (\sim q)]$ in place of $p \wedge q$. These statements use only the symbols $\sim$ and $\vee$. Using these replacements in $p \leftrightarrow q$, written as $(p \rightarrow q) \wedge (q \rightarrow p)$, it becomes $[(\sim p) \vee q] \wedge [p \vee (\sim q)]$, which still has the symbol $\wedge$, and so in turn becomes $\sim\{\sim[(\sim p) \vee q] \vee \sim[p \vee (\sim q)]\}$, using only the symbols $\sim$ and $\vee$. Of course, these statements are quite unwieldy and therefore not suitable for practical purposes, but that the reduction to two symbols can be done at all is of theoretical interest. In fact, it is even possible to get down to only one symbol. The one symbol, called Sheffer's stroke, is not one of those we have used.

EXERCISE 19 (*a*) Use truth tables to show that $q \rightarrow p$ is logically equivalent to $p \vee (\sim q)$. (*b*) Use truth tables to show that $p \wedge q$ is logically equivalent to $\sim[(\sim p) \vee (\sim q)]$. (*c*) Use truth tables to show that $p \leftrightarrow q$ is logically equivalent to $[(\sim p) \vee q] \wedge [p \vee (\sim q)]$. (*d*) Use truth tables to show that $p \leftrightarrow q$ is logically equivalent to $\sim\{\sim[(\sim p) \vee q] \vee \sim[p \vee (\sim q)]\}$.

6

The Law of Detachment

Perhaps of all the remarks about mathematics by mathematicians, the most often quoted is Bertrand Russell's aphorism: "Mathematics may be defined as the subject in which we never know what we are talking about, nor whether what we are saying is true." This curious situation arises because mathematical deduction is of the form "if p then q." We know only that *if* p is true, then q is also true. We never know absolutely that either p or q is true.

Suppose you believe that if you get your feet wet, then you will catch a cold. Suppose further that you do get your feet wet. Then you start taking precautions against a cold. You have reasoned that the two premises "If you get your feet wet, then you will catch a cold" and "You get your feet wet" together lead to the conclusion "You will catch a cold."

We call such reasoning an *argument*, and write the argument in the following form:

Premises If you get your feet wet, then you will catch a cold.
You get your feet wet.

Conclusion You will catch a cold.

Let p be the simple statement "You get your feet wet" and q be "You will catch a cold." Then the argument can be written in symbolic form:

Premises $p \to q$
p

Conclusion q

The symbolic form says that if we have the premises $p \to q$ and p, then we have the conclusion q. Written as a compound statement, the argument is $[(p \to q) \land p] \to q$. The truth table for this statement is constructed by finding the truth values for the part in brackets, in the dotted box, and combining them with the truth values for q, in the other dotted box, using the conditional. Recall that the conditional is true unless it goes from true to false.

p	q	$[(p \to q)$	$\land$	$p]$	$\to$	q
T	T	T	T	T	T	T
T	F	F	F	T	T	F
F	T	T	F	F	T	T
F	F	T	F	F	T	F

The truth values in the final solid box are all T's. The statement is a tautology. The entire compound statement is always true, regardless of the truth values of p and q. When this is the case, we say that the argument is *valid*. Any argument in this form is valid:

Premises $p \to q$
p

Conclusion q

The form is called the *law of detachment*, or *modus ponens.* We can make another example of the law of detachment using the conditional "If you were born in New York City, then you are a United States citizen." The law of detachment becomes

Premises If you were born in New York City, then you are a United States citizen.
You were born in New York City.

Conclusion You are a United States citizen.

You should observe that the validity of the argument is independent of whether the premises and conclusion are true or false. In fact, if we start with false premises, the conclusion of a valid argument can be either true or false. If the premises are true, however, the conclusion of a valid argument will always be true also.

We have agreed that the statement "If you were born in New York City, then you are a United States citizen," allowing for changes of national allegiance, is true. Chances are the statement "You are a United States citizen" is true for you also. Suppose we wrote an argument in the following form:

Premises If you were born in New York City, then you are a United States citizen.
You are a United States citizen.

Conclusion You were born in New York City.

While the premises are true for many people, the conclusion is not necessarily true. This argument should not be valid, because true premises should never lead to a false conclusion. The symbolic form of the argument is

Premises $p \rightarrow q$
q

Conclusion p

In this form, $p \rightarrow q$ and q together imply p. The compound statement is

$[(p \to q) \wedge q] \to p$, and its truth table is

p	q	$[(p \to q)$	$\wedge$	$q]$	$\to$	p
T	T	T	T	T	T	T
T	F	F	F	F	T	T
F	T	T	T	T	F	F
F	F	T	F	F	T	F

The statement is not a tautology, so the argument is indeed not valid.

A closely related form of argument is

Premises $p \to q$
$\sim p$

Conclusion $\sim q$

Its symbolic statement is $[(p \to q) \wedge (\sim p)] \to (\sim q)$. In this form, our example becomes

Premises If you were born in New York City, then you are a United States citizen.
You were not born in New York City.

Conclusion You are not a United States citizen.

According to our previous agreements, the premises are true for all but New Yorkers, but the conclusion is not necessarily true. The argument should not be valid, and the truth table for the compound statement shows that indeed it is not.

There is just one more arrangement left. Its symbolic form is

Premises $p \to q$
$\sim q$

Conclusion $\sim p$

The compound statement is $[(p \to q) \wedge (\sim q)] \to (\sim p)$; our example is

Premises If you were born in New York City, then you are a United States citizen.
You are not a United States citizen.

Conclusion You were not born in New York City.

The second premise is probably not true for you. If it is, however, the conclusion is then true as well. This form of argument is valid. It is a second form of the law of detachment, called *modus tollens.*

EXERCISE 20 Use "Today is Monday" for p and "You go to your math class" for q. (a) Write the first form of the law of detachment in words. (b) Write the symbolic form of this argument. (c) Write this argument as a compound statement.

EXERCISE 21 Use "Today is Monday" for p and "You go to your math class" for q. Write each of these symbolic forms of arguments in words. For each argument, say whether or not it is valid.

(a) $\begin{array}{l} p \rightarrow q \\ \underline{q} \\ p \end{array}$ (b) $\begin{array}{l} p \rightarrow q \\ \underline{\sim p} \\ \sim q \end{array}$ (c) $\begin{array}{l} p \rightarrow q \\ \underline{\sim q} \\ \sim p \end{array}$

EXERCISE 22 Write the symbolic form of each of these arguments. For each argument, say whether or not it is valid.

(a) If you are good at math, then you will pass this course.
You are good at math.
_____
You will pass this course.

(b) If you are good at math, then you will pass this course.
You are not good at math.
_____
You will not pass this course.

(c) If you like English, then you will like this course.
You like this course.
_____
You like English.

(d) If you like English, then you will like this course.
You do not like this course.
_____
You do not like English.

(e) If you are not good at English, then you will fail this course.
You do not fail this course.
_____
You are good at English.

EXERCISE 23 (a) Use a truth table to show that the statement $[(p \rightarrow q) \wedge (\sim p)] \rightarrow (\sim q)$ is not a tautology. (b) Use a truth table to show that the statement $[(p \rightarrow q) \wedge (\sim q)] \rightarrow (\sim p)$ is a tautology.

EXERCISE 24 Some "prestige" publications like to use advertising of the type, "If you are an intelligent, informed person, then you are the type who subscribes to X magazine." Their tactic is for you to subscribe to X magazine in order to infer that you are an intelligent, informed person. (*a*) Write the argument in words. (*b*) Write the argument in symbolic form. (*c*) Is the argument valid?

EXERCISE 25 When fluoride toothpastes were introduced, many companies used advertising of the type, "If you use brand Y toothpaste, then you will get fewer cavities." They assumed that you were not using brand Y toothpaste and hoped you would then conclude that you were not getting fewer cavities. (*a*) Write the argument in words. (*b*) Write the argument in symbolic form. (*c*) Is the argument valid?

EXERCISE 26 A company dealing in a type of frozen food once had the roundabout slogan, "Nobody doesn't like brand Z," meaning that "If you are somebody, then you like brand Z." Suppose you do not like brand Z. They want you to be forced to the conclusion that you are nobody. (*a*) Write the argument in words. (*b*) Write the argument in symbolic form. (*c*) Is the argument valid?

7
Quantified Statements

The most famous argument of all time is most likely the classical syllogism

Premises All men are mortal.
Socrates is a man.

Conclusion Socrates is mortal.

You should feel that this argument is valid. However, it cannot be put in either of the forms of the law of detachment. If you recall the discussion at the beginning of this chapter, you will see that the argument depends on the three types of is. The statement "All men are mortal" is the "is a subset of" type, meaning that the set of all men *is a subset of* the set of all mortals. The statements "Socrates is a man" and "Socrates is mortal" are both the "is an element of" type. The first means that Socrates *is an element of* the set of men, and the second means that Socrates *is an element of* the set of mortals. The Venn-Euler diagrams for the statements are the following.

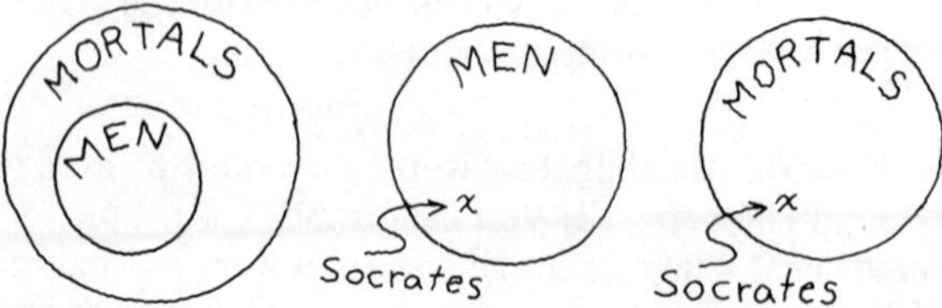

The crux of the argument is the word "all." It is the assurance that *all* men are mortal which makes us believe that Socrates, being a man, must be mortal. If we were assured that only *some* men are mortal, we would have a different Venn-Euler diagram.

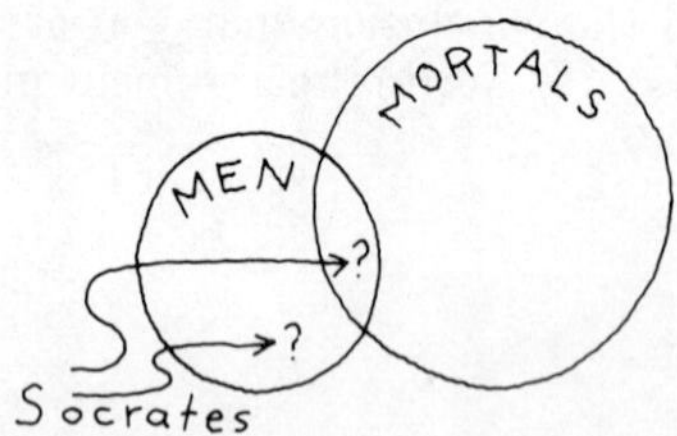

This diagram indicates that there are some men who are mortal, but possibly also some who are not. The premises do not indicate whether to put Socrates in the part of the set of men that are mortals or in the part that are not. We see that a distinction must be made between the words "all" and "some." These two words are called *quantifiers*, and the statements in which they appear are called *quantified statements.*

"All" is called the *universal quantifier*, written $\forall$. The quantified statement "All X's are Y's" is called the *universal-affirmative*, and is written $\forall x{:}p$. The x is any element of X, and p is the simple statement "X is Y." We often use synonyms such as "each" and "every" in place of "all" in the universal-affirmative.

"Some" is called the *existential quantifier*, written $\exists$. The name and symbol come from the word "exists," since the phrase "there exists at least one" is often used for "some." The quantified statement "Some X's are Y's," or "At least one X is a Y," is called the *existential-affirmative,* written $\exists x\!:\!p$. Again x is any element of X, and p is the simple statement "X is Y."

The statement "Some men are not mortal" is also a form of the existential quantifier, called the *existential-negative*. Any statement of the type "Some X's are not Y's," or "At least one X is not a Y," is existential-negative. If p is the simple statement "X is Y," then the statement "X is not Y" is $\sim p$. Thus the existential-negative is written $\exists x\!:\!\sim p$. These are the Venn-Euler diagrams for "Some men are mortal" and "Some men are not mortal."

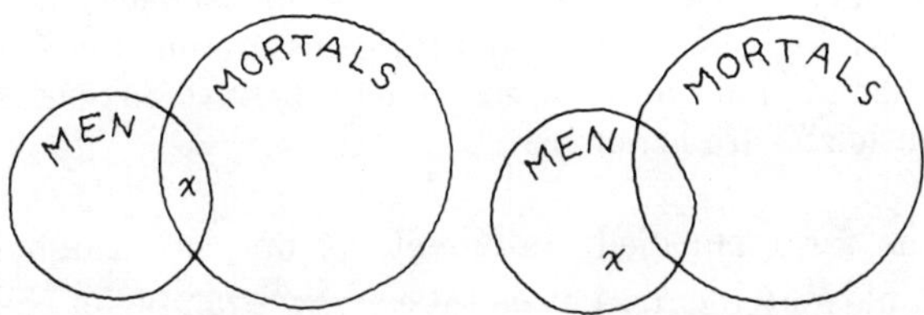

In the first diagram, x indicates that we know of at least one man who is mortal but do not know if there is a man who is not. In the second diagram, x indicates that we know of at least one man who is not mortal, but the statement does not say whether any men are mortal.

There is one more possible arrangement for this example.

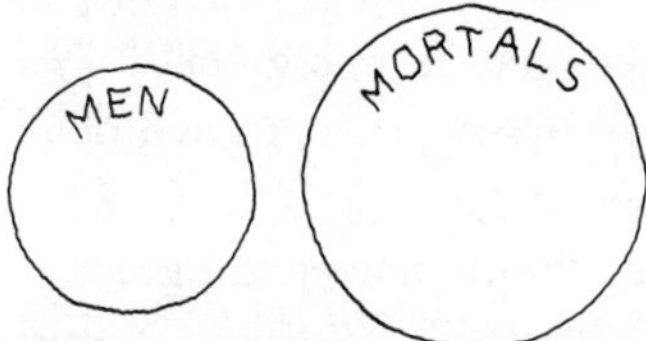

This Venn-Euler diagram indicates that every man is not in the set of mortals, or "All men are not mortal." This is another form of the universal quantifier called the *universal-negative*. Any statement of the type "All X's are not Y's" is universal-negative, written $\forall x\!:\!\sim p$. Another wording of the universal-negative is "No X's are Y's." Although the second wording is probably the more familiar, the first has the advantage that it translates directly into symbolic form.

EXERCISE 27 Draw a Venn-Euler diagram for each of these statements. (*a*) George Boole was a logician. (*b*) Aristotle was not a mathematician.

EXERCISE 28 Write the symbolic form of each of these quantified statements. Draw a Venn-Euler diagram for each statement. (*a*) All logicians are philosophers. (*b*) Some logicians are mathematicians. (*c*) All logicians are not politicians. (*d*) Some logicians are not mathematicians. (*e*) No logicians are politicians.

EXERCISE 29 Draw a Venn-Euler diagram for each statement in these pairs of statements. If the diagrams are identical, the statements mean the same thing. For each pair of statements, say whether or not the statements mean the same thing. (*a*) All logicians are philosophers. All philosophers are logicians. (*b*) Some logicians are mathematicians. Some mathematicians are logicians. (*c*) Some logicians are not mathematicians. Some mathematicians are not logicians. (*d*) No logicians are politicians. No politicians are logicians.

Forming negations for quantified statements is not as straightforward as it is for simple statements. Let us take a new example of the universal-affirmative that is clearly false: "All horses are green." Surely you will want to deny this statement. If you claim, "It is false that all horses are green," your statement is $\sim(\forall x\colon p)$, the negation of the universal-affirmative. To justify your claim you might bring in a purple horse. Then you may say, "There exists at least one horse that is purple." Of course, your horse would not have to be purple. Any horse that is not green will do. To deny the statement "All horses are green," you must be able to say, "There exists at least one horse that is not green." *The negation of the universal-affirmative is the existential-negative.* In symbols, $\sim(\forall x\colon p) \leftrightarrow \exists x\colon \sim p$. Similarly, $\sim(\exists x\colon \sim p) \leftrightarrow \forall x\colon p$.

Now consider the existential-affirmative, "Some horses are green." To deny this statement you must round up every horse there is and demonstrate that none of them is green. Your denial must now be "All horses are not green" or "No horses are green." *The negation of the existential-affirmative is the universal-negative.* In symbols, $\sim(\exists x\colon p) \leftrightarrow \forall x\colon \sim p$. Similarly, $\sim(\forall x\colon \sim p) \leftrightarrow \exists x\colon p$.

As another example, we return to the statement "All men are mortal." Its symbolic form is $\forall x\colon p$, so the symbolic form of its negation is $\sim(\forall x\colon p)$ or $\exists x\colon \sim p$. In words, the negation is "Some men are not mortal." The negation of "Some men are not mortal" is "All men are mortal." Turning to the statement "Some men are mortal," we see that its symbolic form is $\exists x\colon p$. The negation is $\sim(\exists x\colon p)$ or $\forall x\colon \sim p$. In words, this negation is "All men are not mortal," which can also be written as "No men are mortal." The negation of "All men are not mortal" or "No men are mortal" is then "Some men are mortal." The statement "Socrates is a man" is a simple statement. Its negation is simply "Socrates is not a man," and vice versa.

EXERCISE 30 For each of these statements that is a quantified statement, write its symbolic form. Write the symbolic form of its negation. Write the negation of each statement in words, without using the phrase "it is false that." (*a*) George Boole was a logician. (*b*) Aristotle was not a mathematician. (*c*) All logicians are philosophers. (*d*) Some logicians are mathematicians. (*e*) All logicians are not politicians. (*f*) Some logicians are not mathematicians. (*g*) No logicians are politicians.

EXERCISE 31 The following advertisement was seen in the Boston subways early in 1974:

All banks do not insure all deposits in full. We do.

(*a*) What is another way to write the quantified statement in this advertisement, using the phrase "it is false that?" (*b*) Is it likely that this statement is what the bank really meant? (*c*) It may be assumed the bank meant to write the negation of "All banks do insure all deposits in full." Write the correct negation of this statement, using the phrase "it is false that" and also without using the phrase "it is false that." (To the bank's credit, the advertisement disappeared after a short time.)

8

Classical Syllogisms

Venn-Euler diagrams provide a way to discover whether arguments involving quantified statements are valid. For example, at the beginning of the preceding section we drew three Venn-Euler diagrams, one for each of the statements in the famous syllogism

Premises	All men are mortal.
	Socrates is a man.
Conclusion	Socrates is mortal.

If we combine the two diagrams for the premises into one, we have the following diagram:

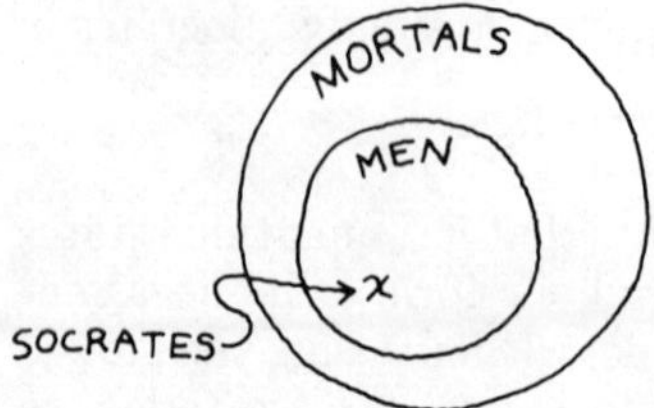

This diagram shows that the x representing Socrates must be in the circle which represents the set of mortals, so the argument is valid.

Aristotle considered only syllogisms in which both the premises and the conclusion are quantified statements. Although he gives few examples, one of those few is

If all broad-leaved plants are deciduous and all vines are broad-leaved plants, then all vines are deciduous.[3]

A classical syllogism consists of two premises and a conclusion. Written in this form, Aristotle's syllogism is

Premises	All broad-leaved plants are deciduous.
	All vines are broad-leaved plants.
Conclusion	All vines are deciduous.

The combined Venn-Euler diagrams for the premises of this argument are

[3] Aristotle, *Analytica Posteriora*, ii, 16, 98^b5; from Richard McKeon (ed.), *The Basic Works of Aristotle* (New York: Random House, 1941).

The circle which represents the set of vines is entirely included in the circle which represents the set of all deciduous plants, so the argument is valid. The diagram suggests the general form for this argument

Premises Y is a subset of Z.
X is a subset of Y.

Conclusion X is a subset of Z.

An argument in this form is called the *transitive law of set inclusion.* The prefix *trans-* is used in the sense of "across," since the argument travels across the set Y from X to Z.

If an argument in the transitive form does not go from X to Y and from Y to Z, then the transitive law of set inclusion does not hold. A common invalid form is

Premises Y is a subset of Z.
X is a subset of Z.

Conclusion X is a subset of Y.

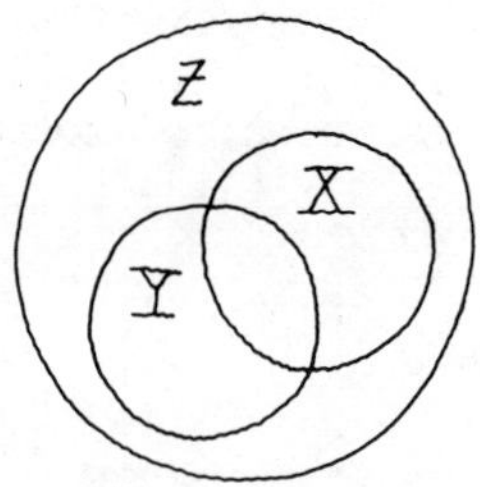

Since it is possible to include both X and Y in Z with no special relationship between X and Y, a careful Venn-Euler diagram will show that the argument is invalid. For example, prejudiced thinking is often based on this type of argument.

Premises Bad people's behavior includes behavior patterns B.
The people in group G exhibit behavior patterns B.

Conclusion The people in group G are bad.

The Venn-Euler diagram is

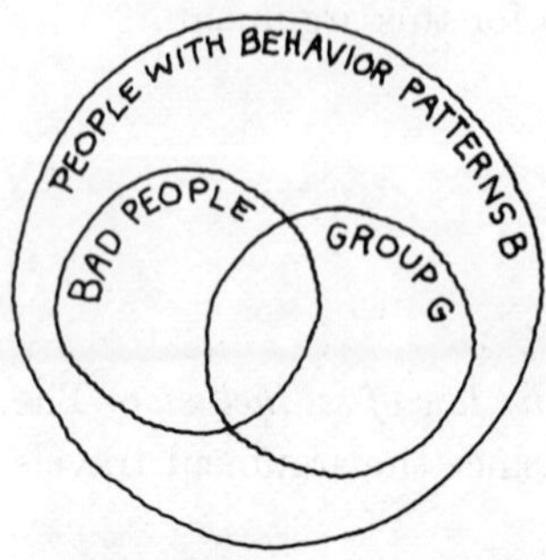

The argument is not valid.

Classical syllogisms can also involve existential quantifiers. For example,

Premises	Some men are mortal.	Some men are mortal.
	Socrates is a man.	Socrates is a man.
Conclusions	Socrates is mortal.	Socrates is not mortal.

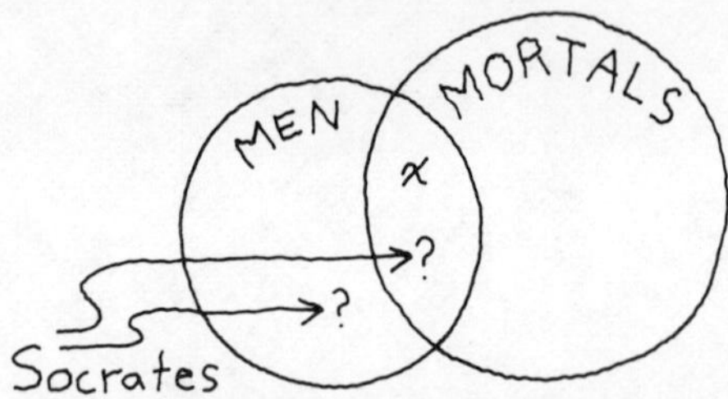

Neither conclusion is valid. The x in the Venn-Euler diagram indicates that there exists at least one man who is mortal. Nothing in the premises, however, says that Socrates is that man, or even that Socrates is in that part of the diagram. The two question marks indicate that we do not know which of the two parts of the set of men Socrates is in.

Finally, a classical syllogism can have both universal and existential quantifiers in its premises or conclusion. A carefully drawn Venn-Euler diagram will usually sort out valid and invalid conclusions. For example,

Premises	Some politicians accept kickbacks. All people who accept kickbacks are deceitful.
Possible conclusions	Some politicians are deceitful. All politicians are deceitful. All politicians accept kickbacks. Some politicians are not deceitful. Some politicians do not accept kickbacks. Some deceitful people are politicians.

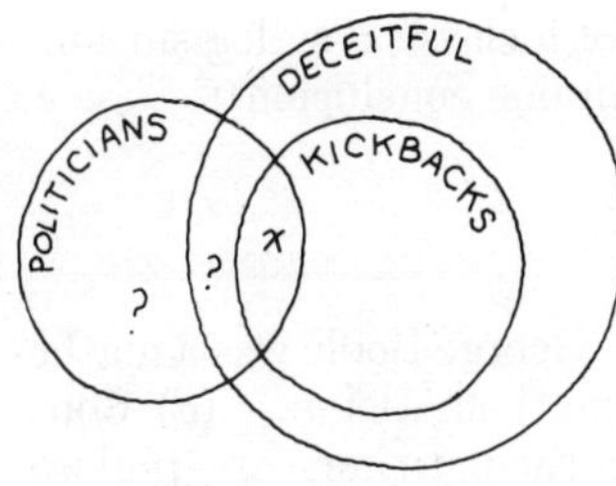

In the diagram, the x indicates that there exists at least one politician who accepts kickbacks, the first premise. The question marks indicate we do not know whether there are elements in the other parts of the set of politicians.

The first conclusion is valid since the x also indicates that there is at least one politician who is deceitful. The next two conclusions are not valid, for the question marks indicate that not all politicians are necessarily inside the set of deceitful people, nor also inside the set of those who accept kickbacks. On the other hand, the next two conclusions are also invalid since neither are we assured that there are politicians outside those sets. The last conclusion is valid since the x indicates that there is at least one deceitful person who is also a politician.

EXERCISE 32 Politician P, wishing to excuse certain small errors of his ways, claims that all politicians are human, and that all humans make mistakes. Thus, he concludes, all politicians make mistakes. (*a*) Write the argument in syllogism form. (*b*) Draw the Venn-Eüler diagram. (*c*) Is the argument valid?

EXERCISE 33 Politician P would like to exhibit his unpretentious nature by claiming that all politicians make mistakes and all humans make mistakes. He wants you to conclude that all politicians are human. (*a*) Write the argument in syllogism form. (*b*) Draw the Venn-Euler diagram. (*c*) Is the argument valid?

EXERCISE 34 Some politicians accept kickbacks for contracts on urban renewal projects. Incumbent politician P, who is the mayor of city C, is planning an urban-renewal project. His opponent would have you believe that therefore politician P is accepting kickbacks. (*a*) Write the argument in syllogism form. (*b*) Draw the Venn-Euler diagram. (*c*) Is the argument valid?

EXERCISE 35 Consider this argument (not a classical syllogism since it has more than two premises and more than one conclusion):

George Boole was a logician.
All logicians are philosophers.
Some logicians are mathematicians.

(*a*) George Boole was a philosopher. (*b*) George Boole was a mathematician. (*c*) George Boole was not a mathematician. (*d*) Some mathematicians are logicians. (*e*) Some mathematicians are philosophers. (*f*) No mathematicians are philosophers. Draw a careful Venn-Euler diagram of the three premises. For each of the conclusions, say whether or not it is valid.

EXERCISE 36 Consider this argument (also not a classical syllogism):

Aristotle was a philosopher.
All logicians are philosophers.
No logicians are politicians.

(*a*) Aristotle was a logician. (*b*) Aristotle was a politician. (*c*) Aristotle was not a politician. (*d*) No politicians are logicians. (*e*) No politicians are philosophers. (*f*) Some politicians are philosophers. Draw a careful Venn-Euler diagram of the three premises. For each of the conclusions, say whether or not it is valid.

9
Many-valued Logics

The classical logic of Aristotle and Boole, in which every statement is either true or false, is called a *two-valued logic*. There are only two truth values, T and F or, in Boole's form, 1 and 0. Toward the end of the nineteenth century logicians began to investigate the possibilities of *three-valued logics*, with truth values true, false, and undecidable, and even *many-valued logics.*

An entertaining case in support of many-valued logics is made by two contemporary mathematicians, J. B. Rosser and A. B. Turquette.[4] It takes the form of a debate between a hypothetical Mr. Rossette, who is interested in the possibility of many-valued logics, and an equally hypothetical Mr. Turquer, who believes that the two-valued logic is the only true logic. As we join them in middebate, it will help to know that Mr. Turquer has defined a "statement form" to be a sentence which can be made into a statement by "binding the variables," that is, providing further information. For example, "It is raining" is not a statement until we declare where and when.

Mr. Rossette: . . . for the sake of argument I will grant your distinction between statements and statement forms. Nevertheless, I believe that there are some actual statements which are possibly neither true nor false.

Mr. Turquer: You know you are being ridiculous. I challenge you to produce such a statement.

Mr. Rossette: Well, suppose that when the janitor arrives we ask him if he is in this room. No doubt, he will reply in the affirmative and with complete assurance, but if the question is repeated as he leaves and while he is passing through the door, then he will certainly be flustered and unable to give an answer.

Mr. Turquer: Oh, that is no problem, for all we need to do is inform the janitor of the necessity of specifying some boundaries to the room which will define precisely when he is in and out of the room.

Mr. Rossette: But what about <u>him</u>? Will it not also be necessary to specify the physical boundaries of the janitor's body?

Mr. Turquer: Yes indeed, or we might just as well use the center of gravity of the janitor's body, specifying that when this is beyond a certain point, the janitor's body is out of the room. In the last analysis, all such difficulties are easily resolved by indicating the presence of certain free variables which must be bound before statement forms can become statements.

Mr. Rossette: If I accept your solution, then I am forced to the conclusion that the janitor's original assurance about being in the room was completely

[4] J. Barkley Rosser and Atwell R. Turquette, *Many-Valued Logics* (Amsterdam: North-Holland Publishing Company, 1952), pp. 4–6.

illusory, for at that time he was as ignorant of the precise meaning of being in and out of the room as he was when we repeated the question at the time he was leaving through the door. Actually, from your point of view, he could only utter statement forms until he became versed in the use of free and bound variables. Thus, it would appear that ordinary discourse must consist entirely of statement forms. If so, it must fail completely ever to convey meaning. But after all, scientific discourse developed from ordinary language.

Mr. Turquer: I would rather say that scientific discourse divorced itself from the vague expressions and ambiguities of ordinary language. In a nutshell, it learned the importance of distinguishing statements from statement forms and, in general, it learned the necessity of precise definition.

Mr. Rosette: I doubt that the transition from ordinary common sense to scientific knowledge is as sharp as you would have it, but again for the sake of argument, let us suppose that it is and further examine the question of the truth or falsity of all statements.

Mr. Turquer: By all means let us get on with a scientific discussion. Ordinary discourse is just so much nonsense to me.

Mr. Rossette: That is another interesting fact about you, but to be scientific let us return to the center of gravity of our janitor. We would like to specify the precise center of gravity of the janitor's body in order to be able to define exactly when he is in or out of the room. How should this be done in an acceptable scientific manner?

Mr. Turquer: Perhaps the best method would be to specify the collection of atoms which constitute the janitor's body, and using their positions, calculate the exact center of gravity of this body.

Mr. Rossette: But the janitor's body is at a temperature which can be estimated, so we can estimate the thermal velocities of the atoms which compose his body. However, does not the principle of indeterminacy assert that if we have any information whatever of the velocities of atoms, then exact information concerning their positions is impossible? . . .

The discussion of the janitor's body is eventually terminated by the arrival of the body in question to clean the room. Rosser and Turquette conclude that, while there is sufficient reason in Rossette's arguments for the investigation of many-valued logics, there is also sufficient reason in Turquer's not to expect any great use of logics other than two-valued in the near future. We might note at this point, however, that in 1944, a physicist, Hans Reichenbach, published a book in which he investigated quantum mechanics with a three-valued logic. In quantum mechanics there are certain phenomena which are known to happen but which cannot be observed or measured. To these, Reichenbach assigned the truth value I, for "interphenomena." We do not claim any particular logic, two- or many-valued, to be true. We only ask, *if* we agree to look at many-valued logics, what might they look like?

Suppose that m is the number of truth values for any logic. For classical logic, $m = 2$. For three-valued logics, the first of the many-valued logics, $m = 3$. The truth values true, false, and undecidable are often represented by T, F, and ?.[5] To construct a truth table for $\sim p$ in a three-valued logic, we have several possibilities. If p is true, $\sim p$ can be undecidable or false. If p is false, $\sim p$ can be undecidable or true. If p is undecidable, $\sim p$ can be true, false, or also undecidable. There are $2 \cdot 2 \cdot 3$, or twelve, possibilities for $\sim p$. The most common of them is

p	$\sim p$
T	F
?	?
F	T

A truth table for $p \wedge q$ in a three-valued logic will have nine lines instead of four. If we agree that $p \wedge q$ is true only when both p and q are true, we begin as follows:

p	q	$p \wedge q$
T	T	T
T	?	
T	F	
?	T	
?	?	
?	F	
F	T	
F	?	
F	F	

There are eight spaces left to be filled, and two choices, ? or F, for each space. There are $2^8 = 256$ choices in all, that is, 256 possible truth tables

[5] This discussion of three-valued logic is suggested by Howard Eves and Carroll V. Newsom, *An Introduction to the Foundations and Fundamental Concepts of Mathematics* (New York: Holt, Rinehart and Winston, Inc., 1958), pp. 278–280.

for $p \wedge q$ in three-valued logics! Here are two possibilities:

p	q	$p \wedge q$
T	T	T
T	?	?
T	F	F
?	T	?
?	?	?
?	F	F
F	T	F
F	?	F
F	F	F

p	q	$p \wedge q$
T	T	T
T	?	?
T	F	F
?	T	?
?	?	?
?	F	?
F	T	F
F	?	?
F	F	F

The first table is constructed by taking the truth value of $p \wedge q$ to be the "lower" of the separate truth values of p and of q, where undecidable is lower than true and false is lower than undecidable. The second is constructed by taking $p \wedge q$ to be undecidable whenever either p or q is undecidable, and to have the same truth value as in two-valued logic otherwise.

EXERCISE 37 Suppose we agree that $p \vee q$ is false only when p and q are both false. (*a*) How many possible truth tables are there for $p \vee q$ in three-valued logics? (*b*) Suggest two truth tables for $p \vee q$. (*c*) Explain how they are constructed.

To extend many-valued logics beyond three values, we must find a new way to indicate the truth values. The usual solution is to use numbers for the truth values. If 0 and 1 are used for false and true, the other truth values are fractions between 0 and 1, in which case many-valued logic is essentially probability theory (Chapter 6). Rosser and Turquette use the counting numbers, with 1 for true and the highest number used in the logic for false. The number M for false is then the same as the total number of values, m, in the logic. For two-valued logic, $m = 2$, T is 1, and F is 2. For three-valued logics, $m = 3$, T is 1, ? is 2, and F is 3.

Having decided what numbers to use for the truth values, we must also determine which of the many tables possible we want to construct. We do this by constructing formulas to calculate the truth values from the truth values of p and q. Rosser and Turquette suggest the following formulas.

Suppose that P and Q are statements with truth values p and q, where p and q are counting numbers no larger than m (the number of values in the logic).

Let $\sim P$ have the truth value $m - p + 1$.

Let $P \wedge Q$ have the truth value which is the larger of p and q.

Let $P \vee Q$ have the truth value which is the smaller of p and q.

Using these formulas, the tables for two-valued logic turn out to be the same as our old tables, with 1 for T and 2 for F:

p	$\sim p$
1	2
2	1

p	q	$p \wedge q$
1	1	1
1	2	2
2	1	2
2	2	2

p	q	$p \vee q$
1	1	1
1	2	1
2	1	1
2	2	2

For $m = 3$, the formulas produce the following tables from among those possible in three-valued logics:

p	$\sim p$
1	3
2	2
3	1

p	q	$p \wedge q$
1	1	1
1	2	2
1	3	3
2	1	2
2	2	2
2	3	3
3	1	3
3	2	3
3	3	3

p	q	$p \vee q$
1	1	1
1	2	1
1	3	1
2	1	1
2	2	2
2	3	2
3	1	1
3	2	2
3	3	3

EXERCISE 38 A truth table for $\sim p$ in a four-valued logic will have four lines, $p = 1, 2, 3$, and 4. The tables for $p \wedge q$ and $p \vee q$ will have sixteen lines, where $p = 1$ and $q = 1, 2, 3, 4$; $p = 2$ and $q = 1, 2, 3, 4$; and so forth. Use Rosser and Turquette's formulas to construct truth tables for $\sim p$, $p \wedge q$, and $p \vee q$ for a four-valued logic.

Looking Back

1 (*a*) Write the definition of logically equivalent. (*b*) Use truth tables to show that each of these pairs of statements are logically equivalent: (*i*) $p \rightarrow q$ and $(\sim q) \rightarrow (\sim p)$; (*ii*) $\sim(p \wedge q)$ and $(\sim p) \vee (\sim q)$; (*iii*) $\sim[(\sim p) \wedge q]$ and $p \vee (\sim q)$; (*iv*) $\sim[(\sim p) \rightarrow q]$ and $(\sim p) \wedge (\sim q)$.

2 (*a*) Write the definition of tautology. (*b*) Use a truth table to show that each of these statements is a tautology: (*i*) $q \rightarrow (p \rightarrow q)$; (*ii*) $\sim(p \vee q) \rightarrow (\sim p)$. (*c*) Is $p \rightarrow (p \rightarrow q)$ a tautology? (*d*) Is $\sim(p \vee q) \rightarrow (\sim q)$ a tautology?

***3** The transitive law of conditional statements is $[(p \rightarrow q) \wedge (q \rightarrow r)] \rightarrow (p \rightarrow r)$. (*a*) Write the transitive law of conditional statements in words. (*b*) Use a truth table to prove that the statement is a tautology. (The truth table for three simple statements p, q, and r has eight lines. Under p write four T's and four F's. Under q write two T's, two F's, two T's, two F's. Under r write T's and F's alternately.)

4 Observe that the negation of a tautology is a contradiction, and the negation of a contradiction is a tautology. (*a*) Use a truth table to show that $\sim[p \vee (\sim p)]$ is a contradiction. (*b*) Use a truth table to show that $\sim[p \wedge (\sim p)]$ is a tautology. (Recall that a truth table with only one statement p has only two lines.)

***5** The Sheffer stroke is $p|q$ ("p is incompatible with q") (Russell and Whitehead, *Principia Mathematica*, 2d. ed., Introduction.) It means p and q are not both true, and is false only when p and q are both true. The truth table for the Sheffer stroke is

p	q	$p \mid q$
T	T	F
T	F	T
F	T	T
F	F	T

Show that (*a*) $\sim p$ is logically equivalent to $p|p$; (*b*) $p \rightarrow q$ is logically equivalent to $p|(q|q)$; (*c*) $p \vee q$ is logically equivalent to $(p|p)|(q|q)$; (*d*) $p \wedge q$ is logically equivalent to $(p|q)|(p|q)$. Thus every compound statement can be written in terms of just one symbol.

6 (*a*) Use truth tables to prove that arguments in the form of the compound statement $[(p \rightarrow q) \wedge p] \rightarrow q$ are valid. (*b*) Use truth tables to prove that arguments in the form of the compound statement $[(p \rightarrow q) \wedge q] \rightarrow p$ are not valid. (*c*) Is an argument with the compound statement

$[(p \to q) \wedge (\sim p)] \to (\sim q)$ valid? (*d*) Is an argument with the compound statement $[(p \to q) \wedge (\sim q)] \to (\sim p)$ valid?

7 Write each of the following arguments in symbolic form. Which ones are valid? (*a*) If two lines are parallel, then a line which meets one also meets the other. The two lines are not parallel. Therefore, a line which meets one does not meet the other. (*b*) If two lines are parallel, then the lines do not meet. The lines do not meet. Therefore, the two lines are parallel. (*c*) If two lines are parallel, then the lines do not meet. The lines do meet. Therefore, the two lines are not parallel.

8 Consider this argument:

Premises I am a music major.
All music majors take English.
Some music majors do not take math.

Conclusions (*a*) I take English.
(*b*) I take math.
(*c*) All music majors take math.
(*d*) Some people who take English take math.
(*e*) I take English or math.
(*f*) Some music majors take English and math.

Draw a careful Venn-Euler diagram of the three premises. For each of the conclusions, say whether or not it is valid.

9 Write in words the negation of each of the conclusions in Exercise 8, without using the phrase "it is false that." For each of the negations of the conclusions, say whether or not it is a valid conclusion for the premises as given in Exercise 8.

10 Draw a Venn-Euler diagram for each of the following arguments. Which ones are valid? (*a*) All parallels are nonmeeting. All nonmeeting lines have no common point. Therefore, all parallels have no common point. (*b*) All parallels are nonmeeting. Some lines in space are parallel. Therefore, some lines in space are nonmeeting. (*c*) All parallels are nonmeeting. Some nonmeeting lines are skew lines. Therefore, some parallels are skew lines. (*d*) All parallels are nonmeeting. Some equidistant lines are not nonmeeting. Therefore, some equidistant lines are not parallels.

***11** Aristotle uses examples frequently in discussing forms of syllogisms which are not valid. These examples show that in the case when all B is A and no C is B, we can have both all C is A and no C is A (*Analytica Priora*, i, 4, 26[a]2). Use a carefully drawn Venn-Euler diagram to show that each of the following examples is, despite appearances, invalid. (*a*) All men are animals. No horses are men. Therefore, all horses are animals. (*b*) All men are animals. No stones are men. Therefore, no stones are animals.

Branching Out

1 Although only part of Aristotle's writings have survived, they still fill several volumes. His subjects range from rules of logic in the *Prior and Posterior Analytics* to rules of tragedy in the *Poetics.* He wrote *Politics* and *Ethics* and even about God (the "unmoved mover") in the *Metaphysics* and about many other subjects, including physics and biology. You will find his work described in books of various levels of difficulty. Most also contain a sketch of his life. Aristotle's own writings, in translation, are very difficult reading. You may prefer to get a general overview of his life and work, using the commentaries. You may, however, wish to tackle a small portion of a subject that interests you and read some of the translations.

2 Not too much material on George Boole's life is available. You will find some isolated stories, and there is a chapter in Bell's *Men of Mathematics.* However, his two main works, *The Mathematical Analysis of Logic* and *An Investigation of the Laws of Thought,* are available exactly as he wrote them. If you have a strong background in logic, you might try to figure out some of his formulas and derivations. Or you can begin with an overview of Boole's work and trace the history of logic through the logistic school to the *Principia Mathematica.* Along the way, you may want to find out about a man named Peano and his postulates for the counting numbers.

3 Gottfried Wilhelm von Leibniz (1646–1716) is probably the first person to have considered a symbolic systemization of logic. Leibniz is credited by Eves in *An Introduction to the History of Mathematics* and Bell in *Men of Mathematics* with stating several principles of symbolic logic, and also some concepts of set theory, including, in our terminology, subsets and the empty set. The systemization Leibniz thought about was not attempted again until modern symbolic logic began with the work of Boole almost 200 years later. Leibniz is far better known for the development of the calculus and for his discovery, coincident with Newton, of the fundamental theorem of the calculus. The controversy over who was the real discoverer is a famous one. Most people are now satisfied that the discoveries were independent, and that each man made important contributions to the study of calculus. The stories of both men may be found in Bell's *Men of Mathematics.*

4 Comments by mathematicians about mathematics can be both entertaining and enlightening. Russell's very famous one at the beginning of Section 6 is from "Mathematics and the Metaphysicians," which you can find in Newman's *The World of Mathematics*. Russell's comments on Boole's work quoted at the beginning of this chapter are also in this essay. Another famous discourse on mathematics by a mathematician is G. H. Hardy's *A Mathematician's Apology*. There are excerpts in Newman, but the book itself is easily available. If you look, you will find many more comments, especially in introductions to, or opening chapters of, books about mathematics. Raymond Wilder has some interesting comments on the teaching of mathematics as well as on the nature of mathematics in the first chapter of *Evolution of Mathematical Concepts*.

5 Fallacies are misuses of forms of arguments, or of their subject matter. "If this is good, then I will like it; and I like it, therefore it is good" is a common fallacy of everyday thought. It is a misuse of the law of detachment. "All presidents of the United States (through 1974) were men, and all men are equal; therefore all presidents of the United States were equal" is a fallacy because it uses the word "men" in two different ways. Almost any elementary book on logic will have a section on fallacies. (Look in the philosophy section of your library or bookstore; there are no elementary books devoted to mathematics logic.) Look for examples of fallacies in advertising, in news reports, and even in general conversation.

6 Quantum theory is one of the fascinating concepts of modern physics. First you will need to know the roles of wave theory and quantum theory in explaining phenomena of light. Then from elementary physics texts you should try to find out about the phenomena of quantum theory which are not observable. Reichenbach's book is only for physicists, but you might be able to find out enough about quantum mechanics to guess what his interphenomena meant. *Warning:* Attempt this subject only if you have a strong background in mathematics, and also some physics or physical science.

7 You can read about the astonishing work in foundations of mathematics done by Kurt Gödel in Nagel and Newman's *Gödel's Proof*. It touches on much of what we have done in Chapters 2 and 3, and more. The symbols are different from ours, but they are explained fully in the book. You will meet David Hilbert, who worked on the question of consistency and, along with Cantor and Russell, created the field of foundations. *Gödel's Proof* is summarized in *Mathematics in the Modern World*, but you should try to read the complete explanations in the book itself. *Beware:* Although written for nonspecialists, it is difficult reading and requires knowledge of algebra and geometry.

I'll meet you
at 12:00

4

Finite Arithmetics

1

A Clock Arithmetic

Suppose it is three o'clock and you want to know what time it will be in 5 hours. You figure that 3 hours plus 5 hours is 8 hours and conclude that it will be eight o'clock. Now, suppose again it is three o'clock and you want to know what time it will be in 10 hours: 3 hours plus 10 is 13. Would you conclude that it will be thirteen o'clock?

You would use the *clock arithmetic of the 12-hour clock.* Every time you go past 12, you subtract 12 hours and start counting over again. In clock arithmetic, 3 hours plus 10 is 13, but 13 minus 12 is 1. You would conclude that it will be one o'clock. In the clock arithmetic of the 12-hour clock, $3 + 5 = 8$ as in ordinary arithmetic but $3 + 10 = 1$.

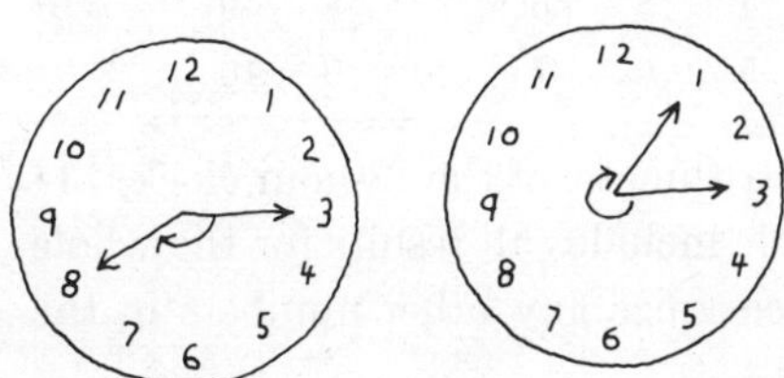

To find the time 9 hours after three o'clock, you would figure 3 hours plus 9 is 12, so it will be exactly twelve o'clock. However, we could also say that it is 0 hours after twelve o'clock. In the clock arithmetic of the 12-hour clock, $3 + 9 = 0$.

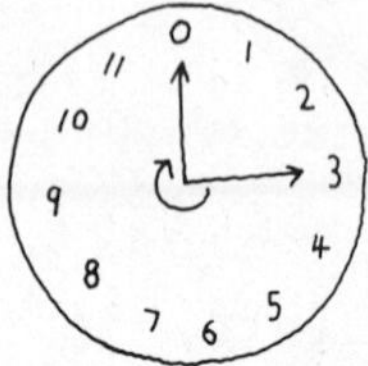

We use 0 in place of 12 on the clock. There are still twelve numbers on the clock, however, and we subtract 12 every time a result is 12 or more. Thus every result must be one of the numbers 0, 1, 2, 3, 4, 5, 6, 7, 8, 9, 10, or 11. When a result is more than 11, we reduce it by subtracting 12.

Using the results of the 12-hour clock, we can create an addition on the set of numbers {0, 1, 2, 3, 4, 5, 6, 7, 8, 9, 10, 11}. An *addition table* for the 12-hour clock shows the sums of all the numbers in the set for the clock arithmetic of the 12-hour clock:

+	0	1	2	3	4	5	6	7	8	9	10	11
0	0	1	2	3	4	5	6	7	8	9	10	11
1	1	2	3	4	5	6	7	8	9	10	11	0
2	2	3	4	5	6	7	8	9	10	11	0	1
3	3	4	5	6	7	8	9	10	11	0	1	2
4	4	5	6	7	8	9	10	11	0	1	2	3
5	5	6	7	8	9	10	11	0	1	2	3	4
6	6	7	8	9	10	11	0	1	2	3	4	5
7	7	8	9	10	11	0	1	2	3	4	5	6
8	8	9	10	11	0	1	2	3	4	5	6	7
9	9	10	11	0	1	2	3	4	5	6	7	8
10	10	11	0	1	2	3	4	5	6	7	8	9
11	11	0	1	2	3	4	5	6	7	8	9	10

Every possible addition for the clock arithmetic of the 12-hour clock, 144 results in all, is in this table. The table includes all results for the whole numbers from 0 to 11. We do not recognize any other numbers in the clock arithmetic of the 12-hour clock.

Library of Congress

Karl Friedrich Gauss is one of the greatest mathematicians of all time. He ignored no branch of mathematics, but he especially loved numbers: discovering their properties and doing arithmetic calculations no one else would undertake. "Mathematics is the queen of the sciences," said Gauss, "and the theory of numbers is the queen of mathematics."

Appropriately enough it is said that Gauss first demonstrated his arithmetical abilities at the age of three when, while watching his father work out a long and difficult payroll computation, Gauss corrected an error in the calculations. According to another story, when at the age of ten Gauss began to study arithmetic in school, on the first day the class was told to add together the numbers from 1 through 100. Gauss handed in his slate almost immediately and sat quietly during the hour the others worked. When the slates were checked, only Gauss had the right answer. He had added $1 + 100$, $2 + 99$, $3 + 98$, and so on to $50 + 51$, giving fifty pairs of numbers each totaling 101, so the answer is $50 \cdot 101 = 5{,}050$.

We can also do multiplication in the clock arithmetic of the 12-hour clock. As in addition, we subtract 12 whenever a result is 12 or more. For example, $2 \cdot 5 = 10$, but $2 \cdot 6 = 0$ and $2 \cdot 7 = 2$ in the clock arithmetic of the 12-hour clock.

EXERCISE 1 The first three rows of the multiplication table for the clock arithmetic of the 12-hour clock are

·	0	1	2	3	4	5	6	7	8	9	10	11
0	0	0	0	0	0	0	0	0	0	0	0	0
1	0	1	2	3	4	5	6	7	8	9	10	11
2	0	2	4	6	8	10	0	2	4	6	8	10
.	.	.	.	.	.	.	.	.	.	.	.	.
.	.	.	.	.	.	.	.	.	.	.	.	.
.	.	.	.	.	.	.	.	.	.	.	.	.

Complete the multiplication table for the clock arithmetic of the 12-hour clock.

We will use the symbol I_{12} ("I sub 12") for the clock arithmetic of the 12-hour clock. In the symbol I_{12}, I stands for the integers and sub 12 for the subscript 12 which, in this context, will mean to subtract twelves. When one integer can be reduced to another by subtracting twelves, we say that the integers are *congruent modulo 12*, with 12 being the *modulus*. For example, 13, 25, 37, . . . are all congruent to 1 modulo 12, because each can be reduced to 1 by subtracting twelves.

The symbol $\equiv$ is used for "congruent." Also, it is conventional to abbreviate the word "modulo" to "mod," and to set off the phrase "mod 12" in parentheses. Thus "13 is congruent to 1 modulo 12" is written $13 \equiv 1 \pmod{12}$. The notation for congruences was introduced by the great German mathematician Carl Friedrich Gauss (1777–1855), although he used congruences to develop a different and far more difficult theory than we will attempt in this chapter.

2
Modular Arithmetics

In the preceding section, we constructed an arithmetic suggested by the 12-hour clock. The arithmetic consists of the set of numbers $I_{12} = \{0, 1, 2, \ldots, 11\}$, with addition and multiplication tables. The tables are constructed by reducing ordinary arithmetic results modulo 12, which means subtracting twelves. For any counting number n, the *modular arithmetic modulo n* is the set $I_n = \{0, 1, 2, \ldots, n - 1\}$, with addition and multiplication tables. The tables are constructed by reducing ordinary arithmetic results modulo n, which means subtracting the counting number n as many times as necessary to get a number in the set I_n. We observe that for each counting number n, I_n has exactly n elements. The number of elements is called the *order* of the arithmetic.

For each I_n, there are $n \cdot n$ results in the addition table, n across times n down, and the same number of results in the multiplication table. Since the set I_n is finite, it is possible to construct addition and multiplication tables for any I_n. For small counting numbers, the tables are easy to construct. As an example, we will construct the addition and multiplication tables for I_5. The set of numbers in I_5 is $\{0, 1, 2, 3, 4\}$ and all numbers are reduced to elements of this set modulo 5; that is, by subtracting fives. We will use the symbol $(I_5,+)$ to denote the addition table for I_5, and $(I_5,\cdot)$ for the multiplication table. The tables are:

$(I_5,+)$	0	1	2	3	4
0	0	1	2	3	4
1	1	2	3	4	0
2	2	3	4	0	1
3	3	4	0	1	2
4	4	0	1	2	3

$(I_5,\cdot)$	0	1	2	3	4
0	0	0	0	0	0
1	0	1	2	3	4
2	0	2	4	1	3
3	0	3	1	4	2
4	0	4	3	2	1

There are twenty-five computations for each of these tables. It is useful, once, to observe all the computations. We list them on the following page, recalling that mod 5 means to subtract as many fives as possible.

The tables $(I_4,+)$ and $(I_4,\cdot)$ are constructed from the set $I_4 =$

$(I_5, +)$

First row:

$0 + 0 = 0$

$0 + 1 = 1$

$0 + 2 = 2$

$0 + 3 = 3$

$0 + 4 = 4$

Second row:

$1 + 0 = 1$

$1 + 1 = 2$

$1 + 2 = 3$

$1 + 3 = 4$

$1 + 4 = 5 \equiv 0 \pmod 5$

Third row:

$2 + 0 = 2$

$2 + 1 = 3$

$2 + 2 = 4$

$2 + 3 = 5 \equiv 0 \pmod 5$

$2 + 4 = 6 \equiv 1 \pmod 5$

Fourth row:

$3 + 0 = 3$

$3 + 1 = 4$

$3 + 2 = 5 \equiv 0 \pmod 5$

$3 + 3 = 6 \equiv 1 \pmod 5$

$3 + 4 = 7 \equiv 2 \pmod 5$

Fifth row:

$4 + 0 = 4$

$4 + 1 = 5 \equiv 0 \pmod 5$

$4 + 2 = 6 \equiv 1 \pmod 5$

$4 + 3 = 7 \equiv 2 \pmod 5$

$4 + 4 = 8 \equiv 3 \pmod 5$

$(I_5, \cdot)$

First row:

$0 \cdot 0 = 0$

$0 \cdot 1 = 0$

$0 \cdot 2 = 0$

$0 \cdot 3 = 0$

$0 \cdot 4 = 0$

Second row:

$1 \cdot 0 = 0$

$1 \cdot 1 = 1$

$1 \cdot 2 = 2$

$1 \cdot 3 = 3$

$1 \cdot 4 = 4$

Third row:

$2 \cdot 0 = 0$

$2 \cdot 1 = 2$

$2 \cdot 2 = 4$

$2 \cdot 3 = 6 \equiv 1 \pmod 5$

$2 \cdot 4 = 8 \equiv 3 \pmod 5$

Fourth row:

$3 \cdot 0 = 0$

$3 \cdot 1 = 3$

$3 \cdot 2 = 6 \equiv 1 \pmod 5$

$3 \cdot 3 = 9 \equiv 4 \pmod 5$

$3 \cdot 4 = 12 \equiv 2 \pmod 5$

Fifth row:

$4 \cdot 0 = 0$

$4 \cdot 1 = 4$

$4 \cdot 2 = 8 \equiv 3 \pmod 5$

$4 \cdot 3 = 12 \equiv 2 \pmod 5$

$4 \cdot 4 = 16 \equiv 1 \pmod 5$

{0, 1, 2, 3}, reducing all results modulo 4:

$(I_4,+)$	0	1	2	3
0	0	1	2	3
1	1	2	3	0
2	2	3	0	1
3	3	0	1	2

$(I_4,\cdot)$	0	1	2	3
0	0	0	0	0
1	0	1	2	3
2	0	2	0	2
3	0	3	2	1

The tables for I_2 are particularly easy to construct. The numbers are simply $I_2 = \{0, 1\}$, and we reduce modulo 2 by subtracting twos. The only result that we must reduce is then $1 + 1 = 2 \equiv 0 \pmod 2$. The tables $(I_2,+)$ and $(I_2,\cdot)$ are

$(I_2,+)$	0	1
0	0	1
1	1	0

$(I_2,\cdot)$	0	1
0	0	0
1	0	1

EXERCISE 2 (*a*) What is the order of each of these arithmetics:

I_6 I_5 I_4 I_3 I_2 I_1

(*b*) Write the set of numbers in each of the arithmetics above. (*c*) Why is there no arithmetic I_0?

EXERCISE 3 Complete each of these calculations used to construct the tables $(I_4,+)$ and $(I_4,\cdot)$:

(*a*) $1 + 3 = 4 \equiv$ ____ (mod 4)
(*b*) $2 + 2 = 4 \equiv$ ____ (mod 4)
(*c*) $2 + 3 = 5 \equiv$ ____ (mod 4)
(*d*) $3 + 3 = 6 \equiv$ ____ (mod 4)
(*e*) $2 \cdot 2 = 4 \equiv$ ____ (mod 4)
(*f*) $2 \cdot 3 = 6 \equiv$ ____ (mod 4)
(*g*) $3 \cdot 3 = 9 \equiv$ ____ (mod 4)

EXERCISE 4 Construct each of these tables for the modular arithmetics I_6, I_3, and I_1: (*a*) $(I_6,+)$ and $(I_6,\cdot)$. (*b*) $(I_3,+)$ and $(I_3,\cdot)$. (*c*) $(I_1,+)$ and $(I_1,\cdot)$.

Each table $(I_n,+)$ and $(I_n,\cdot)$ for a modular arithmetic I_n consists of two parts. The numbers across the top and down the left edge are called the *headings*. They are simply the numbers in I_n and, as we have observed, there are always exactly n of them. The numbers which are the results of adding or multiplying heading numbers reduced modulo n are the *entries in the table*. There are always $n \cdot n$ entries in each table.

Each of the entries in the table is the result of adding or multiplying two heading numbers, one from the left edge and one from the top of the table. A process in which two (not necessarily different) numbers are combined to produce a (not necessarily different) number is called a *binary operation*. Binary means "two by two"; and "operation" indicates that the two numbers are combined by some process.

We associate two words, "existence" and "uniqueness," with any operation. For a binary operation such as an operation on two numbers in a modular arithmetic, *there must exist a unique result*. For the addition and multiplication tables of a modular arithmetic, *existence* means that there is an entry in each place of the table. If a result did not exist for some two numbers in the arithmetic, then there would be a blank space in the table. *Uniqueness* means that there is not more than one entry in each place of the table. If for some two numbers the result were not unique, then there would be two or more entries in some place of the table. Together, existence and uniqueness guarantee that there is one and only one result of each binary operation for every two numbers of the arithmetic.

If you examine the tables of $(I_5,+)$ and $(I_5,\cdot)$, you will observe that every entry in each table is an element of $I_5 = \{0, 1, 2, 3, 4\}$; that is, every time we add or multiply two elements of I_5 and reduce modulo 5, we get an element of I_5:

$(I_5,+)$	0	1	2	3	4
0	0	1	2	3	4
1	1	2	3	4	0
2	2	3	4	0	1
3	3	4	0	1	2
4	4	0	1	2	3

$(I_5,\cdot)$	0	1	2	3	4
0	0	0	0	0	0
1	0	1	2	3	4
2	0	2	4	1	3
3	0	3	1	4	2
4	0	4	3	2	1

This observation provides us with an important property of the binary operations of addition and multiplication for modular arithmetics.

DEFINITION An arithmetic has the *closure property for addition* (or *is closed under addition*) if $a + b$ is an element of the arithmetic for every two elements a and b in the arithmetic. An arithmetic has the *closure property for multiplication* (or *is closed under multiplication*) if $a \cdot b$ is an element of the arithmetic for every two elements a and b in the arithmetic.

We say that I_5 is closed under both its binary operations, addition and multiplication, since the entries in each of the tables $(I_5,+)$ and $(I_5,\cdot)$ are all elements of I_5.

Because the closure property is so basic, it is sometimes included along with existence and uniqueness as a requirement for a binary operation. Using modular arithmetics, however, it is possible to construct an arithmetic with a binary operation which does not have the closure property.

First we will construct a new arithmetic which does have the closure property. Suppose we take the table $(I_5,\cdot)$ and on observing that the first row and column of the entries are all zeros, we decide simply to delete zero. We have a new arithmetic on the set $I_5 - 0$ ("I sub 5 without 0"), where $I_5 - 0 = \{1, 2, 3, 4\}$, and there is just one operation, multiplication. The table for this arithmetic is

$(I_5 - 0,\cdot)$	1	2	3	4
1	1	2	3	4
2	2	4	1	3
3	3	1	4	2
4	4	3	2	1

We observe that every entry in the table is an element of the set $I_5 - 0 = \{1, 2, 3, 4\}$, so the arithmetic is closed under multiplication.

Now we try to construct a similar arithmetic for I_4. We delete zero, so $I_4 - 0 = \{1, 2, 3\}$. The table for multiplication in $I_4 - 0$ is

$(I_4 - 0,\cdot)$	1	2	3
1	1	2	3
2	2	0	2
3	3	2	1

One of the entries in the table for $(I_4 - 0,\cdot)$ is 0, which is not an element of $I_4 - 0$, so $(I_4 - 0,\cdot)$ is not closed under multiplication.

EXERCISE 5 Refer to the tables you have made for

$(I_6,+)$ $(I_6,\cdot)$ $(I_3,+)$ $(I_3,\cdot)$ $(I_1,+)$ $(I_1,\cdot)$

(*a*) What does "existence" mean in terms of the entries in the table? (*b*) What does "uniqueness" mean in terms of the entries in the table? (*c*) Does each table have the closure property for its binary operation, addition or multiplication? (If not, you have probably forgotten to reduce every result by subtracting as many sixes, threes, or ones as possible.)

EXERCISE 6 (*a*) Construct each of these tables:

$(I_6 - 0,\cdot)$ $(I_3 - 0,\cdot)$ $(I_2 - 0,\cdot)$

(*b*) Which ones of the arithmetics are closed under multiplication? (*c*) For each of the arithmetics which is not closed under multiplication, explain why it is not. (*d*) Is there an arithmetic $(I_1 - 0,\cdot)$? Why or why not?

EXERCISE 7 This design is a reproduction of a Math/Art poster. The parts that appear white are red (r) on the poster, the black parts are dark green (dg), the lighter gray are green (g), and the darker gray are blue (b). The poster is in four sections; each section consists of sixteen squares and each square is divided into two triangles of different colors.

The pattern for the first four squares is:

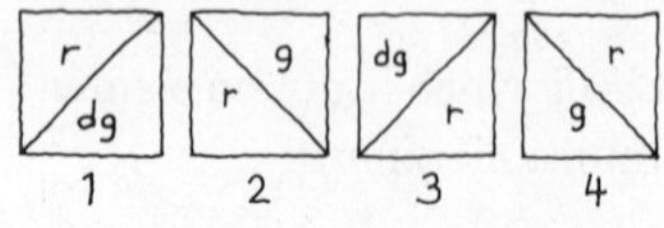

The remaining squares of the first section are filled in using $\{1, 2, 3, 4\}$ according to the table for $(I_5 - 0, \cdot)$. (*a*) Check this pattern for the first section. (*b*) Observe that the table for $(I_5 - 0, \cdot)$ is repeated in each of the other three sections. (*c*) The manual which accompanies the poster says that

There are two "errors" made by the artist in this Math/Art. Can you find them? An ancient bit of oriental philosophy says that the artist or craftsman should always have at least one error or imperfection in each piece of work so as not to offend the gods by attempting to produce a perfect work. In any event such small "errors" seldom distract from the overall visual pleasure of the art and in fact often enhance its interest.[1]

3
The Associative Property

There are many occasions when we must combine three numbers using a binary operation, but a binary operation can combine only two numbers at once. We ask, for example, what does $1 + 3 + 4 \pmod 5$ mean? There are two possible answers. We can add $1 + 3$ and then add the result to $4 \pmod 5$, as follows:

$$(1 + 3) + 4 = 4 + 4 \equiv 3 \pmod 5$$

(You are probably familiar with the use of parentheses to group the numbers which are to be combined first.) Alternatively, we can add $3 + 4 \pmod 5$ and then add 1 to the result, as follows:

$$1 + (3 + 4) \equiv 1 + 2 = 3 \pmod 5$$

We observe that the result, $3 \pmod 5$, is the same in either computation, so we conclude that the two computations are equal:

$$(1 + 3) + 4 = 1 + (3 + 4) \pmod 5$$

Similarly, to multiply three numbers we can write

$$(1 \cdot 3) \cdot 4 = 3 \cdot 4 \equiv 2 \pmod 5$$

[1] From *Math/Art Posters Teacher's Guide* by Andria Troutman and Sonia Forseth. Published by Creative Publications, Palo Alto, California.

or

$1 \cdot (3 \cdot 4) \equiv 1 \cdot 2 = 2 \pmod 5$

Again, observing that the results are the same, we conclude that the computations are equal:

$(1 \cdot 3) \cdot 4 = 1 \cdot (3 \cdot 4) \pmod 5$

DEFINITION An arithmetic has the *associative property for addition* (or *is associative for addition*) if $(a + b) + c = a + (b + c)$ for every three elements a, b, and c in the arithmetic. An arithmetic has the *associative property for multiplication* (or *is associative for multiplication*) if $(a \cdot b) \cdot c = a \cdot (b \cdot c)$ for every three elements a, b, and c in the arithmetic.

The associative property is often called the *associative law* because it always holds in ordinary arithmetic.

Although we cannot simply look at the tables for a modular arithmetic and tell whether it is associative for addition and multiplication, we can construct new tables which serve this purpose.[2] We observe that the letter b plays a special role in the associative properties. It is inside the parentheses in both the formulas $(a + b) + c = a + (b + c)$ and $(a \cdot b) \cdot c = a \cdot (b \cdot c)$.

First, we note that if $b = 0$, the associative properties will hold for all elements a and c in an arithmetic. For addition, if $b = 0$, we have $(a + 0) + c = a + c$, and also $a + (0 + c) = a + c$, so $(a + 0) + c = a + (0 + c)$. For multiplication, if $b = 0$, $(a \cdot 0) \cdot c = 0 \cdot c = 0$ and $a \cdot (0 \cdot c) = a \cdot 0 = 0$, so again $(a \cdot 0) \cdot c = a \cdot (0 \cdot c)$.

Now, suppose that $b = 1$. We will show that $(I_5,+)$ has the associative property for all elements a and c in I_5 when $b = 1$. We must show that $(a + 1) + c = a + (1 + c)$ for all a and c in I_5. To demonstrate this equality, we construct two special tables. The first has $a + 1$, for each element a in I_5, down the left-hand side. The elements of I_5 are across the top as usual. The entries in this table then represent $(a + 1) + c$ for all a and c in I_5. The second table has the elements of I_5 down the left-hand side, but $1 + c$ for each element c in I_5 across the top. It represents $a + (1 + c)$.

[2] The method used here is suggested by A. H. Clifford and G. B. Preston, *The Algebraic Theory of Semigroups*, Vol. I (Providence, R.I.: American Mathematical Society, 1961), pp. 7–8.

a	$a+1$	0	1	2	3	4
0	1	1	2	3	4	0
1	2	2	3	4	0	1
2	3	3	4	0	1	2
3	4	4	0	1	2	3
4	0	0	1	2	3	4

c	0	1	2	3	4
$1+c$	1	2	3	4	0
0	1	2	3	4	0
1	2	3	4	0	1
2	3	4	0	1	2
3	4	0	1	2	3
4	0	1	2	3	4

Comparing the entries in these tables, we see that they are identical; therefore $(a + 1) + c = a + (1 + c)$ for all elements a and c in I_5.

The tables below represent $(a + 2) + c$ and $a + (2 + c)$; and since their entries are identical, they show that $(a + 2) + c = a + (2 + c)$ for all elements a and c in I_5.

a	$a+2$	0	1	2	3	4
0	2	2	3	4	0	1
1	3	3	4	0	1	2
2	4	4	0	1	2	3
3	0	0	1	2	3	4
4	1	1	2	3	4	0

c	0	1	2	3	4
$2+c$	2	3	4	0	1
0	2	3	4	0	1
1	3	4	0	1	2
2	4	0	1	2	3
3	0	1	2	3	4
4	1	2	3	4	0

By constructing similar pairs of tables for $b = 3$ and for $b = 4$, we can prove that $(a + b) + c = a + (b + c)$ for all elements a and c in I_5 for each b in I_5; that is, we can prove the associative property for $(I_5, +)$.

For multiplication, if $b = 2$, we construct tables in which the left-hand side of the first table has $a \cdot 2$ for each element a in I_5 and the top of the second table has $2 \cdot c$ for each element c in I_5.

a	$a \cdot 2$	0	1	2	3	4
0	0	0	0	0	0	0
1	2	0	2	4	1	3
2	4	0	4	3	2	1
3	1	0	1	2	3	4
4	3	0	3	1	4	2

c	0	1	2	3	4
$2 \cdot c$	0	2	4	1	3
0	0	0	0	0	0
1	0	2	4	1	3
2	0	4	3	2	1
3	0	1	2	3	4
4	0	3	1	4	2

Since the entries in these tables are identical, we have proved that $(a \cdot 2) \cdot c = a \cdot (2 \cdot c)$ for all a and c in I_5. By constructing similar tables for $b = 3$ and for $b = 4$, we can prove the associative property for $(I_5, \cdot)$.

EXERCISE 8 Verify the associative properties for addition and multiplication in I_5 by computing each side of each of these examples:

(*a*) $(1 + 2) + 3 = 1 + (2 + 3) \pmod 5$
(*b*) $(1 + 2) + 4 = 1 + (2 + 4) \pmod 5$
(*c*) $(2 + 3) + 4 = 2 + (3 + 4) \pmod 5$
(*d*) $(3 + 3) + 4 = 3 + (3 + 4) \pmod 5$
(*e*) $(1 \cdot 2) \cdot 3 = 1 \cdot (2 \cdot 3) \pmod 5$
(*f*) $(1 \cdot 2) \cdot 4 = 1 \cdot (2 \cdot 4) \pmod 5$
(*g*) $(2 \cdot 3) \cdot 4 = 2 \cdot (3 \cdot 4) \pmod 5$
(*h*) $(3 \cdot 3) \cdot 4 = 3 \cdot (3 \cdot 4) \pmod 5$

EXERCISE 9 Complete the proof of the associative property for $(I_5, +)$ by constructing these tables for I_5:

(*a*) $(a + 3) + c$ and $a + (3 + c)$
(*b*) $(a + 4) + c$ and $a + (4 + c)$

EXERCISE 10 Show that $(a \cdot 1) \cdot c = a \cdot (1 \cdot c)$ in any arithmetic, so that it is not necessary to construct tables for $b = 1$ for multiplication.

EXERCISE 11 Complete the proof of the associative property for $(I_5, \cdot)$ by constructing these tables for I_5:

(*a*) $(a \cdot 3) \cdot c$ and $a \cdot (3 \cdot c)$
(*b*) $(a \cdot 4) \cdot c$ and $a \cdot (4 \cdot c)$

EXERCISE 12 Prove the associative properties for addition and multiplication for I_4 by constructing these pairs of tables in I_4. (*a*) Addition tables for $b = 1$. (*b*) Addition tables for $b = 2$. (*c*) Addition tables for $b = 3$. (*d*) Multiplication tables for $b = 2$. (*e*) Multiplication tables for $b = 3$.

EXERCISE 13 Prove the associative properties for addition and multiplication for I_3 by constructing these pairs of tables in I_3. (*a*) Addition tables for $b = 1$. (*b*) Addition tables for $b = 2$. (*c*) Multiplication tables for $b = 2$.

EXERCISE 14 (*a*) What one pair of tables must be constructed to prove both associative properties for I_2? (*b*) Construct the one pair of tables. (*c*) Explain why the associative properties for I_1 can be proved without constructing any tables.

4
Identities

You have probably noticed many patterns in the addition and multiplication tables for the modular arithmetics. In this table for $(I_5,+)$, the first row and column of entries are circled:

$(I_5,+)$	0	1	2	3	4
0	0	1	2	3	4
1	1	2	3	4	0
2	2	3	4	0	1
3	3	4	0	1	2
4	4	0	1	2	3

The reason the first row and column are special is clear. They are exactly the same as the headings. The entries in the first row are the results

$0 + 0 = 0$
$0 + 1 = 1$
$0 + 2 = 2$
$0 + 3 = 3$
$0 + 4 = 4$

The entries in the first column are the results

$0 + 0 = 0$
$1 + 0 = 1$
$2 + 0 = 2$
$3 + 0 = 3$
$4 + 0 = 4$

We see that 0 plays a special role in $(I_5,+)$. Whenever 0 is added to an element of I_5, or an element of I_5 is added to 0, the result is that element.

DEFINITION An arithmetic has the *identity property for addition* (or has *an additive identity*) if it contains an element 0 such that $a + 0 = a$ and $0 + a = a$ for every element a in the arithmetic.

For any modular arithmetic I_n, the first row and column of the addition table $(I_n, +)$ repeat the headings. This row and column are next to and under the heading element 0, which indicates that 0 is the additive identity for any I_n.

In $(I_5, \cdot)$, the first row and column are all zeros. It is the second row and column which repeat the headings.

$(I_5, \cdot)$	0	1	2	3	4
0	0	0	0	0	0
1	0	1	2	3	4
2	0	2	4	1	3
3	0	3	1	4	2
4	0	4	3	2	1

The entries in the second row of $(I_5, \cdot)$ are the results

$1 \cdot 0 = 0$
$1 \cdot 1 = 1$
$1 \cdot 2 = 2$
$1 \cdot 3 = 3$
$1 \cdot 4 = 4$

The entries in the second column of $(I_5, \cdot)$ are the results

$0 \cdot 1 = 0$
$1 \cdot 1 = 1$
$2 \cdot 1 = 2$
$3 \cdot 1 = 3$
$4 \cdot 1 = 4$

Whenever 1 is multiplied by an element of I_5, or an element of I_5 is multiplied by 1, the result is that element. Thus 1 plays a role in $(I_5, \cdot)$ similar to the role of 0 in $(I_5, +)$.

DEFINITION An arithmetic has the *identity property for multiplication* (or has *a multiplicative identity*) if it contains an element 1 such that $a \cdot 1 = a$ and $1 \cdot a = a$ for every element a in the arithmetic.

In each multiplication table $(I_n, \cdot)$, with one exception, the second row and column of entries repeat the headings. The row and column are next to and under the heading element 1, which indicates that 1 is the multiplicative identity for all I_n which have a multiplicative identity. There is one modular arithmetic which has no multiplicative identity.

These are the addition and multiplication tables for I_1:

$(I_1,+)$	0
0	0

$(I_1,\cdot)$	0
0	0

Zero is the only number in I_1; thus the first row and column of $(I_1,+)$ consist of only zero. Zero is the additive identity, so I_1 has the additive identity. The number 1, however, is not in I_1, and there is no second row and column in the multiplication table $(I_1,\cdot)$. I_1 is thus the one modular arithmetic which does not have the multiplicative identity.

EXERCISE 15 (*a*) List the additions which give each of the results in the circled row and column of the following table for $(I_4,+)$:

$(I_4,+)$	0	1	2	3
0	0	1	2	3
1	1	2	3	0
2	2	3	0	1
3	3	0	1	2

(*b*) List the multiplications which give each of the results in the circled row and column of the following table for $(I_4,\cdot)$:

$(I_4,\cdot)$	0	1	2	3
0	0	0	0	0
1	0	1	2	3
2	0	2	0	2
3	0	3	2	1

EXERCISE 16 (*a*) Circle the first row and column of each of these addition tables:

$(I_2,+)$ $(I_3,+)$ $(I_6,+)$

Observe that 0 is the additive identity for each of these arithmetics.
(*b*) Circle the second row and column of each of these multiplication tables:

$(I_2, \cdot)$ $(I_3, \cdot)$ $(I_6, \cdot)$

Observe that 1 is the multiplicative identity for each of these arithmetics.

5
Inverses and Divisors of Zero

In any modular arithmetic I_n, the additive identity 0 always follows a distinctive pattern in the addition table $(I_n, +)$. After its first appearance in the upper left corner, 0 moves to the far right of the second row and then back to the left one place in each row after that. This pattern indicates that there is always a zero in every row and every column of an addition table $(I_n, +)$. For I_5, the pattern in the table $(I_5, +)$ is

$(I_5, +)$	0	1	2	3	4
0	(0)	1	2	3	4
1	1	2	3	4	(0)
2	2	3	4	(0)	1
3	3	4	(0)	1	2
4	4	(0)	1	2	3

The circled entries are the results

$0 + 0 = 0$
$1 + 4 = 0$
$2 + 3 = 0$
$3 + 2 = 0$
$4 + 1 = 0$

For every element in I_5, there is an element (possibly the same one) which we can add to get the additive identity.

DEFINITION An arithmetic has the *inverse property for addition* (or has *an additive inverse for each element*) if for every element a in the arithmetic there exists an element a' such that $a + a' = 0$ and $a' + a = 0$.

For I_5, the pattern of zeros in the addition table $(I_5, +)$ shows us that 0 is its own additive inverse, 1 and 4 are additive inverses of one another, and also 2 and 3 are additive inverses of one another.

In $(I_5, \cdot)$, there is not so distinct a pattern for the multiplicative identity 1 among the entries in the table. However, there is a 1 in every row and every column, except of course the first row and column, which have only zeros. The ones in the table $(I_5, \cdot)$ are

$(I_5, \cdot)$	0	1	2	3	4
0	0	0	0	0	0
1	0	(1)	2	3	4
2	0	2	4	(1)	3
3	0	3	(1)	4	2
4	0	4	3	2	(1)

The circled entries are the results

$1 \cdot 1 = 1$
$2 \cdot 3 = 1$
$3 \cdot 2 = 1$
$4 \cdot 4 = 1$

For every element in I_5, except 0, there is an element (possibly the same one) by which we can multiply to get the multiplicative identity.

DEFINITION An arithmetic has the *inverse property for multiplication* (or *has a multiplicative inverse for each element*) if for every element a in the arithmetic there exists an element a' such that $a \cdot a' = 1$ and $a' \cdot a = 1$.

In I_5, 0 has no multiplicative inverse, 1 is its own multiplicative inverse, 2 and 3 are multiplicative inverses of one another, and 4 is again its own multiplicative inverse. We do not say that $(I_5, \cdot)$ has the multiplicative inverse property because there is an element, 0, which has no multiplicative inverse in I_5. There is a multiplicative inverse for every nonzero element in I_5, however.

In general, 0 never has a multiplicative inverse. If 0 did have a multiplicative inverse, it would be an element $0'$ such that $0 \cdot 0' = 1$ and

$0' \cdot 0 = 1$. But for any element a in a modular arithmetic, $a \cdot 0 = 0$ and $0 \cdot a = 0$, so there cannot be an element $a = 0'$ such that $a \cdot 0 = 1$ and $0 \cdot a = 1$. Thus some modular arithmetics have a multiplicative inverse for each nonzero element, but no modular arithmetic has the inverse property for multiplication.

EXERCISE 17 (*a*) List the additions which give each of the results circled in this table for $(I_4,+)$:

$(I_4,+)$	0	1	2	3
0	(0)	1	2	3
1	1	2	3	(0)
2	2	3	(0)	1
3	3	(0)	1	2

(*b*) List the additive inverse of each element in I_4.
(*c*) List the multiplications which give each of the results circled in this table for $(I_4,\cdot)$:

$(I_4,\cdot)$	0	1	2	3
0	0	0	0	0
1	0	(1)	2	3
2	0	2	0	2
3	0	3	2	(1)

(*d*) List the multiplicative inverse of each element in I_4 which has a multiplicative inverse.

EXERCISE 18 (*a*) Circle each 0 among the entries in each of these addition tables:

$(I_2,+)$ $\quad$ $(I_3,+)$ $\quad$ $(I_6,+)$

(*b*) For each arithmetic, list the additive inverse of each element.

EXERCISE 19 (*a*) Circle each 1 among the entries in each of these multiplication tables:

$(I_2,\cdot)$ $\quad$ $(I_3,\cdot)$ $\quad$ $(I_6,\cdot)$

(*b*) For each arithmetic, list the multiplicative inverse of each element which has a multiplicative inverse. (*c*) For each arithmetic list the elements which have no multiplicative inverse. (*d*) Which of the arithmetics have a multiplicative inverse for every nonzero element?

EXERCISE 20 Recall these arithmetics without zero, constructed in Section 2 and Exercise 6:

$(I_2 - 0,\cdot)$ $(I_3 - 0,\cdot)$ $(I_4 - 0,\cdot)$ $(I_5 - 0,\cdot)$ $(I_6 - 0,\cdot)$

(*a*) For each arithmetic, list the multiplicative inverse of each element which has a multiplicative inverse. (*b*) For each arithmetic list the elements which have no multiplicative inverse. (*c*) Which of the arithmetics have the multiplicative inverse property?

There may be elements besides 0 in modular arithmetics which do not have multiplicative inverses. You may remember an unusual result involving 0 in $(I_4,\cdot)$:

$(I_4,\cdot)$	0	1	2	3
0	0	0	0	0
1	0	1	2	3
2	0	2	(0)	2
3	0	3	2	1

A 0 appears in the middle of the table $(I_4,\cdot)$. The multiplication which produces this 0 is $2 \cdot 2 \equiv 0 \pmod 4$.

DEFINITION An arithmetic has *divisors of zero* if it has elements a and b (possibly the same element) such that $a \neq 0$ and $b \neq 0$, but $a \cdot b = 0$.

Two is a divisor of zero in I_4, since $2 \neq 0$, but $2 \cdot 2 = 0$ in I_4. Moreover, we observe that there is no 1 in the row next to 2 or in the column under 2 in $(I_4,\cdot)$. The absence of a 1 indicates that there is no element $2'$ in I_4 such that $2 \cdot 2' = 1$ and $2' \cdot 2 = 1$. Thus 2 has no multiplicative inverse in I_4.

We can prove that for any element a in a modular arithmetic, if a is a divisor of zero, then a has no multiplicative inverse in the arithmetic. We suppose that a is a divisor of zero; thus there is an element b such that $a \neq 0$ and $b \neq 0$, but $a \cdot b = 0$. We also suppose that a has a multiplicative inverse, so that there is an element a' such that $a \cdot a' = 1$ and

$a' \cdot a = 1$. Then,

$$\begin{aligned} a \cdot b &= 0 \\ a' \cdot (a \cdot b) &= a' \cdot 0 \\ (a' \cdot a) \cdot b &= a' \cdot 0 \\ 1 \cdot b &= a' \cdot 0 \\ b &= a' \cdot 0 \\ b &= 0 \end{aligned}$$

But $b \neq 0$ since a is a divisor of zero; therefore, the supposed multiplicative inverse a', by which we multiplied both sides of the equation in the second line, must not exist. It is also true for any modular arithmetic that if $a \neq 0$ and a is not a divisor of zero, then a does have a multiplicative inverse.

EXERCISE 21 (*a*) Circle all the zeros, other than those in the first row and column, among the entries in each of these tables:

$(I_2,\cdot)$ $(I_3,\cdot)$ $(I_5,\cdot)$

(*b*) List all the divisors of zero in each of the arithmetics. (*c*) Does every nonzero element in each of the arithmetics have a multiplicative inverse?

EXERCISE 22 (*a*) Circle all the zeros, other than those in the first row and column, among the entries in $(I_6,\cdot)$
(*b*) List all the divisors of zero in $(I_6,\cdot)$ (*c*) Does any divisor of zero in $(I_6,\cdot)$ have a multiplicative inverse?

EXERCISE 23 (*a*) Construct each of these multiplication tables:

$(I_7,\cdot)$ $(I_8,\cdot)$ $(I_9,\cdot)$

(*b*) List all the divisors of zero in each of the arithmetics which have any. (*c*) Can you find a way to predict, for any counting number n, if I_n will have any divisors of zero?

EXERCISE 24 (*a*) Circle all the zeros among the entries in each of these tables:

$(I_2 - 0,\cdot)$ $(I_3 - 0,\cdot)$ $(I_4 - 0,\cdot)$ $(I_5 - 0,\cdot)$ $(I_6 - 0,\cdot)$

(*b*) Which of the tables have the closure property? (*c*) Which of the tables have the multiplicative inverse property? (*d*) Which of the tables have divisors of zero? (*e*) Can you find a way to predict, for any counting number n greater than 1, if $(I_n - 0, \cdot)$ will have the closure property? the multiplicative inverse property? divisors of zero?

6

The Commutative Property

In this table for $(I_5,+)$, we have circled the second row and column next to and under 1:

$(I_5,+)$	0	1	2	3	4
0	0	1	2	3	4
1	1	2	3	4	0
2	2	3	4	0	1
3	3	4	0	1	2
4	4	0	1	2	3

This row and column do not repeat the headings in the original order. You may have observed, however, that for any $(I_n,+)$ or $(I_n,\cdot)$, each pair of corresponding rows and columns is identical. In fact, you may have used this observation to construct tables more quickly. The row and column circled on the table for $(I_5,+)$ both are, in order, 1, 2, 3, 4, 0. These entries are the results

Row	*Column*
$1 + 0 = 1$	$0 + 1 = 1$
$1 + 1 = 2$	$1 + 1 = 2$
$1 + 2 = 3$	$2 + 1 = 3$
$1 + 3 = 4$	$3 + 1 = 4$
$1 + 4 = 0$	$4 + 1 = 0$

Since the results are identical, we have

$1 + 0 = 0 + 1$
$1 + 1 = 1 + 1$
$1 + 2 = 2 + 1$
$1 + 3 = 3 + 1$
$1 + 4 = 4 + 1$

Similarly, we can circle any corresponding row and column of the multiplication table $(I_5, \cdot)$. Here we have chosen the fourth row and column, next to and under 3:

$(I_5,\cdot)$	0	1	2	3	4
0	0	0	0	0	0
1	0	1	2	3	4
2	0	2	4	1	3
3	0	3	1	4	2
4	0	4	3	2	1

This row and column both are, in order, 0, 3, 1, 4, 2. The entries in the row and column are the results

Row	*Column*
$3 \cdot 0 = 0$	$0 \cdot 3 = 0$
$3 \cdot 1 = 3$	$1 \cdot 3 = 3$
$3 \cdot 2 = 1$	$2 \cdot 3 = 1$
$3 \cdot 3 = 4$	$3 \cdot 3 = 4$
$3 \cdot 4 = 2$	$4 \cdot 3 = 2$

Since the results are identical, we have

$$3 \cdot 0 = 0 \cdot 3$$
$$3 \cdot 1 = 1 \cdot 3$$
$$3 \cdot 2 = 2 \cdot 3$$
$$3 \cdot 3 = 3 \cdot 3$$
$$3 \cdot 4 = 4 \cdot 3$$

We see that the entries in corresponding rows and columns of $(I_5,+)$ represent the addition of the same two elements in either order. The corresponding rows and columns of $(I_5,\cdot)$ represent the multiplication of the same two elements in either order. Since in every case the results are the same, we can add or multiply two elements of I_5 in either order.

DEFINITION An arithmetic has the *commutative property for addition* (or *is commutative for addition*) if $a + b = b + a$ for every two elements a and b in the arithmetic. An arithmetic has the *commutative property for multiplication* (or *is commutative for multiplication*) if $a \cdot b = b \cdot a$ for every two elements a and b in the arithmetic.

Like the associative property, the commutative property was once thought to be so common for addition and multiplication that it was called the *commutative law*. It has turned out not to be so common a property, as we shall see in Section 9. However, all the modular arithmetics do have the commutative property for both addition and multiplication.

EXERCISE 25 (*a*) List the additions which give the results in the third row and column of $(I_5,+)$ next to and under 2. (*b*) List the additions which give the results of the fourth row and column of $(I_5,+)$ next to and under 3. (*c*) List the additions which give the results in the fifth row and column of $(I_5,+)$ next to and under 4. (*d*) List the multiplications which give the results in the third row and column of $(I_5,\cdot)$ next to and under 2. (*e*) List the multiplications which give the results in the fifth row and column of $(I_5,\cdot)$ next to and under 4.

EXERCISE 26 Compare the corresponding pairs of rows and columns for each of these tables:

$(I_2,+)$ $(I_3,+)$ $(I_4,+)$ $(I_6,+)$
$(I_2,\cdot)$ $(I_3,\cdot)$ $(I_4,\cdot)$ $(I_6,\cdot)$

Observe that each arithmetic has the commutative property for its operation.

7
Semigroups and Groups

Algebraic structures are sets with one or more operations, where the operations have certain combinations of properties such as those discussed in Sections 2 to 6. Using the tables we constructed in those sections as examples, we can identify two basic types of algebraic structures for sets with one operation. Of all our tables, $(I_1,\cdot)$ has the fewest properties:

$(I_1,\cdot)$	0
0	0

$(I_1,\cdot)$ has the closure and associative properties for multiplication. It does not have the multiplicative identity 1, and it has no element with a multiplicative inverse. It is commutative for multiplication, but the commutative property is listed last, for special reasons.

DEFINITION An arithmetic with one operation is a *semigroup* if it has at least the properties

1 Closure
2 Associative

for the operation.

The arithmetic $(I_1, \cdot)$ is a semigroup. Also, all the tables $(I_n, +)$ and $(I_n, \cdot)$ are semigroups, since all have at least the closure and associative properties.

We also know arithmetics which are not semigroups. $(I_4 - 0, \cdot)$ does not have the closure property, so it is not a semigroup. $(I_5 - 0, \cdot)$ does have at least the closure and associative properties, however, so it is a semigroup.

$(I_4 - 0, \cdot)$	1	2	3
1	1	2	3
2	2	0	2
3	3	2	1

$(I_5 - 0, \cdot)$	1	2	3	4
1	1	2	3	4
2	2	4	1	3
3	3	1	4	2
4	4	3	2	1

We see that some arithmetics of the type $(I_n - 0, \cdot)$ are semigroups, and some are not semigroups.

Any arithmetic $(I_n, +)$ always has the identity and inverse properties for addition as well as the closure and associative properties. We still reserve the commutative property.

DEFINITION An arithmetic with one operation is a *group* if it has at least the properties

1 Closure
2 Associative
3 Identity
4 Inverse

for the operation.

For any counting number n, $(I_n, +)$ is a group.

For multiplication, however, the situation is again more complicated. $(I_1,\cdot)$ cannot be a group because it has neither the multiplicative identity nor multiplicative inverse property. $(I_4,\cdot)$ cannot be a group because it has divisors of zero which have no multiplicative inverses. $(I_5,\cdot)$ has no divisors of zero, but it has the element 0 which has no multiplicative inverse. We see that no arithmetic $(I_n,\cdot)$ is a group.

$(I_5 - 0,\cdot)$ is, however, a group as well as a semigroup. It has the closure and associative properties, and also the identity and inverse properties. There is a multiplicative inverse for every element because zero has been deleted and there are no divisors of zero in $(I_5 - 0,\cdot)$. If there are no divisors of zero, then an arithmetic $(I_n - 0,\cdot)$ will be a group as well as a semigroup. If there are divisors of zero, then $(I_n - 0,\cdot)$ is not even a semigroup.

EXERCISE 27 (*a*) List all the properties of each of these arithmetics:

$(I_2 - 0,\cdot)$ $(I_3 - 0,\cdot)$ $(I_6 - 0,\cdot)$
$(I_7 - 0,\cdot)$ $(I_8 - 0,\cdot)$ $(I_9 - 0,\cdot)$

(*b*) Which of the arithmetics are semigroups? (*c*) Can you find a way to predict, for any counting number n greater than 1, if $(I_n - 0,\cdot)$ will be a semigroup?

EXERCISE 28 (*a*) Which of these arithmetics are groups:

$(I_2 - 0,\cdot)$ $(I_3 - 0,\cdot)$ $(I_6 - 0,\cdot)$
$(I_7 - 0,\cdot)$ $(I_8 - 0,\cdot)$ $(I_9 - 0,\cdot)$

(*b*) Can you find a way to predict, for any counting number n greater than 1, if $(I_n - 0,\cdot)$ will be a group?

You are probably already familiar with the following definition.

DEFINITION A counting number greater than 1 is *prime* if the only counting numbers which divide it are itself and 1. A counting number greater than 1 is *composite* if it has divisors which are counting numbers other than itself and 1.

A modular arithmetic I_n has no divisors of zero whenever n is prime. Among our tables, $(I_2 - 0,\cdot)$, $(I_3 - 0,\cdot)$, $(I_5 - 0,\cdot)$, and $(I_7 - 0,\cdot)$ have no divisors of zero. When n is composite, the modular arithmetic I_n does have divisors of zero. $(I_4 - 0,\cdot)$, $(I_6 - 0,\cdot)$, $(I_8 - 0,\cdot)$, and $(I_9 - 0,\cdot)$

do have divisors of zero. Thus an arithmetic $(I_n - 0, \cdot)$ will be a group if n is prime.

Last of all, we consider an algebraic structure with the commutative property.

DEFINITION A *commutative*, or *Abelian*, group is a group which has the commutative property for its operation.

The name "Abelian," often used for commutative groups, is in honor of Niels Henrik Abel (1802–1829), a Norwegian mathematician. The word "group" was first used by a French mathematician, Évariste Galois (1811–1832), who flourished briefly but brilliantly (a whole field of modern algebra, Galois theory, is named for him) just after the Napoleonic era. Both Abel and Galois died at a very early age, Abel at twenty-six and Galois at twenty, and both died directly or indirectly as a consequence of the indifference of their contemporaries. Either, had he lived, might have equalled or even surpassed the great Gauss.

Commutative groups have all of the properties we have discussed:

1 Closure
2 Associative
3 Identity
4 Inverse
5 Commutative

Since all the modular arithmetics are commutative for both addition and multiplication, all the tables which we have found to be groups are also commutative groups. These tables are $(I_n, +)$ for any counting number n, and $(I_n - 0, \cdot)$ when n is prime.

EXERCISE 29 Consider each of these arithmetics:

$(I_1, +)$ $(I_2, +)$ $(I_3, +)$ $(I_4, +)$ $(I_5, +)$ $(I_6, +)$
$(I_1, \cdot)$ $(I_2, \cdot)$ $(I_3, \cdot)$ $(I_4, \cdot)$ $(I_5, \cdot)$ $(I_6, \cdot)$
$(I_7, \cdot)$ $(I_8, \cdot)$ $(I_9, \cdot)$
$(I_2 - 0, \cdot)$ $(I_3 - 0, \cdot)$ $(I_4 - 0, \cdot)$ $(I_5 - 0, \cdot)$
$(I_6 - 0, \cdot)$ $(I_7 - 0, \cdot)$ $(I_8 - 0, \cdot)$ $(I_9 - 0, \cdot)$

(*a*) Which ones are semigroups? (*b*) Which ones are groups? (*c*) Which ones are commutative groups? (*d*) Which ones are neither semigroups nor groups?

E. T. Bell calls his chapter on Niels Henrik Abel "Genius and Poverty." Born in Oslo (then Christiania), Norway, this cheerful and optimistic mathematician is the only Scandinavian to have achieved a prominent place in the history of mathematics. When his father died, leaving him at the age of eighteen to provide for his mother and six sisters and brothers, he managed somehow to survive by tutoring. His friends finally persuaded the Norwegian government to give Abel a small grant with which to travel in France and Germany. He had his first important paper, which eventually led Galois to the invention of group theory, printed as a letter of introduction to the mathematicians of Europe. It is said that, in Germany, Gauss threw it aside without reading it. In Paris, France, subsisting on almost nothing, Abel wrote another important paper and left it with Cauchy for the Académie. Cauchy mislaid it. Meanwhile, Abel had returned to Norway, with both himself and his funds exhausted. There he sustained himself for two years, again mostly by tutoring, before he died of tuberculosis. Just two days after his death, news arrived that he was to be appointed professor at the prestigious University of Berlin.

Photo: David Eugene Smith Collection, Columbia University Library

Courtesy of IBM

E. T. Bell calls his chapter on Évariste Galois "Genius and Stupidity." Galois' dream was to study at the École Polytechnique, center of both mathematical and revolutionary activity during the turbulent times following the Napoleonic era in France. Twice he failed the entrance examinations, being too far ahead of his examiners to be understood.

Galois published four papers in journals during his brief career. His major work, however, he sent to the Académie des Sciences. Twice the work was lost, once through the negligence of Cauchy, a second time when Fourier died leaving it among his papers. Galois submitted a third paper, but it was rejected by Poisson as incomprehensible.

Galois submerged himself in politics, and twice he was arrested as a dangerous radical. While on parole he was challenged to a duel, whether over his politics or a woman is not clear. The whole night before the duel he wrote—three letters, the third of which outlines whole fields of mathematics he had not had time to develop. The next day he was mortally wounded, and the day following he died.

EXERCISE 30 Without constructing their tables, predict which of these arithmetics will be groups:

$(I_{10} - 0,\cdot)$ $(I_{11} - 0,\cdot)$ $(I_{12} - 0,\cdot)$
$(I_{13} - 0,\cdot)$ $(I_{14} - 0,\cdot)$ $(I_{15} - 0,\cdot)$

8
Rings and Fields

So far we have considered addition and multiplication separately for the modular arithmetics. We should be able to study the structure of an arithmetic with both addition and multiplication. Before we study algebraic structures with two operations, we need one more property which connects the two operations. Given three numbers in an arithmetic, can we add two of them and then multiply by the third, or should we multiply each of the first two by the third and then add the results? For example, in I_5,

$$2 \cdot (3 + 4) \equiv 2 \cdot 2 = 4 \pmod 5$$

and

$$(2 \cdot 3) + (2 \cdot 4) \equiv 1 + 3 = 4 \pmod 5$$

and so

$$2 \cdot (3 + 4) = (2 \cdot 3) + (2 \cdot 4) \pmod 5$$

DEFINITION An arithmetic has the *distributive property* (or *is distributive) for multiplication over addition* if $a \cdot (b + c) = (a \cdot b) + (a \cdot c)$ and $(b + c) \cdot a = (b \cdot a) + (c \cdot a)$ for every three elements a, b, and c in the arithmetic.

We observe that if an arithmetic has the commutative property for multiplication, then $a \cdot (b + c) = (b + c) \cdot a$ and $a \cdot b = b \cdot a$ and $a \cdot c = c \cdot a$, so only the first formula of the distributive property must be demonstrated.

The distributive property for multiplication over addition cannot be demonstrated for modular arithmetics by considering their addition and multiplication tables like, for example, the commutative properties. Nor can it be demonstrated by the construction of new tables like the associative properties. It is true, however, that all modular arithmetics I_n have the distributive property for multiplication over addition. Furthermore, like the commutative and associative properties, the distributive property for multiplication over addition is true in ordinary arithmetic, so it is often called the *distributive law.*

EXERCISE 31 Verify the distributive property for multiplication over addition in I_5 by computing each side of each of these examples:

(*a*) $1 \cdot (2 + 3) = (1 \cdot 2) + (1 \cdot 3) \pmod 5$
(*b*) $2 \cdot (1 + 3) = (2 \cdot 1) + (2 \cdot 3) \pmod 5$
(*c*) $3 \cdot (1 + 2) = (3 \cdot 1) + (3 \cdot 2) \pmod 5$
(*d*) $2 \cdot (1 + 4) = (2 \cdot 1) + (2 \cdot 4) \pmod 5$
(*e*) $4 \cdot (1 + 2) = (4 \cdot 1) + (4 \cdot 2) \pmod 5$
(*f*) $4 \cdot (1 + 3) = (4 \cdot 1) + (4 \cdot 3) \pmod 5$
(*g*) $2 \cdot (3 + 4) = (2 \cdot 3) + (2 \cdot 4) \pmod 5$
(*h*) $3 \cdot (2 + 4) = (3 \cdot 2) + (3 \cdot 4) \pmod 5$
(*i*) $4 \cdot (2 + 3) = (4 \cdot 2) + (4 \cdot 3) \pmod 5$
(*j*) Are these all the examples of the distributive property for multiplication over addition in I_5?

EXERCISE 32 Verify the distributive property for multiplication over addition in I_4 by computing each side of each of these examples:

(*a*) $1 \cdot (2 + 3) = (1 \cdot 2) + (1 \cdot 3) \pmod 4$
(*b*) $2 \cdot (1 + 3) = (2 \cdot 1) + (2 \cdot 3) \pmod 4$
(*c*) $3 \cdot (1 + 2) = (3 \cdot 1) + (3 \cdot 2) \pmod 4$
(*d*) $2 \cdot (2 + 3) = (2 \cdot 2) + (2 \cdot 3) \pmod 4$
(*e*) $3 \cdot (2 + 3) = (3 \cdot 2) + (3 \cdot 3) \pmod 4$
(*f*) $2 \cdot (3 + 3) = (2 \cdot 3) + (2 \cdot 3) \pmod 4$
(*g*) Are these all the examples of the distributive property for multiplication over addition in I_4?

EXERCISE 33 (*a*) Compute all the examples of the distributive property for multiplication over addition in I_1. (*b*) How many examples are there in I_1?

The distributive property for multiplication over addition makes it possible to consider algebraic structures with two operations. Again we start with I_1, the arithmetic with the fewest elements and also the fewest properties. Since I_1 is the only modular arithmetic with no identity for multiplication, $(I_1,+)$ is a commutative group but $(I_1,\cdot)$ is only a semigroup.

DEFINITION An arithmetic with addition and multiplication is a *ring* if it is a commutative group for addition, at least a semigroup for multiplication, and has the distributive property for multiplication over addition.

We may describe the structure of a modular arithmetic which is a ring by listing its properties for addition and multiplication:

A1	Closure for addition
A2	Associative for addition
A3	Identity for addition
A4	Inverse for addition
A5	Commutative for addition
M1	Closure for multiplication
M2	Associative for multiplication
D	Distributive for multiplication over addition

I_1 with both addition and multiplication, written $(I_1,+,\cdot)$, is a ring. In fact, since I_1 also has the commutative property for multiplication, we call it a *commutative ring*. All other modular arithmetics $(I_n,+,\cdot)$, since they have at least the properties of $(I_1,+,\cdot)$, are also commutative rings. I_4, for example, has the multiplicative identity, so we call $(I_4,+,\cdot)$ a *commutative ring with identity*, or *with unity*. All modular arithmetics $(I_n,+,\cdot)$ except $(I_1,+,\cdot)$ are commutative rings with unity.

I_5 is distinguished from I_4 by the fact that $(I_5 - 0,\cdot)$ is a commutative group, but since $(I_4 - 0,\cdot)$ has divisors of zero it is not a group.

DEFINITION An arithmetic with addition and multiplication is a *field* if it is a commutative group for addition and, without 0, a commutative group for multiplication, and has the distributive property for multiplication over addition.

We may describe the structure of a modular arithmetic which is a field by listing its properties for addition and multiplication. We note that the presence of zero affects only the inverse property of the commutative group for multiplication:

A1	Closure for addition
A2	Associative for addition
A3	Identity for addition
A4	Inverse for addition
A5	Commutative for addition
M1	Closure for multiplication
M2	Associative for multiplication
M3	Identity for multiplication
M4	Inverse for multiplication (without 0)
M5	Commutative for multiplication
D	Distributive for multiplication over addition

We have studied five properties separately for addition and for multiplication and one which combines multiplication with addition, making eleven separate properties. The field is the algebraic structure which has all eleven properties.

EXERCISE 34 Consider each of these arithmetics with addition and multiplication:

$(I_1,+,\cdot)$ $(I_2,+,\cdot)$ $(I_3,+,\cdot)$ $(I_4,+,\cdot)$ $(I_5,+,\cdot)$
$(I_6,+,\cdot)$ $(I_7,+,\cdot)$ $(I_8,+,\cdot)$ $(I_9,+,\cdot)$ $(I_{10},+,\cdot)$
$(I_{11},+,\cdot)$ $(I_{12},+,\cdot)$ $(I_{13},+,\cdot)$ $(I_{14},+,\cdot)$ $(I_{15},+,\cdot)$

(*a*) Which of the arithmetics are rings? (*b*) Which ones are commutative rings with unity? (*c*) Which ones are fields? (*d*) Find a way to predict, for any counting number n, whether $(I_n,+,\cdot)$ will be a field.

Until the nineteenth century, algebra was merely a generalization of ordinary arithmetic. The properties, or "laws," of algebra were simply the properties of ordinary numbers. In the nineteenth, and especially the current, century algebra became the study of algebraic structures with various properties. Although the names of the algebraic structures we have studied are familiar to all mathematicians, the names of those who invented them are less well known. The beginnings of modern algebra are generally credited to the British school of mathematicians of the nineteenth century; in particular, to a British mathematician, George Peacock (1791–1858).

We have named four basic algebraic structures—the semigroup, group, ring, and field—all with properties of addition or multiplication or both, and all with properties included among the field properties. By changing the combinations of properties, or even the set itself and its operations and thus the types of properties, many more types of algebraic structures can be produced. Eves and Newsom say that mathematicians have constructed some 200 algebraic structures (*An Introduction to the Foundations and Fundamental Concepts of Mathematics*, p. 127). E. T. Bell claims that up to 1,152 might be constructed from various combinations of properties like the field properties (*Mathematics, Queen and Servant of Science*, p. 35).

9

A Noncommutative Group

An arithmetic was not required to have the commutative property to be a group. Since all our modular arithmetics, and also ordinary arithmetic, do have the commutative properties for both addition and multiplication, you might wonder why the commutative property is set apart from the other properties. The reason is that the commutative property is not so universal as it was once thought to be.

The first noncommutative type of number was discovered by an Irish mathematician, Sir William Rowan Hamilton (1805–1865) in 1843. Until then, the notion that all arithmetic was associative and commutative was so strong that Hamilton struggled with a new kind of number for fifteen years before he realized he would have to give up one property or the other. Using what we now know of finite arithmetic tables and their properties, we can reconstruct Hamilton's arithmetic in a very different way from his first efforts. Thus, with the advantage of hindsight, we may reduce his fifteen years to a few pages.

We begin with a special number i. The number i is defined to be the square root of -1. Since i is the square root of -1, we can also define $i \cdot i = -1$. Then $-i \cdot i = -(i \cdot i) = -(-1) = 1$, and $-i \cdot -i =$

$i \cdot i = -1$. Using the numbers 1, -1, i, and $-i$, we can construct this arithmetic of order 4:

(C, ·)	1	-1	i	$-i$
1	1	-1	i	$-i$
-1	-1	1	$-i$	i
i	i	$-i$	-1	1
$-i$	$-i$	i	1	-1

If a is a real number, then a times itself is not negative. If b is a real number and $b \neq 0$, then bi times itself is negative. The combination $a + bi$ is called a *complex number* (Chapter 1, Section 5).

EXERCISE 35 (*a*) Explain why table $(C, \cdot)$ has the closure property. (*b*) Construct tables which show that $(C, \cdot)$ has the associative property for the case $(a \cdot i) \cdot c = a \cdot (i \cdot c)$. (*c*) Circle the row and column of $(C, \cdot)$ which indicate the identity for the table. What element is the identity for multiplication for $(C, \cdot)$? (*d*) Circle the identity for $(C, \cdot)$ each time it appears among the entries of the table for $(C, \cdot)$. For each element of $(C, \cdot)$, what is the inverse of the element? (*e*) Explain why $(C, \cdot)$ has the commutative property. (*f*) Explain why the arithmetic $(C, \cdot)$ is a commutative group.

You might feel, as Hamilton did, that the next type of number after the complex numbers should be of the form $a + bi + cj$, where $j \cdot j = -1$. If this were the case, we should be able to construct a table for an arithmetic of order 6 using the numbers 1, -1, i, $-i$, j, and $-j$, where the arithmetic is a group:

(Q_b, ·)	1	-1	i	$-i$	j	$-j$
1	1	-1	i	$-i$	j	$-j$
-1	-1	1	$-i$	i	$-j$	j
i	i	$-i$	-1	1		
$-i$	$-i$	i	1	-1		
j	j	$-j$			-1	1
$-j$	$-j$	j			1	-1

(The symbol Q_b has no special mathematical significance. The b is meant to indicate that this table is a bad attempt at constructing a table we will eventually call Q.)

What should the result be for $i \cdot j$? If we repeat an entry within any row or column of the table, we lose the associative property and the table will not be a group. For example, suppose we decide that we will have $i \cdot j = i$; then,

$$(i \cdot j) \cdot j = i \cdot j = i$$

but

$$i \cdot (j \cdot j) = i \cdot -1 = -i$$

Therefore, $(i \cdot j) \cdot j \neq i \cdot (j \cdot j)$, so the table is not associative. The results are similar for other choices of $i \cdot j$. There is no group for the six elements 1, -1, i, $-i$, j, and $-j$.

We find that we must introduce a fourth symbol k, where $i \cdot j = k$. We define $i \cdot i = -1$, $j \cdot j = -1$, and also $k \cdot k = -1$. Our new type of number is of the form $a + bi + cj + dk$, called a *quarternion.* Suppose we assume that the quaternions, being merely an extension of the real and complex numbers, must be commutative for multiplication as are real and complex numbers. Then we also have $j \cdot i = k$. We can construct a table for an arithmetic of order 8 using the numbers 1, -1, i, $-i$, j, $-j$, k, and $-k$:

$(Q_a, \cdot)$	1	-1	i	$-i$	j	$-j$	k	$-k$
1	1	-1	i	$-i$	j	$-j$	k	$-k$
-1	-1	1	$-i$	i	$-j$	j	$-k$	k
i	i	$-i$	-1	1	k	$-k$	j	$-j$
$-i$	$-i$	i	1	-1	$-k$	k	$-j$	j
j	j	$-j$	k	$-k$	-1	1	i	$-i$
$-j$	$-j$	j	$-k$	k	1	-1	$-i$	i
k	k	$-k$	j	$-j$	i	$-i$	-1	1
$-k$	$-k$	k	$-j$	j	$-i$	i	1	-1

The associative property does not hold for table $(Q_a, \cdot)$. We know that $(a \cdot b) \cdot c = a \cdot (b \cdot c)$ for $b = 1$. The associative property also holds for $b = -1$. But suppose that $b = i$. We construct tables where the left-hand side of the first table has the elements $a \cdot i$ and the top of the second table has the elements $i \cdot c$:

a	$a \cdot i$	1	-1	i	$-i$	j	$-j$	k	$-k$
1	i	i	$-i$	-1	1	k	$-k$	j	$-j$
-1	$-i$	$-i$	i	1	-1	$-k$	k	$-j$	j
i	-1	-1	1	$-i$	i	$-j$	j	$-k$	k
$-i$	1	1	-1	i	$-i$	j	$-j$	k	$-k$
j	k	k	$-k$	j	$-j$	i	$-i$	-1	1
$-j$	$-k$	$-k$	k	$-j$	j	$-i$	i	1	-1
k	j	j	$-j$	k	$-k$	-1	1	i	$-i$
$-k$	$-j$	$-j$	j	$-k$	k	1	-1	$-i$	i

c	1	-1	i	$-i$	j	$-j$	k	$-k$
$i \cdot c$	i	$-i$	-1	1	k	$-k$	j	$-j$
1	i	$-i$	-1	1	k	$-k$	j	$-j$
-1	$-i$	i	1	-1	$-k$	k	$-j$	j
i	-1	1	$-i$	i	j	$-j$	k	$-k$
$-i$	1	-1	i	$-i$	$-j$	j	$-k$	k
j	k	$-k$	$-j$	j	i	$-i$	-1	1
$-j$	$-k$	k	j	$-j$	$-i$	i	1	-1
k	j	$-j$	$-k$	k	-1	1	i	$-i$
$-k$	$-j$	j	k	$-k$	1	-1	$-i$	i

We see that the entries in these tables for $(a \cdot i) \cdot c$ and $a \cdot (i \cdot c)$ are not identical. There are different entries in the third and fourth rows and third and fourth columns. The arithmetic is not associative for $b = i$, so it does not have the associative property.

Although our approach has been very different from Hamilton's, we have been forced to the same conclusion. The quaternions cannot have both the associative and the commutative properties for multiplication. We give up the commutative property, and define $i \cdot j = k$ but $j \cdot i = -k$. There is an easy diagram which indicates the multiplications for i, j, and k:

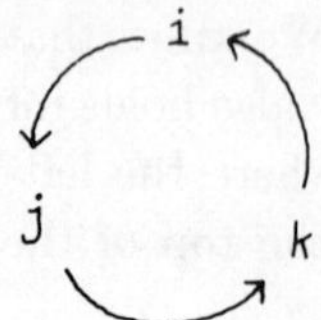

When we multiply counterclockwise, in the direction of the arrows, the result is positive. However, when we multiply clockwise, against the direction of the arrows, the result is negative. The table for the quaternions, which has the associative property for multiplication and is a group, is

$(Q, \cdot)$	1	-1	i	$-i$	j	$-j$	k	$-k$
1	1	-1	i	$-i$	j	$-j$	k	$-k$
-1	-1	1	$-i$	i	$-j$	j	$-k$	k
i	i	$-i$	-1	1	k	$-k$	$-j$	j
$-i$	$-i$	i	1	-1	$-k$	k	j	$-j$
j	j	$-j$	$-k$	k	-1	1	i	$-i$
$-j$	$-j$	j	k	$-k$	1	-1	$-i$	i
k	k	$-k$	j	$-j$	$-i$	i	-1	1
$-k$	$-k$	k	$-j$	j	i	$-i$	1	-1

The arithmetic of table $(Q, \cdot)$ is a group, but not a commutative group.

EXERCISE 36 Consider table $(Q_b, \cdot)$. Compute $(i \cdot j) \cdot j$ and $i \cdot (j \cdot j)$ for each of these choices for the result of $i \cdot j$: (*a*) $i \cdot j = j$. (*b*) $i \cdot j = 1$. (*c*) $i \cdot j = -i$. (*d*) $i \cdot j = -j$. (*e*) $i \cdot j = -1$. (*f*) Is it possible to complete table $(Q_b, \cdot)$ so that it has the associative property?

EXERCISE 37 Consider table $(Q_a, \cdot)$. (*a*) Explain why table $(Q_a, \cdot)$ has the closure property. (*b*) Circle the row and column of $(Q_a, \cdot)$ which indicate the identity for the table. What element is the identity for multiplication for $(Q_a, \cdot)$? (*c*) Circle the identity for $(Q_a, \cdot)$ each time it appears among the entries of the table for $(Q_a, \cdot)$. For each element of $(Q_a, \cdot)$, what is the inverse of the element? (*d*) Explain why table $(Q_a, \cdot)$ has the commutative property.

EXERCISE 38 Consider table $(Q_a, \cdot)$. Construct the two tables which show that $(Q_a, \cdot)$ does not have the associative property for each of these values of b. (*a*) $b = j$. (*b*) $b = k$.

EXERCISE 39 Suppose it is given that for any numbers x and y, $x \cdot -y = -x \cdot y = -(x \cdot y)$. (*a*) Show that $(a \cdot b) \cdot c = a \cdot (b \cdot c)$ for $b = -1$. (*b*) Show that if $(a \cdot b) \cdot c = a \cdot (b \cdot c)$ is true for a number b, then the associative property is also true for $-b$.

National Library of Ireland

One of the most famous of mathematical stories is about William Rowan Hamilton. Having struggled with the problem of quaternion multiplication for fifteen years, he suddenly saw the solution while crossing a bridge, and scratched the formulas in its stone with his pocket-knife. The bridge is supposed to have been the Brougham Bridge across the Royal Canal in Dublin. Joseph Ayton, a present-day teacher of mathematics, located the bridge and photographed a commemorative plaque he found there. His photograph appears below. You can read about his search for the bridge in his article "A Visit to a Mathematical Shrine."

Photograph by Joseph Ayton, "A Visit to a Mathematical Shrine." Reprinted from the *Mathematics Teacher*, October 1969 (Vol. LXII, p. 479), copyright 1969 by the National Council of Teachers of Mathematics. Reprinted by permission.

Photograph by Joseph Ayton

EXERCISE 40 Consider table $(Q,\cdot)$. (*a*) Explain why table $(Q,\cdot)$ has the closure property. (*b*) Construct tables which show that $(Q,\cdot)$ has the associative property for $b = i$, $b = j$, and $b = k$. (*c*) Circle the row and column of $(Q,\cdot)$ which indicate the identity for the table. What element is the identity for multiplication for $(Q,\cdot)$? (*d*) Circle the identity for $(Q,\cdot)$ each time it appears among the entries of the table for $(Q,\cdot)$. For each element of $(Q,\cdot)$, what is the inverse of the element? (*e*) Explain why the arithmetic $(Q,\cdot)$ does not have the commutative property. (*f*) Explain why the arithmetic $(Q,\cdot)$ is a group but not a commutative group.

Looking Back

1 Fill in the blank with a nonnegative integer less than the modulus for each of these congruences:

(*a*) $3 \equiv$ _____ (mod 7) (*b*) $10 \equiv$ _____ (mod 7)
(*c*) $14 \equiv$ _____ (mod 7) (*d*) $5 \cdot 6 \equiv$ _____ (mod 7)
(*e*) $10 \equiv$ _____ (mod 8) (*f*) $4 + 4 \equiv$ _____ (mod 8)
(*g*) $4 \cdot 4 \equiv$ _____ (mod 8) (*h*) $7 \cdot 7 \equiv$ _____ (mod 8)

***2** The symbol I_n stands for the integers modulo n. To apply congruences to negative integers, we add the modulus until we obtain an element of the set $I_n = \{0, 1, 2, \cdots, n - 1\}$. For example, $-1 \equiv 11$ (mod 12), $-13 \equiv 11$ (mod 12), and so on. Fill in the blank with a nonnegative integer less than the modulus for each of these congruences;

(*a*) $-1 \equiv$ _____ (mod 7) (*b*) $-6 \equiv$ _____ (mod 7)
(*c*) $-8 \equiv$ _____ (mod 7) (*d*) $-12 \equiv$ _____ (mod 7)
(*e*) $-1 \equiv$ _____ (mod 8) (*f*) $-7 \equiv$ _____ (mod 8)
(*g*) $-12 \equiv$ _____ (mod 8) (*h*) $-8 \equiv$ _____ (mod 8)

***3** If $a \equiv b$ (mod m), then $a + c \equiv b + c$ (mod m), $a - c \equiv b - c$ (mod m), and $a \cdot c \equiv b \cdot c$ (mod m). Thus congruences can be "solved" like equations, by adding, subtracting, or multiplying the same number on each side. Solve each of these congruences for x, where x is a nonnegative integer less than the modulus:

(*a*) $x + 2 \equiv 4$ (mod 7) (*b*) $x + 10 \equiv 4$ (mod 7)
(*c*) $x + 12 \equiv 4$ (mod 8) (*d*) $\frac{1}{2}x + 2 \equiv 4$ (mod 8)
(*e*) $\frac{1}{2}x + 2 \equiv 6$ (mod 8) (*f*) $\frac{1}{2}x + 9 \equiv 2$ (mod 8)
(*g*) $x - 7 \equiv 2$ (mod 8) (*h*) $\frac{1}{2}x - 9 \equiv 2$ (mod 8)

(*i*) If c is a factor of both a and b, and $a \equiv b$ (mod m), is it true that $a/c \equiv b/c \pmod{m}$? [*Hint*: Use an example such as $10 \equiv 2 \pmod{8}$ and divide by 2. Are the results congruent?]

4 Write the definition of each of these properties. (*a*) The closure property for addition. (*b*) The associative property for addition. (*c*) The identity property for addition. (*d*) The inverse property for addition. (*e*) The identity property for multiplication. (*f*) The inverse property for multiplication. (*g*) The commutative property for addition. (*h*) The distributive property for multiplication over addition.

5 (*a*) Construct the table for the arithmetic $(I_7,+)$. (*b*) How do you tell that the table is closed for addition? (*c*) How do you find the identity for addition? What element is the identity for addition? (*d*) How do you find the inverse for addition of each element? List the inverse for addition of each element. (*e*) How do you tell that the table is commutative for addition? (*f*) Is $(I_7,+)$ a commutative group?

6 Compute each side of each of these examples of the associative properties for $(I_7,+)$ and $(I_7,\cdot)$:

(*a*) $(2 + 3) + 4 = 2 + (3 + 4) \pmod{7}$
(*b*) $(3 + 2) + 5 = 3 + (2 + 5) \pmod{7}$
(*c*) $(4 + 6) + 5 = 4 + (6 + 5) \pmod{7}$
(*d*) $(4 + 4) + 6 = 4 + (4 + 6) \pmod{7}$
(*e*) $(2 \cdot 3) \cdot 4 = 2 \cdot (3 \cdot 4) \pmod{7}$
(*f*) $(3 \cdot 2) \cdot 5 = 3 \cdot (2 \cdot 5) \pmod{7}$
(*g*) $(4 \cdot 6) \cdot 5 = 4 \cdot (6 \cdot 5) \pmod{7}$
(*h*) $(4 \cdot 4) \cdot 6 = 4 \cdot (4 \cdot 6) \pmod{7}$

7 Construct pairs of tables which show that (*a*) $(I_7,+)$ is associative for $b = 2$ and (*b*) $(I_7,\cdot)$ is associative for $b = 2$.

8 (*a*) Construct the table for the arithmetic $(I_{10},\cdot)$. (*b*) What are the divisors of zero of $(I_{10},\cdot)$? (*c*) What is the inverse for each element of $(I_{10},\cdot)$ which has an inverse? (*d*) Write the definition of a semigroup. (*e*) Is $(I_{10},\cdot)$ a semigroup? (*f*) Is $(I_{10} - 0,\cdot)$ a semigroup? (*g*) Write the definition of a group. (*h*) Is $(I_{10},\cdot)$ a group? (*i*) Is $(I_{10} - 0,\cdot)$ a group?

9 Compute each side of each of these examples of the distributive property for multiplication over addition for $(I_7,+,\cdot)$:

(*a*) $2 \cdot (3 + 4) = (2 \cdot 3) + (2 \cdot 4) \pmod{7}$
(*b*) $3 \cdot (2 + 5) = (3 \cdot 2) + (3 \cdot 5) \pmod{7}$
(*c*) $4 \cdot (6 + 5) = (4 \cdot 6) + (4 \cdot 5) \pmod{7}$
(*d*) $4 \cdot (4 + 6) = (4 \cdot 4) + (4 \cdot 6) \pmod{7}$

10 (*a*) Write the definition of a ring. (*b*) Is $(I_{10},+,\cdot)$ a ring? (*c*) Is $(I_7,+,\cdot)$ a ring? (*d*) Write the definition of a field. (*e*) Is $(I_{10},+,\cdot)$ a field? (*f*) Is $(I_7,+,\cdot)$ a field?

***11** *Two-by-two matrices* are "arrays" of the form $\begin{pmatrix} a & b \\ c & d \end{pmatrix}$.
Matrices are multiplied by this formula:

$$\begin{pmatrix} a_1 & b_1 \\ c_1 & d_1 \end{pmatrix} \cdot \begin{pmatrix} a_2 & b_2 \\ c_2 & d_2 \end{pmatrix} = \begin{pmatrix} a_1 \cdot a_2 + b_1 \cdot c_2 & a_1 \cdot b_2 + b_1 \cdot d_2 \\ c_1 \cdot a_2 + d_1 \cdot c_2 & c_1 \cdot b_2 + d_1 \cdot d_2 \end{pmatrix}$$

(Observe that this pattern goes across the rows of the first matrix and down the columns of the second.) Let

$$M_1 = \begin{pmatrix} 1 & 0 \\ 0 & 1 \end{pmatrix} \quad M_2 = \begin{pmatrix} -1 & 0 \\ 0 & -1 \end{pmatrix} \quad M_3 = \begin{pmatrix} 0 & 1 \\ 1 & 0 \end{pmatrix}$$

and $M_4 = \begin{pmatrix} 0 & -1 \\ -1 & 0 \end{pmatrix}$

(*a*) Construct a multiplication table for the four elements M_1, M_2, M_3, and M_4. (*b*) What element is the identity for the table? (*c*) What element is the inverse for each element of the table? (*d*) Is the arithmetic of the table a group? (*e*) Is the arithmetic of the table a commutative group? Let

$$N_1 = \begin{pmatrix} i & 0 \\ 0 & i \end{pmatrix} \quad N_2 = \begin{pmatrix} -i & 0 \\ 0 & -i \end{pmatrix} \quad N_3 = \begin{pmatrix} 0 & i \\ i & 0 \end{pmatrix}$$

and $N_4 = \begin{pmatrix} 0 & -i \\ -i & 0 \end{pmatrix}$

(*f*) Construct a multiplication table for the eight elements M_1, M_2, M_3, M_4, N_1, N_2, N_3, and N_4. (*g*) What element is the identity for the table? (*h*) What element is the inverse for each element of the table? (*i*) Is the arithmetic of the table a group? (*j*) Is the arithmetic of the table a commutative group?

Branching Out

1 Interest in algebraic solutions to algebraic equations reached two peaks in the history of mathematics before the question was settled by Abel and Galois. The first was in the mathematics of the ancient Babylonians (see Chapter 1). The second occurred in Italy in the fifteenth and sixteenth centuries. The Italian mathematicians of these centuries were a boisterous group, given to a curious custom of challenging one another to contests in equation-solving, and also occasionally to stealing one anothers methods of solution. A mathematics history text such as Howard Eves' *An Introduction to the History of Mathematics* will introduce you to such men as Tartaglia and Cardano (or Cardan, whom we met in Chapter 1). Once acquainted with the cast of characters, you can find out more about them in other books.

2 The biographies of Abel, Galois, and Hamilton are among the most appealing in the history of mathematics. You can find Sarton's essay "Évariste Galois" in his *The Life of Science*, and also in Rapport and Wright's *Mathematics*. There is an excellent article on Hamilton by Sir Edmund Whittaker in Rapport and Wright which also appears in Kline's *Mathematics in the Modern World*. E. T. Bell has a chapter devoted to each of Abel, Galois, and Hamilton in *Men of Mathematics*. You may wish to search out additional material on one or more of these three, or you may find other mathematicians whose lives or work interest you in the books you consult.

3 There is relatively little material readily available on the British school of mathematicians of the nineteenth century responsible for the development of modern algebra. You will find Peacock usually mentioned, but rarely with information on his life or work, in mathematics histories. Augustus De Morgan, known also for work in the development of logic, is often discussed more fully. Perhaps the best known of this group, besides Hamilton, is Arthur Cayley. You will find Cayley, as well as Hamilton, in E. T. Bell's *Men of Mathematics*. There is also a German mathematician, Hermann Grassmann, known for contributions to the study of algebraic structures. If you enjoy the search for information, you might see what you can find out about any or all of these mathematicians.

4 Hamilton's original concept of the quaternions has its roots in physics, in particular in vectors in three-dimensional space. A *vector* is a directed line segment. In two-dimensional space, for example, if $a = 2$ and $b = 1$, the vectors for the complex numbers $a + bi$ and $a - bi$ are

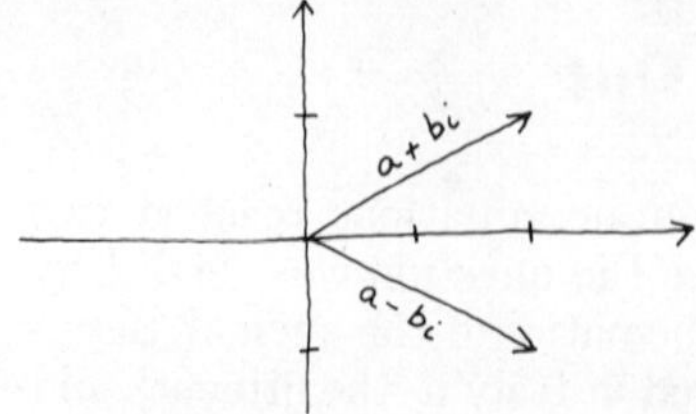

Whittaker's article on Hamilton in Rapport and Wright's *Mathematics*, also reproduced in Kline's *Mathematics in the Modern World*, gives a readable summary of Hamilton's approach. You should have a fairly strong background in algebra to pursue this subject further. In choosing books, beware of those which are meant for advanced students of mathematics.

5 There are several easily understood algebraic structures besides the semigroup, group, ring, and field. Two of them are the division ring, or skew field, which is a ring with the inverse property but not the commutative property for multiplication (the quaternions are an example); and the integral domain, which is a commutative ring with unity and with no divisors of zero. The integers are an integral domain—hence its name. Other algebraic structures have such intriguing names as ideals and lattices. E. T. Bell outlines several algebraic structures in *Mathematics, Queen and Servant of Science.* If you choose this subject, beware of abstract algebra books meant for advanced students of mathematics.

6 A kind of algebraic structure called a Boolean algebra is named for George Boole. It can be derived from three of the operations for statements, negation, conjunction, and disjunction, in Chapter 3. Besides familiar properties such as closure, associative, and commutative, Boolean algebra has other properties with names such as the idempotent laws, and De Morgan's Laws, named for Augustus De Morgan. Boolean algebra also has two distributive properties, one for conjunction over disjunction, and one for disjunction over conjunction. There are interesting applications to electrical circuits. You will find an excellent chapter on Boolean algebra in Allendoerfer and Oakley's *Principles of Mathematics.* A reference that emphasizes the algebraic structure of Boolean algebra but not the applications is E. T. Bell's *Mathematics, Queen and Servant of Science.* Be sure to heed Bell's warning concerning an unusual use of the symbols $\vee$ and $\wedge$. Eves and Newsom's *An Introduction to the Foundations and Fundamental Concepts of Mathematics* has a very sophisticated presentation, including many theorems.

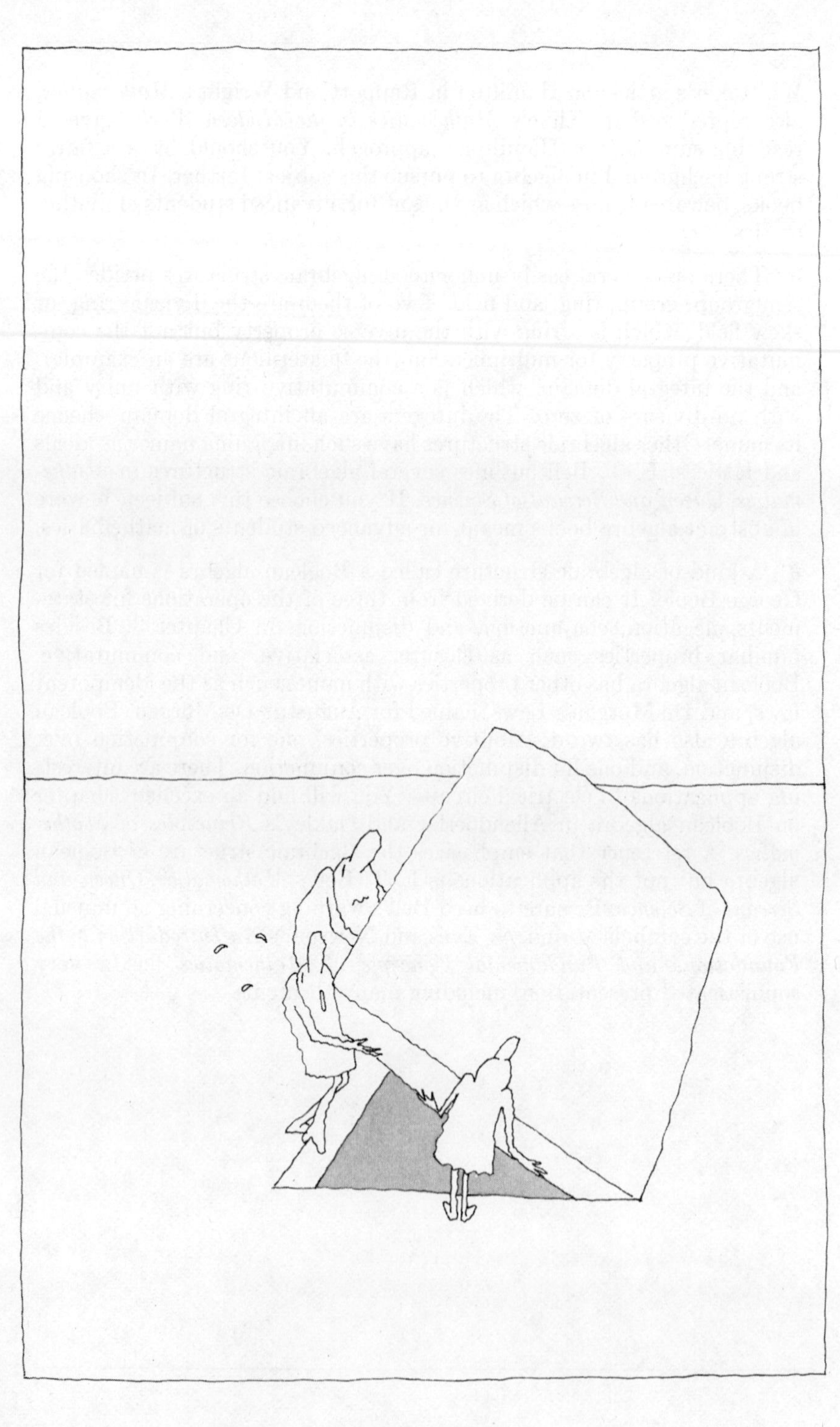

5

Arithmetics Without Numbers

1

The First Groups

The study of groups grew out of a seemingly unrelated area of mathematics. If you have studied a little algebra, you know that the linear (first-degree) equation $ax + b = 0$ has a solution $x = -b/a$. If you have had more algebra, you may know a formula for finding solutions to the quadratic (second-degree) equation $ax^2 + bx + c = 0$. There are also formulas which give solutions to the cubic and quartic (third- and fourth-degree) equations. The types of solutions given by such formulas are called *algebraic solutions*. Some special cases of algebraic solutions were known even in ancient times. The general algebraic solutions to the cubic and quartic equations were published in the sixteenth and seventeenth centuries.

Mathematicians of three centuries, the sixteenth, seventeenth, and eighteenth, attempted to solve the problem of the quintic (fifth-degree) equation. The first real progress toward its solution, which turned out to be that it has no algebraic solution, was made by one of several great French mathematicians of the eighteenth century, Joseph-Louis Lagrange (1736–1813). Lagrange saw a connection between the problem of constructing algebraic solutions to equations and that of arranging letters *a*, *b*, *c*, and so forth, in all possible different ways.

The problem of arranging letters in all possible ways was systematized by Augustin-Louis Cauchy (1789–1857), a French mathematician of the nineteenth century. You will see that Cauchy's life span overlapped the short careers of Abel, who proved that the quintic equation has no algebraic solution, and Galois, who generalized the theory of algebraic solutions and coined the word "group." Some historians attribute to Cauchy, however, because of his study of arrangements of letters, the beginning of the concept of a group. E. T. Bell,[1] who says of Cauchy he was "the first of the great French mathematicians whose thought belongs definitely to the modern age," describes the nature of Cauchy's contribution to the creation of the theory of groups.

Modern mathematics is indebted to Cauchy for two of its major interests, each of which marks a sharp break with the mathematics of the eighteenth century. The first was the introduction of rigor into mathematical analysis. It is difficult to find an adequate simile for the magnitude of this advance; perhaps the following will do. Suppose that for centuries an entire people has been worshipping false gods and that suddenly their error is revealed to them. Before the introduction of rigor mathematical analysis was a whole pantheon of false gods. In this Cauchy was one of the great pioneers with Gauss and Abel. Gauss might have taken the lead long before Cauchy entered the field, but did not, and it was Cauchy's habit of rapid publication and his gift for effective teaching which really got rigor in mathematical analysis accepted.

The second thing of fundamental importance which Cauchy added to mathematics was on the opposite side—the combinatorial. Seizing on the heart of Lagrange's method in the theory of equations, Cauchy made it abstract and began the systematic creation of the theory of groups. The nature of this will be described later; for the moment we note only the modernity of Cauchy's outlook.

Without enquiring whether the thing he had invented had any application or not, even to other branches of mathematics, Cauchy developed it on its own merits as an abstract system. His predecessors, with the exception of the

[1] *Men of Mathematics*. Copyright © 1937 by E. T. Bell. Reprinted by permission of Simon and Schuster, Inc.

David Eugene Smith Collection, Columbia University Library

Augustin-Louis Cauchy was either the second or the third most prolific mathematician of all time (the first was, without dispute, Euler, and the other possible runner-up is Arthur Cayley). Cauchy's name appears on theorems and formulas throughout mathematics.

Cauchy, one of many great mathematicians of the Napoleonic era, was loyal to the Bourbons. He apparently survived the reign of Napoleon only because of his useful talents as an engineer. The Bourbons returned to France after Napoleon's defeat; and when they were deposed again, Cauchy refused to take the oath of allegiance to the new government. He left France for eight years and returned to spend another ten years in Paris in positions for which the oath was not required. Finally he was allowed to return to his professorship at the École Polytechnique when another government abolished the oath in 1848.

universal Euler who was as willing to write a memoir on a puzzle in numbers as on hydraulics or the "system of the world," had found their inspiration growing out of the applications of mathematics. This statement of course has numerous exceptions, notably in arithmetic; but before the time of Cauchy few if any sought profitable discoveries in the mere manipulations of algebra. Cauchy looked deeper, saw the operations and their laws of combination beneath the symmetries of algebraic formulas, isolated them, and was led to the theory of groups.

The important point of the preceding passage is in the phrase "an abstract system." In Chapter 4, we used specific numbers and the specific operations addition and multiplication. One stage of abstraction is to use letters such as a and b to stand for any number in an arithmetic, and to write formulas such as $a + b = b + a$ for any two numbers a and b in the arithmetic. When we use letters to stand for any number, you are probably used to changing from the name arithmetic to the name algebra. It is at this point that "Cauchy looked deeper" than ordinary algebra and into what is now called *abstract algebra.*

The next stage of abstraction is to let the letters a and b stand not just for numbers but for elements of any given set. At this stage the operation is not necessarily addition, multiplication, or one of the other arithmetic operations. For example, if a and b are elements of the set of colors, the operation might be "mix." Yellow mixed with blue gives green and also blue mixed with yellow gives green; that is, a "mix" b = b "mix" a.

The final stage of abstraction for our purposes is to use some general symbol to stand for the operation, whatever the operation is. Suppose $\circ$ (circle) stands for any operation. Then $a + b = b + a$ and a "mix" b = b "mix" a both can be written as the general formula $a \circ b = b \circ a$.

DEFINITION A *group* is a set S with an operation $\circ$, where $\circ$ has the following properties over the set S:

1 $a \circ b$ is an element of S for every two elements a and b of S (closure)
2 $(a \circ b) \circ c = a \circ (b \circ c)$ for every three elements a, b, and c of S (associative)
3 There is an element u in S such that $a \circ u = a$ and $u \circ a = a$ for every element a of S (identity)
4 For every element a of S there is an element a' in S such that $a \circ a' = u$ and $a' \circ a = u$ (inverse)

The group is a *commutative* or *Abelian* group if it also has the property that $a \circ b = b \circ a$ for every two elements a and b in S.

In the next two sections, we will construct groups in which the set is not a set of numbers and the operation is not addition or multiplication. The groups are different from those constructed by Cauchy, but the type of construction is similar. Instead of rearranging letters, we will turn polygons around their center points and across their center lines.

2

Rotation Groups

You are probably familiar with the basic polygons: the triangle (three-sided), the quadrilateral (four-sided), the pentagon (five-sided), the hexagon (six-sided), and perhaps a few others. A *regular* polygon is a polygon with all its sides equal and all its angles equal. The regular triangle is the *equilateral triangle*; and the regular quadrilateral is the *square*.

Suppose we label the four corners of a square with the letters a, b, c, and d:

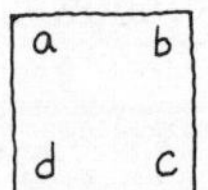

This labeling is for the purpose of keeping track of the different corners of the square. It does not have to be done in any particular order, but once a labeling is chosen we must stick to it throughout the discussion. Thus we will use the order of labeling above throughout this chapter.

Now imagine there is a pin in the center of the square so that we can turn it around its center point like a pinwheel. If the square is turned one-quarter of a rotation in a clockwise direction, its outline will be the same as before but the letters will be in new locations which indicate the new location of each corner:

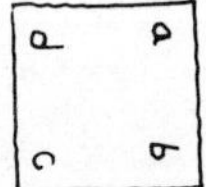

For a half turn, these are the new locations,

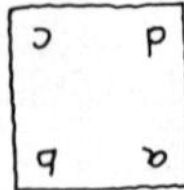

and for a three-quarter turn,

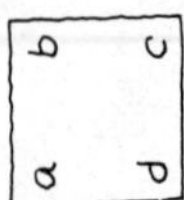

Again the outline is the same but the letters are in new locations. A full turn, of course, is the same as if the square were left alone.

You will find that these four turns are the only turns around its center point which do not change the outline of the square. Of course, we could have done them all in a counterclockwise direction instead, but the results would have been the same with only the order in which they are derived reversed. Once we decide on the direction we will turn the square, we must stick with that decision. Throughout this chapter, turns will be in the clockwise direction. The four turns of the square around its center point are called the *rotations* of the square. Writing the letters in the appropriate corners, but right side up, we name the rotations as follows:

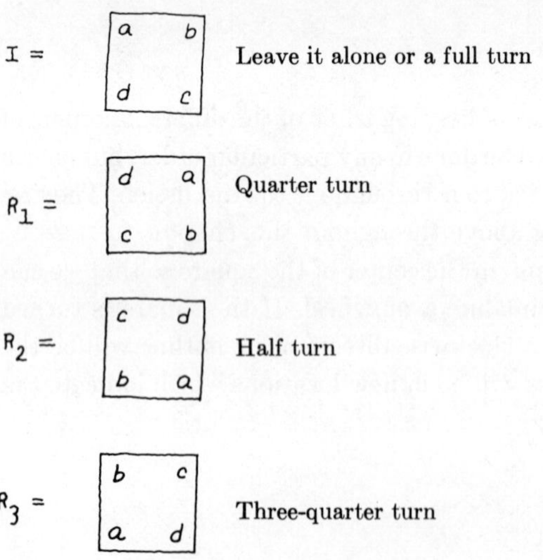

We now define a binary operation for the rotations of the square. Two rotations can be combined by doing one rotation *followed by* another. The operation "followed by" is indicated by an asterisk. For example, $R_1 * R_2$ means the rotation R_1 followed by the rotation R_2. Clearly, a quarter turn followed by a half turn gives a three-quarter turn, so $R_1 * R_2 = R_3$. You may find it helpful to cut a square out of paper and label it like square I above. By turning your square the amounts of the different rotations, you can verify the entries in table $(R_4,*)$ for the set of rotations of the square with the operation "followed by":

$(\boldsymbol{R_4},*)$	I	R_1	R_2	R_3
I	I	R_1	R_2	R_3
R_1	R_1	R_2	R_3	I
R_2	R_2	R_3	I	R_1
R_3	R_3	I	R_1	R_2

$(R_4,*)$ is a group of order 4, where the set is the set $R_4 = \{I, R_1, R_2, R_3\}$ and the operation is the operation "followed by."

A similar group can be constructed for any regular polygon. Here are the rotations of the regular pentagon:

$I =$ 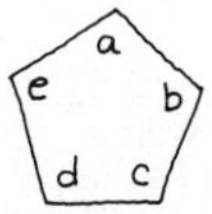 Leave it alone or a full turn

$R_1 =$ 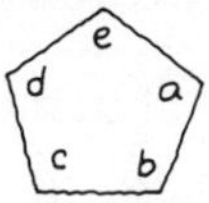 One-fifth turn

$R_2 =$ 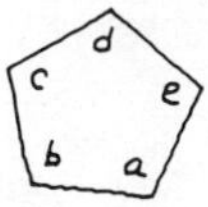 Two-fifths turn

$R_3 =$ 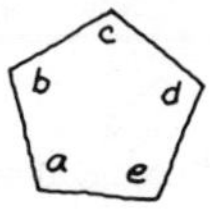 Three-fifths turn

$R_4 =$ 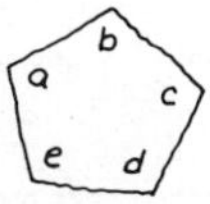 Four-fifths turn

Table $(R_5,*)$ for the rotations of the regular pentagon with the operation "followed by" is

$(R_5,*)$	I	R_1	R_2	R_3	R_4
I	I	R_1	R_2	R_3	R_4
R_1	R_1	R_2	R_3	R_4	I
R_2	R_2	R_3	R_4	I	R_1
R_3	R_3	R_4	I	R_1	R_2
R_4	R_4	I	R_1	R_2	R_3

EXERCISE 1 (*a*) Explain why table $(R_4,*)$ has the closure property. (*b*) Construct tables which show that $(R_4,*)$ has the associative property for $b = R_1$, $b = R_2$, and $b = R_3$. (*c*) Circle the row and column of $(R_4,*)$ which indicate the identity for the table. What element is the identity for "followed by" for $(R_4,*)$? (*d*) Circle the identity for $(R_4,*)$ each time it appears among the entries of the table for $(R_4,*)$. For each element of $(R_4,*)$, what is the inverse of the element? (*e*) Explain why table $(R_4,*)$ has the commutative property. (*f*) Explain why $(R_4,*)$ is a commutative group.

EXERCISE 2 (*a*) Label an equilateral triangle as follows:

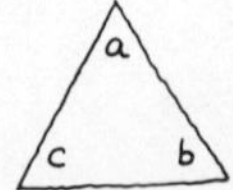

Construct a table $(R_3,*)$ for the rotations of the equilateral triangle with the operation "followed by."
(*b*) Label a regular hexagon as follows:

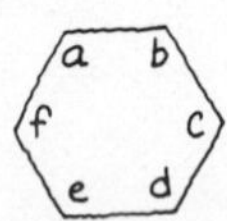

Construct a table $(R_6,*)$ for the rotations of the regular hexagon with the operation "followed by."
(*c*) Consider these tables:

$(R_3,*)$ $(R_5,*)$ $(R_6,*)$

Observe that each table has the properties of a commutative group.

3
Symmetry Groups

Any motion of a figure that leaves the outline of the figure unchanged is a *symmetry* of the figure. Rotations of regular polygons are symmetries of the polygons. Each n-sided regular polygon has n rotations. Besides these, each n-sided regular polygon has n other symmetries called *reflections.*

Let us return to the square in its original position:

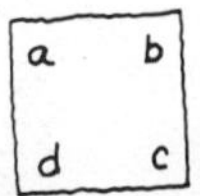

Now take the two left-hand corners, a and d in the original position, and turn the square over its center line:

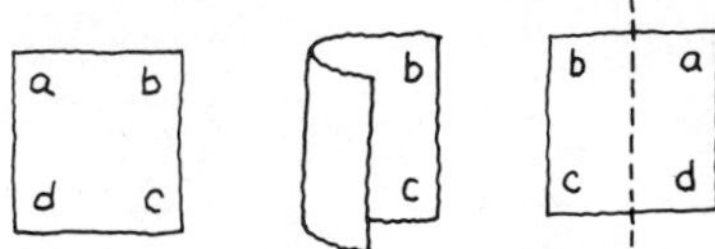

The back side of the square is now toward you. We can identify the new position by labeling each corner of the back of the square with the same letter which is on the front. The outline of the square is unchanged, so this motion is a symmetry of the square. The symmetry is the *vertical reflection* V. Observe that it interchanges the letters to the left of the square with the letters to the right.

Now take the two top corners, a and b in the original position, and turn the square over horizontally:

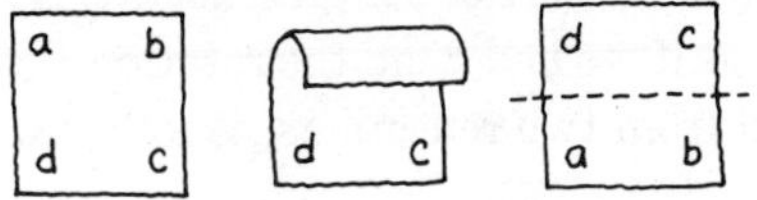

This symmetry is the *horizontal reflection* H. Observe that it interchanges the letters at the top of the square with the letters at the bottom.

If we take the square by two opposite corners, a and c in the original position, and turn it over across a diagonal, the other two opposite corners are interchanged:

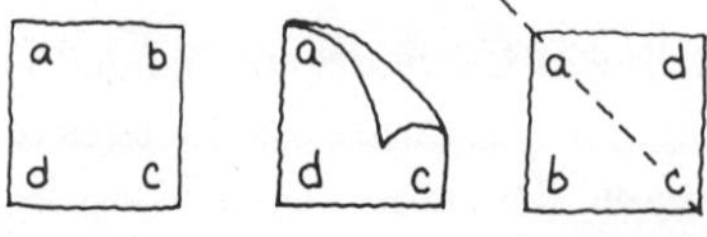

This symmetry is one of two *diagonal reflections* D_1. For the other diagonal reflection D_2, we take the other two opposite corners, b and d in the original position, and turn the square over across its other diagonal:

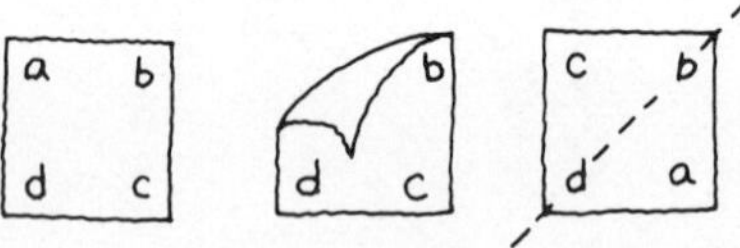

These are the four reflections of the square:

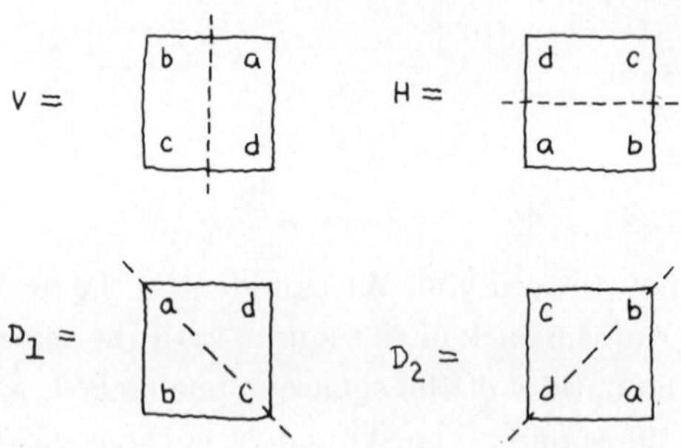

As with rotations, the operation $*$ for two reflections means "one followed by the other." For example, $V * H = R_2$. To show this result, we start with the square in the original position and do the vertical reflection V. Then, without returning to the original position, we turn the square horizontally across its center line to do the horizontal reflection H. The square is now in the same position as if we had done the rotation R_2. Observe that although it is constructed from two reflections, V followed by H, gives a rotation.

We can also do rotations followed by reflections and reflections followed by rotations. As an example of the latter, $V * R_1 = D_2$. We start with the square in the original position and do V. Then, without returning to the original position, we rotate the square through one quarter turn. The square is now in the same position as if we had done the reflection D_2.

Table $(S_4,*)$ gives the group of the symmetries of the square, the four rotations and the four reflections, with the operation "followed by." You can verify the entries in the table by turning the square you have labeled around or over as indicated by the headings. Be sure to do the symmetry to the left first, followed by the symmetry in the heading at the top.

$(S_4,*)$	I	R_1	R_2	R_3	V	H	D_1	D_2
I	I	R_1	R_2	R_3	V	H	D_1	D_2
R_1	R_1	R_2	R_3	I	D_1	D_2	H	V
R_2	R_2	R_3	I	R_1	H	V	D_2	D_1
R_3	R_3	I	R_1	R_2	D_2	D_1	V	H
V	V	D_2	H	D_1	I	R_2	R_3	R_1
H	H	D_1	V	D_2	R_2	I	R_1	R_3
D_1	D_1	V	D_2	H	R_1	R_3	I	R_2
D_2	D_2	H	D_1	V	R_3	R_1	R_2	I

EXERCISE 3 (*a*) Explain why table $(S_4,*)$ has the closure property. (*b*) Construct tables which show that $(S_4,*)$ has the associative property for $b = R_1$ and for $b = V$. (*c*) Circle the row and column of $(S_4,*)$ which indicate the identity for the table. What element is the identity for "followed by" for $(S_4,*)$? (*d*) Circle the identity for $(S_4,*)$ each time it appears among the entries of the table for $(S_4,*)$. For each element of $(S_4,*)$, what is the inverse of the element? (*e*) Does table $(S_4,*)$ have the commutative property? Explain. (*f*) Is $(S_4,*)$ a group? Is it a commutative group? Explain.

EXERCISE 4 Use these symmetries for the equilateral triangle:

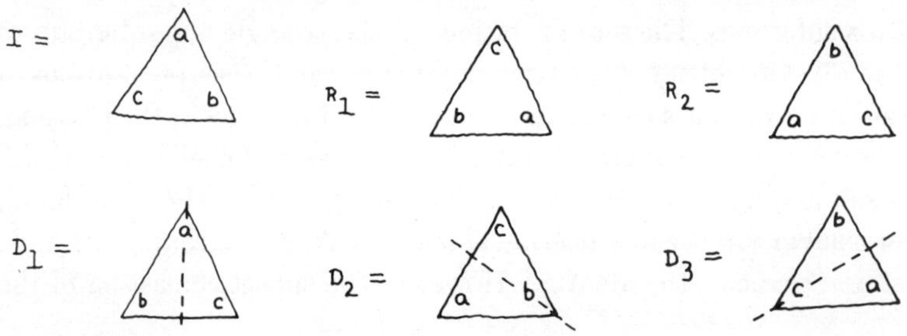

(*a*) Construct table $(S_3,*)$ for the symmetries of the equilateral triangle with the operation "followed by." (*b*) Is $(S_3,*)$ a group? Is it a commutative group? Explain.

4
Subgroups

Perhaps you have observed that the upper-left quarter of table $(S_4,*)$ for the symmetry group of the square is identical to table $(R_4,*)$ for the rotation group of the square.

$(S_4,*)$	I	R_1	R_2	R_3	V	H	D_1	D_2
I	I	R_1	R_2	R_3	V	H	D_1	D_2
R_1	R_1	R_2	R_3	I	D_1	D_2	H	V
R_2	R_2	R_3	I	R_1	H	V	D_2	D_1
R_3	R_3	I	R_1	R_2	D_2	D_1	V	H
V	V	D_2	H	D_1	I	R_2	R_3	R_1
H	H	D_1	V	D_2	R_2	I	R_1	R_3
D_1	D_1	V	D_2	H	R_1	R_3	I	R_2
D_2	D_2	H	D_1	V	R_3	R_1	R_2	I

Similarly, the upper-left quarter of $(S_3,*)$ is identical to $(R_3,*)$ for the triangle. In each case, one group has another group as a subset.

DEFINITION A subset of a group is a *subgroup* if it is itself a group under the same operation.

The set of rotations of the square $\{I, R_1, R_2, R_3\}$ is a subset of the set of all symmetries of the square and is itself a group with the operation $*$, so it is also a subgroup. The set of rotations of the triangle is a subgroup of the set of all symmetries of the triangle. In general, the set of rotations of any regular polygon is a subgroup of the set of symmetries of the polygon.

Every group is a subgroup of itself, since its set of elements is a subset of itself and is, of course, a group. The empty set, on the other hand, is never a subgroup because it is never a group. Any subgroup must have at least one element, the identity. However, the subset consisting of the identity alone is always a subgroup. For symmetry groups such as $(S_4,*)$ and $(S_3,*)$, the subgroup of only the identity is

*	I
I	I

Any symmetry group $(S_n,*)$ of a regular polygon has at least two subgroups: itself and the subgroup of the identity alone. Each must also have a third subgroup, its rotation group $(R_n,*)$. We recall that the *order* of any finite arithmetic is the number of elements in the arithmetic. The orders of these three subgroups, the identity alone, the rotation group, and the symmetry group itself, are 1, n, and $2n$.

A group of order 2 must have as its two elements the identity and one other element. Since the identity is its own inverse, the other element must also be its own inverse in order to have an inverse. Thus, to construct a subgroup of order 2, we must choose the identity and any other element which is its own inverse. In $(S_4,*)$, for example, the elements which are their own inverses, besides I, are R_2, V, H, D_1, and D_2. $(S_4,*)$ has five subgroups of order 2:

*	I	R_2
I	I	R_2
R_2	R_2	I

*	I	V
I	I	V
V	V	I

*	I	H
I	I	H
H	H	I

*	I	D_1
I	I	D_1
D_1	D_1	I

*	I	D_2
I	I	D_2
D_2	D_2	I

$(R_4,*)$ is one subgroup of order 4 of $(S_4,*)$. There are two other subgroups of order 4:

*	I	R_2	V	H
I	I	R_2	V	H
R_2	R_2	I	H	V
V	V	H	I	R_2
H	H	V	R_2	I

*	I	R_2	D_1	D_2
I	I	R_2	D_1	D_2
R_2	R_2	I	D_2	D_1
D_1	D_1	D_2	I	R_2
D_2	D_2	D_1	R_2	I

RULE FOR SUBGROUPS For any group of finite order n, the order of any subgroup is a factor of the order n of the group.

This rule is called *Lagrange's theorem* (although Bell attributes its proof to Cauchy). It tells us, for example, that $(S_4,*)$, which has order 8, can have only subgroups whose orders are factors of 8; that is, 1, 2, 4, and 8. $(S_4,*)$ has no subgroups of orders 3, 5, 6, or 7. Unfortunately, Lagrange's theorem does not tell us how many subgroups, if any, of each possible order there are.

EXERCISE 5 Construct the table for each of these subsets of $(S_4,*)$. (*a*) $\{I, R_1, R_3\}$. (*b*) $\{I, V, D_1\}$. (*c*) $\{I, R_1, R_3, V, D_1\}$. (*d*) $\{I, R_2, V, H, D_1, D_2)$. (*e*) What property of a group is missing for each of these tables?

EXERCISE 6 (*a*) Construct the table for the set of reflections of $(S_4,*)$: $\{V, H, D_1, D_2\}$. (*b*) What two properties of a group are missing for this table? (*c*) Is the set of reflections of a regular polygon a subgroup of the group of symmetries of the polygon?

EXERCISE 7 (*a*) What are all the possible orders of subgroups of $(S_3,*)$? (*b*) Construct the table for each subgroup of $(S_3,*)$.

For every counting number n, the modular arithmetic with addition $(I_n,+)$ is a group. All except one have at least two subgroups, the identity alone and the group itself. For $(I_n,+)$, the table for the identity alone is

+	0
0	0

For $(I_1,+)$, the identity alone and the group itself are identical, so there is just one subgroup, $(I_1,+)$ itself.

If n is a counting number greater than 1, $(I_n,+)$ has two or more distinct subgroups. When n is prime, its only factors are itself and 1, so Lagrange's theorem tells us that the only possible orders for subgroups are n and 1. The subgroup of order n is the group itself, and the subgroup of order 1 is the identity alone. If n is prime, then $(I_n,+)$ has only these two subgroups. The first and only even prime is 2. The two subgroups of $(I_2,+)$ are

+	0
0	0

+	0	1
0	0	1
1	1	0

The next prime is 3, and $(I_3,+)$ has the two subgroups

+	0
0	0

+	0	1	2
0	0	1	2
1	1	2	0
2	2	0	1

Four is a composite number with factors 1, 2, and 4. For a subgroup of order 2, we must choose an element which is its own inverse in $(I_4,+)$. Since 2 is the only such element, there is only one subgroup of order 2 of $(I_4,+)$. Thus $(I_4,+)$ has the three subgroups

+	0
0	0

+	0	2
0	0	2
2	2	0

+	0	1	2	3
0	0	1	2	3
1	1	2	3	0
2	2	3	0	1
3	3	0	1	2

When we turn to modular arithmetics with multiplication, the situation is more complicated. Many of the modular arithmetics are not groups for multiplication, even with zero deleted. To begin with, $(I_1 - 0,\cdot)$ is the empty set and thus is not a group. $(I_2 - 0,\cdot)$ is a group but has order 1, so it has only one subgroup, itself.

·	1
1	1

$(I_3 - 0,\cdot)$ is also a group. Since zero is deleted, it has order 2 and thus has only the two subgroups, the identity alone and the group itself.

·	1
1	1

·	1	2
1	1	2
2	2	1

$(I_4 - 0,\cdot)$ is not a group. We know that $(I_n - 0,\cdot)$ will fail to be a group whenever there are divisors of zero. When n is composite, $(I_n - 0,\cdot)$ has divisors of zero. When n is prime, $(I_n - 0,\cdot)$ is a group. However, since zero is deleted, its order is $n - 1$, which is composite if n is prime. Thus whenever n is prime $(I_n - 0,\cdot)$ has at least two subgroups, the identity alone and $(I_n - 0,\cdot)$ itself, and possibly subgroups of other orders as well.

EXERCISE 8 (*a*) List all the possible orders of subgroups for each of these groups:

$(I_5,+)$ $(I_6,+)$

(*b*) Construct the table for each subgroup of each of the groups. (*Hint:* There is only one subgroup of each possible order for each group.)

EXERCISE 9 (*a*) What is the order of each of these groups:

$(I_5 - 0,\cdot)$ $(I_7 - 0,\cdot)$

(*b*) List all the possible orders of subgroups for each of the groups. (*c*) Construct the table for each subgroup of each of the groups. (*Hint:* There is only one subgroup of each possible order for each group.)

5
Isomorphism

We know three different groups, each consisting of just one element. Each group consists of the identity alone. The subgroup of the identity alone of any modular arithmetic with addition $(I_n,+)$ is

+	0
0	0

For any modular arithmetic with multiplication $(I_n - 0,\cdot)$, where n is prime and zero is deleted, the subgroup consisting of the identity alone is

·	1
1	1

For the symmetry group of any regular polygon with the operation "followed by" $(S_n,*)$, the subgroup consisting of the identity alone is

*	I
I	I

Comparing these three tables, which we may call the addition group of one element, the multiplication group of one element, and the symmetry group of one element, we see that they all have the same pattern. Suppose $\circ$ represents any of the operations addition, multiplication, or followed by. Suppose also that u represents the identity that corresponds to the operation $\circ$: 0 for addition, 1 for multiplication, and I for followed by. Then each of the three tables can be written in the form

$\circ$	u
u	u

In the preceding section, we found that there is only one way to form a group of two elements. One element must be the identity, which is its own inverse; the other element must therefore be an element which is its own inverse. An addition group of two elements is, for example,

$(I_2,+)$	0	1
0	0	1
1	1	0

An example of a multiplication group of two elements is

$(I_3 - 0,\cdot)$	1	2
1	1	2
2	2	1

There are many subgroups of two elements for other addition and multiplication groups. We have also found many subgroups of two elements for symmetry groups. For example, if D_1 is the first reflection for any regular polygon, a symmetry group of two elements is

*	I	D_1
I	I	D_1
D_1	D_1	I

Comparing all our groups of two elements, we find that if $\circ$ is the operation, u is the identity, and a is the second element, each table has the form:

$\circ$	u	a
u	u	a
a	a	u

DEFINITION Two finite groups are *isomorphic* if their elements can be put in one-to-one correspondence in such a way that the entries in their tables follow the same correspondence; that is, the results of corresponding operations also correspond.

All groups of order 1 are isomorphic to one another since they all correspond to the table

$\circ$	u
u	u

Also, all groups of order 2 are isomorphic to one another since they all correspond to the table

$\circ$	u	a
u	u	a
a	a	u

You have probably observed that the rotation groups of regular polygons have a familiar pattern. The groups $(R_3,*)$ of the equilateral triangle, $(R_4,*)$ of the square, $(R_5,*)$ of the regular pentagon, and $(R_6,*)$ of the regular hexagon have the same orders as the addition groups $(I_3,+)$, $(I_4,+)$, $(I_5,+)$, and $(I_6,+)$. The identity element I of each group $(R_n,*)$ follows the same pattern as the identity 0 of each $(I_n,+)$. Moreover, R_1 in any $(R_n,*)$ follows the same pattern as 1 in $(I_n,+)$, R_2 follows the same pattern as 2, and so forth.

Consider, for example, the rotation group of the square. If we make the one-to-one correspondence,

$$\begin{array}{cccc} 0 & 1 & 2 & 3 \\ \updownarrow & \updownarrow & \updownarrow & \updownarrow \\ I & R_1 & R_2 & R_3 \end{array}$$

then the entries in the tables for $(R_4,*)$ and $(I_4,+)$ will follow the same correspondence. Thus the groups $(R_4,*)$ and $(I_4,+)$ are isomorphic.

$(R_4,*)$	I	R_1	R_2	R_3
I	I	R_1	R_2	R_3
R_1	R_1	R_2	R_3	I
R_2	R_2	R_3	I	R_1
R_3	R_3	I	R_1	R_2

$(I_4,+)$	0	1	2	3
0	0	1	2	3
1	1	2	3	0
2	2	3	0	1
3	3	0	1	2

EXERCISE 10 (*a*) What is the order of each of these groups?

$(I_3,+)$ $(I_4,+)$ $(I_5,+)$ $(I_6,+)$
$(I_5 - 0,\cdot)$ $(I_7 - 0,\cdot)$
$(R_3,*)$ $(S_3,*)$ $(R_4,*)$ $(S_4,*)$

(*b*) Construct the table for each subgroup of two elements, when there are any, for each of the groups. (*c*) Which of the groups have no subgroup of two elements? (*d*) Does a group with an odd order ever have a subgroup of two elements? Explain.

EXERCISE 11 (*a*) Show the one-to-one correspondence between the sets of elements of each of these pairs of groups:

$(R_3,*)$ and $(I_3,+)$
$(R_5,*)$ and $(I_5,+)$
$(R_6,*)$ and $(I_6,+)$

(*b*) By comparing the patterns of their tables, observe that each pair of groups is isomorphic.

Groups which have the pattern of any rotation group $(R_n,*)$ or any addition group $(I_n,+)$ are called *cyclic groups*. You can observe that the pattern of the table of a cyclic group is circular. The first row follows the order of the headings. The second row starts with the second element, goes in order to the last element, and then circles back to the first element. The third row starts with the third element, goes in order to the last element, then circles back to the first and second elements, and so on.

The groups of order 1 and order 2 are considered to be cyclic groups. We have two examples of groups of order 3, $(I_3,+)$ and $(R_3,*)$. There is no multiplication group of order 3 since $(I_4 - 0,\cdot)$ is not a group. The two groups $(I_3,+)$ and $(R_3,*)$ are both cyclic.

$(R_3,*)$	I	R_1	R_2
I	I	R_1	R_2
R_1	R_1	R_2	I
R_2	R_2	I	R_1

$(I_3,+)$	0	1	2
0	0	1	2
1	1	2	0
2	2	0	1

In these two tables, if $\circ$ is the operation, u the identity, and a and b the other elements, we observe that a and b are one another's inverses; and the general table is

$\circ$	u	a	b
u	u	a	b
a	a	b	u
b	b	u	a

Even though we have no example for it, there is another possibility for a group with three elements. We should investigate the case where a and b are each their own inverse. In this case, the table would begin as follows:

$\circ$	u	a	b
u	u	a	b
a	a	u	
b	b		u

Now we must fill in the two blank spaces with one of the elements u, a, or b. Suppose we try an a in each blank, so that $a \circ b = a$ and $b \circ a = a$. The table becomes:

$\circ$	u	a	b
u	u	a	b
a	a	u	a.
b	b	a	u

Now suppose we use a pair of tables to test the associative property. We will write the associative property for $\circ$: $(x \circ y) \circ z = x \circ (y \circ z)$, using x, y, and z so as not to confuse the letters in the formula with the letters on the table. The pair of tables for $y = a$ is

x	$x \circ a$	u	a	b
u	a	a	u	a
a	u	u	a	b
b	a	a	u	a

z	u	a	b
$a \circ z$	a	u	a
u	a	u	a
a	u	a	u
b	a	b	a

Since the entries in these tables are not identical, the table we made for $\circ$ does not have the associative property. In general, if we fill in a table in such a way that the same element appears twice in any row or column, the table will not be a group. Thus the only possible group of order 3 is the cyclic group, and all groups of order 3 are isomorphic to one another.

For order 4, our examples are $(I_4,+)$ and $(R_4,*)$, and also a multiplication group $(I_5 - 0,\cdot)$. We have looked at the first two. They are isomorphic to each other, both are cyclic, and their tables have the pattern

$\circ$	u	a	b	c
u	u	a	b	c
a	a	b	c	u
b	b	c	u	a
c	c	u	a	b

In this table, b is its own inverse, and a and c are one another's inverses. Again, we must also try the case in which a and c are their own inverses. This time it is possible to complete the table without repeating an element in any row or column:

$\circ$	u	a	b	c
u	u	a	b	c
a	a	u	c	b
b	b	c	u	a
c	c	b	a	u

Groups with the pattern of this table are called *Klein four groups.* Klein four groups are of order 4, and each of the four elements is its own inverse.

There is a cyclic group for each counting number n. For $n = 1, 2,$ and 3, these are the only groups, and all other groups of these orders are isomorphic to them. For $n = 4$, there are two groups, the cyclic group of order 4 and the Klein four group.

EXERCISE 12 Which ones of these groups are cyclic groups:

$(I_3,+)$ $(I_4,+)$ $(I_5,+)$ $(I_6,+)$
$(R_3,*)$ $(S_3,*)$ $(R_4,*)$ $(S_4,*)$

EXERCISE 13 Consider this table:

$\circ$	u	a	b
u	u	a	b
a	a	u	
b	b		u

(*a*) Suppose that $a \circ b = u$ and $b \circ a = u$. Fill in the table. Construct a pair of tables for $y = a$ which shows that the table does not have the associative property. (*b*) Suppose that $a \circ b = b$ and $b \circ a = b$. Fill in the table. Construct a pair of tables for $y = a$ and also a pair of tables for $y = b$ which shows that the table does not have the associative property. (*c*) Suppose that $a \circ b = a$ but $b \circ a = b$. Fill in the table. Construct a pair of tables for $y = a$ which shows that the table does not have the associative property.

EXERCISE 14 Construct the tables for two subgroups of $(S_4,*)$ which are Klein four groups.

EXERCISE 15 (*a*) List the inverse of each element of $(I_5 - 0,\cdot)$. (*b*) Is $(I_5 - 0,\cdot)$ a cyclic or Klein four group? Explain. (*c*) Construct an arrangement of the table for $(I_5 - 0,\cdot)$ which matches the pattern of the cyclic group of order 4. (*Hint:* Rearrange the order in which the elements are listed in the headings.) (*d*) Similarly, construct an arrangement of the table for $(I_7 - 0,\cdot)$ which matches the pattern of the cyclic group of order 6.

EXERCISE 16 (*a*) List the elements of $(I_6,+)$ which are their own inverse. (*b*) List the elements of $(S_3,*)$ which are their own inverse. (*c*) Can $(I_6,+)$ and $(S_3,*)$ be isomorphic? Give two reasons why not. [There are at least two types of groups of order 6, those isomorphic to $(I_6,+)$ which are cyclic and those isomorphic to $(S_3,*)$.] (*d*) Explain why neither $(S_4,*)$ nor $(Q,\cdot)$ is isomorphic to $(I_8,+)$, and also why $(S_4,*)$ and $(Q,\cdot)$ are not isomorphic. (There are at least three types of groups of order 8.)

6

Kaleidoscopes and Tessellations

In every cyclic group there is a particular element, called the *generator*, from which all the other elements of the group can be derived using the operation of the group. For example, every element of $(R_3,*)$, the rotation group of the equilateral triangle, can be derived from one particular element of $(R_3,*)$ and the operation "followed by." The element which generates the group is the one-third turn R_1. Since $R_1 * R_1 = R_2$ and $(R_1 * R_1) * R_1 = R_2 * R_1 = I$, the other two elements can be constructed from R_1, so that $(R_3,*)$ is generated by R_1.

Suppose we draw a line OA from the center of the equilateral triangle to its top vertex. In doing the rotation R_1, we move the line OA to the position OB:

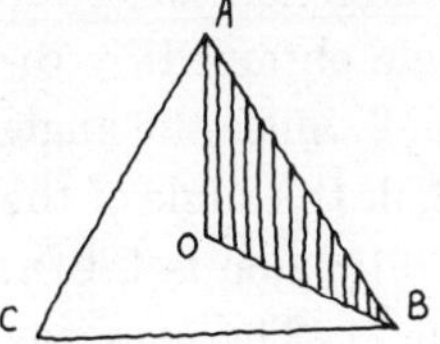

The shaded area of the triangle, bounded by OA and OB, is the *generator* of the triangle since the triangle can be constructed by rotations of this generator about its center O.

EXERCISE 17 (*a*) Show that each element of $(R_4,*)$, the rotation group of the square, can be derived from the one element R_1, the one-quarter turn, and the operation "followed by." (*b*) Draw a square, and shade an area of the square which is the generator of the square.

The *kaleidoscope* was named by the British physicist Sir David Brewster, who patented it in 1817. The kaleidoscope is based on two principles: the construction of a regular polygon from its generator, and the construction of a rotation by two reflections. Now generally considered to be a children's toy, the kaleidoscope was once used to create patterns for weaving and other crafts. The word "kaleidoscope" is made up of words meaning "beautiful," "form," and "to see." The kaleidoscope forms patterns by reflecting bits of colored glass or other objects in two mirrors placed at a certain angle to one another. The mirrors are encased in a tube, with the bits of glass in a container at one end and a hole to view through at the other. Reflections of the bits of glass in the mirrors cause a polygon of images to appear to the viewer.

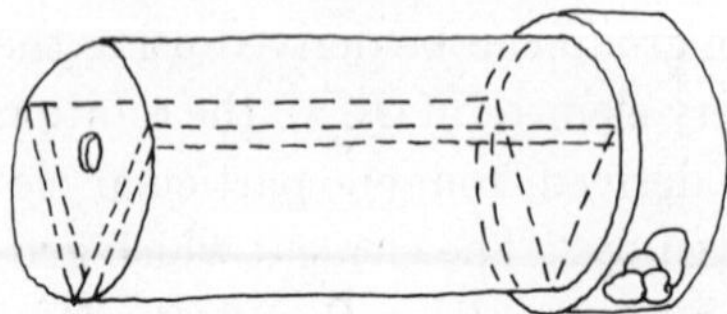

It is the angle at which the mirrors are placed that determines the polygon formed by the kaleidoscope. The region formed by the mirrors is the *fundamental region* of the kaleidoscope. Since two reflections form a rotation, the angle of the fundamental region of a kaleidoscope is one-half the angle of the generator of the polygon formed by the kaleidoscope.

We recall that there are 360 degrees in a circle. The generator of the polygon formed by a kaleidoscope can have any angle obtained by dividing 360 degrees by a counting number greater than 2. Since the angle of the fundamental region of the kaleidoscope is one-half the angle of the generator of the polygon, the angle of the fundamental region is $180/n$, where n is a counting number (usually greater than 2). The counting number n, if n is greater than 2, is the number of sides of the polygon formed by the kaleidoscope. The total number of reflected images formed by the kaleidoscope is $2n$.

An angle often used as the angle of the fundamental region of a kaleidoscope is 60 degrees. Since $180/3 = 60$, n is 3, and the polygon formed is a triangle. Twice 3 is 6, so the total number of images formed is 6. Suppose there is only one object in the kaleidoscope.

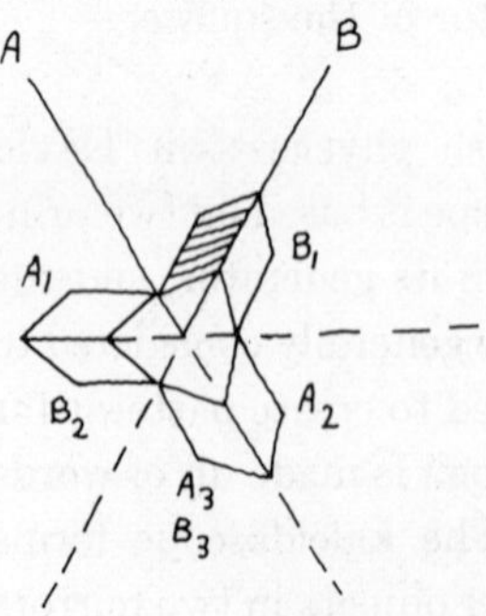

The *first images* of the object are its reflections in each mirror. The reflection in mirror A is marked A_1 and the reflection in mirror B is marked B_1. Also, the horizontal dotted lines represent the reflection of each mirror in the other. Each first image is then reflected in the other mirror to form *second images* A_2 and B_2. The first image A_1 is reflected in mirror B to form the second image A_2, and the first image B_1 is reflected in mirror A to form the second image B_2. Finally, each second image is reflected again in its own mirror to form *third images* A_3 and B_3. The extensions of mirrors A and B which create the third images are shown by dotted lines. If the mirrors are placed at an angle of exactly 60 degrees, the third images A_3 and B_3 will coincide.

There are exactly five distinct reflections of the original object. These reflections, together with the original object, form six images called a *sixfold* pattern. Furthermore, the object together with one first reflection of it, or two reflections together, form a pattern which corresponds to a generator of an equilateral triangle.

EXERCISE 18 How many fold is the reflection pattern formed by the fundamental region of the kaleidoscope with its mirrors at each of these angles: (*a*) 90 degrees; (*b*) 45 degrees; (*c*) 36 degrees. (A toy kaleidoscope I bought in a five-and-dime store has a fundamental region with an angle of 36 degrees.) What regular polygon has the generator corresponding to the pattern formed by a fundamental region with one reflection, of the kaleidoscope with its mirrors at each of these angles: (*d*) 45 degrees; (*e*) 36 degrees. (*f*) Describe the pattern which corresponds to the fundamental region with one reflection, of the kaleidoscope with its mirrors at an angle of 90 degrees.

EXERCISE 19 Using only one object, draw the first and second images formed by the kaleidoscope in which the angle of the fundamental region is 90 degrees.

EXERCISE 20 Consider a kaleidoscope in which the angle of the fundamental region is 180 degrees. (*a*) How many fold is the reflection pattern formed by the kaleidoscope? (*b*) Using only one object, draw the images formed by the kaleidoscope.

Suppose now that a kaleidoscope has three mirrors of equal width, each placed at an angle of 60 degrees to the other two. An object placed in this kaleidoscope has three first images, one reflected in each of the three

Photograph by Manfred Kage

This unusual design is a photograph of a kaleidoscopic image. The object reflected in the mirrors of the kaleidoscope is a cut agate, a translucent stone. The colors in the original of the photograph are beautiful shades of pink.

You can see that the generator of the polygon to which the pattern corresponds appears six times, corresponding to a hexagon. The reflection is therefore twelvefold, and the mirrors of the kaleidoscope are placed at an angle of 30 degrees. With so small an angle, besides the original object there are first, second, third, fourth, fifth, and sixth images. The two sixth images coincide.

You can be sure that such a photograph is not made simply by putting an ordinary camera to an ordinary kaleidoscope. Manfred Kage is world-renown for his photographs of kaleidoscopic images and of very small objects, using microscopes, special lenses, and other special equipment, much of which he makes himself. There are a great many beautiful kaleidoscopic photographs by Manfred Kage in an article "The Kaleidoscope—Vehicle of Beautiful Images," which appeared in the November 1968 issue of Camera. Life Library of Photography, Photography as a Tool, a Time-Life Book, includes two articles on Kage. One is about the photography of the very small. The other, about kaleidoscopic photography, has a few of Kage's kaleidoscopic photographs and a section on how to do it yourself (but you will need some special mirrors and a single-lens reflex camera).

mirrors. These first images each give two second images, one in each of the other two mirrors. The second images give third images, some of which coincide as before, but some of which begin new sixfold patterns and give fourth images, and so on, spreading out into the plane. The solid triangle represents the original mirrors; and the solid circle, the original object. The images of the circle are represented by open circles; and the images of the mirrors, by dotted lines. Rather than try to keep track of which mirror creates each image, we simply mark all first images 1, all second images 2, all third images 3, and so on.

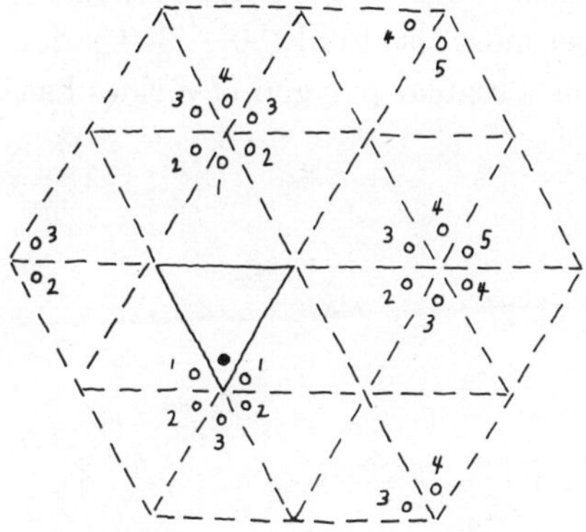

If we ignore the circles and draw only the triangles which represent the mirrors and their reflections, we have a *tessellation of the plane.* This tessellation is a *regular tessellation* because it is made up of identical regular polygons, in this case equilateral triangles. Although it is possible to construct a kaleidoscope which gives a regular polygon with n sides for every counting number n greater than 2, there are only three regular tessellations of the plane!

The proof of this fact involves some observations about regular polygons and a bit of algebra. First we observe that a regular polygon with n sides can be constructed from n isosceles triangles (triangles with two sides equal).

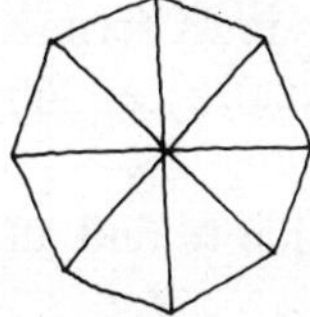

The lines from the center of the polygon to its vertices (points where its sides meet) must all be equal, so the two sides of each triangle formed by the lines must be equal.

We use two theorems from geometry (see Chapter 8). If the triangles are isosceles, then the angles opposite the equal sides are equal. Also, the sum of the three angles of a triangle is 180 degrees. Our regular polygon with n sides has n angles formed at the center, so each isosceles triangle has $360/n$ degrees in its angle at the center. Thus, its two other angles, which are equal, total $180 - 360/n$ degrees. Since two adjacent angles which together make up an outside angle of the polygon are also equal, the two adjacent angles together must also total $180 - 360/n$ degrees. Therefore, each outside angle of a regular polygon of n sides has $180 - 360/n$ degrees.

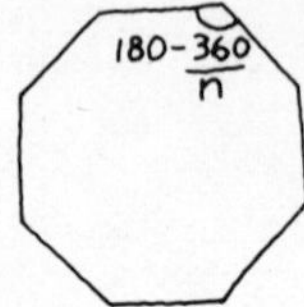

Now we look at the polygons around one vertex (point where the polygons meet) of the tessellation:

180-360/n

The ends of the 1-inch segement and the new half-inch segment are joined to form a triangle.

We must place a certain number of polygons around the vertex so that they touch without overlapping. Suppose there are p polygons around each vertex, where each polygon has n sides. Each polygon has $180 - 360/n$ degrees in its outside angle, and together the angles must form a full circle of 360 degrees; so we must have $p \cdot (180 - 360/n) = 360$ degrees.

Finally, we do some elementary algebra on this equation to find all the possible values of p and n:

$$\begin{aligned}
p \cdot \left(180 - \frac{360}{n}\right) &= 360 \\
180p - \frac{360p}{n} &= 360 \\
180np - 360p &= 360n \\
180np - 360p - 360n &= 0 \\
np - 2p - 2n &= 0 \\
np - 2p - 2n + 4 &= 4 \\
(n - 2)(p - 2) &= 4
\end{aligned}$$

Since n, the number of sides on a polygon, and p, the number of polygons around each vertex, are each at least 3, $n - 2$ and $p - 2$ must be counting numbers. Their product is 4. There are exactly three such products:

$1 \cdot 4 = 4 \qquad 2 \cdot 2 = 4 \qquad 4 \cdot 1 = 4$

If $n - 2 = 1$ and $p - 2 = 4$, then $n = 3$ and $p = 6$. This tessellation is the one formed by the three mirrors, since $n = 3$ means equilateral triangles and $p = 6$ means six triangles at each vertex.

If $n - 2 = 2$ and $p - 2 = 2$, then $n = 4$ and $p = 4$. This tessellation consists of squares placed four around each vertex. It is the pattern of a chessboard, as well as of plain tile walls and floors.

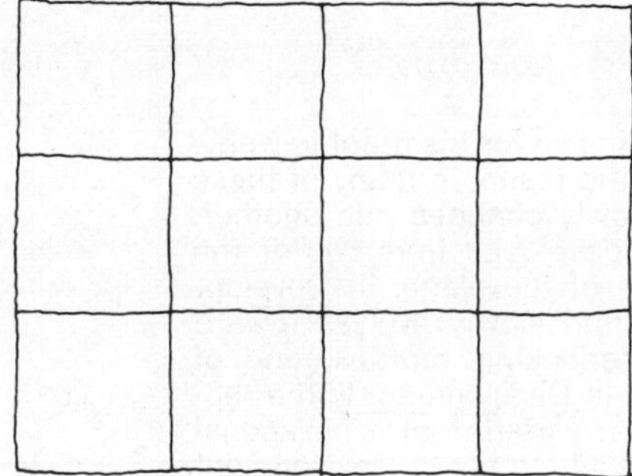

If $n - 2 = 4$ and $p - 2 = 1$, then $n = 6$ and $p = 3$. This tessellation consists of regular hexagons with three around each vertex. Bees build their honeycombs in this pattern, and by so doing, it turns out, use the least possible amount of wax.

Escher Foundation, Haags Gemeentemuseum, The Hague

DEVELOPMENT I

Maurits C. Escher is best known for his graphic works based on tessellations of the plane. In many of these works, one tessellation slowly changes into another. The two shown here are developed from two of the three regular tessellations of the plane. In Development I, a square tessellation slowly evolves inward into a tessellation of interlocking reptiles, one of Escher's favorite subjects. In Development II, the reptiles evolve outward from a variation of a hexagonal tessellation. The hexagons which begin Development

Escher Foundation, Haags Gemeentemuseum, The Hague

DEVELOPMENT II

II, do not actually form a regular tessellation because the spiral pattern of the evolution causes them to increase in size so that they are not identical. For the same reason, the hexagons are also not quite regular, since the edges toward the outside of the spiral are slightly larger than the edges toward the inside. It would have been possible, however, to make the development from a regular hexagonal tessellation (see the lithograph Reptiles, Chapter 7).

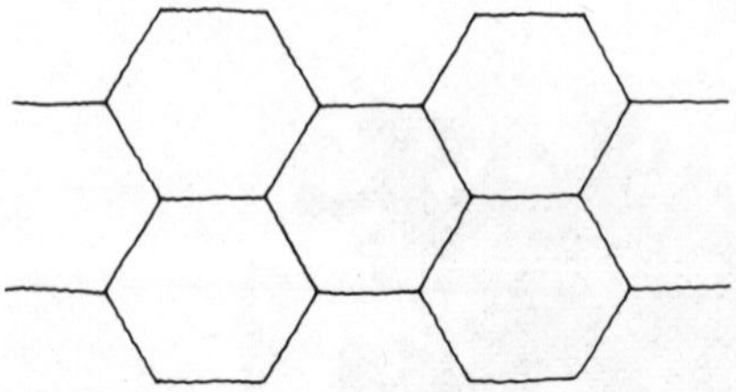

EXERCISE 21 Try the construction suggested below.

To put a chicken-wire pattern of creases into a small sheet of paper, first roll the sheet into a tube about half an inch in diameter. With the thumb and forefinger of your left hand pinch one end of the tube flat. Keeping pressure on the pinch with your left hand, your right thumb and forefinger, now pinch the tube flat at a spot as close as possible to the first pinch, making the pinch at right angles to the first one. Press firmly with both hands, at the same time pushing the two pinches tightly against each other to make the creases as sharp as possible. Now the right hand retains its pinch, while the left hand makes a third pinch adjacent to and perpendicular to the second one. Continue in this way, alternating hands as you move along the tube, until the entire tube has been pinched. (Children often do this with soda straws to make "chains.") Unroll the paper. You will find it hexagonally tessellated in a manner that is most puzzling to the uninitiated.[2]

Looking Back

1 The rectangle is a quadrilateral with its four angles equal and its opposite pairs of sides equal. It has four symmetries, I, R_2, V, and H. (*a*) Use a rectangle marked

$$I = \begin{array}{|cc|} a & b \\ d & c \end{array}$$

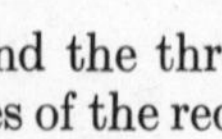

and find the three other symmetries. (*b*) Make the table for the symmetries of the rectangle with the operation "followed by." (*c*) Show that the table is a commutative group. The rhombus is a quadrilateral with its four sides equal and its opposite pairs of angles equal. It also has four

[2] Martin Gardner, "Mathematical Games," *Scientific American*, June 1973, p. 110.

symmetries. (*d*) Use a rhombus marked

I =

and find the three other symmetries. (*e*) Make the table for the symmetries of the rhombus with the operation "followed by." (*f*) Show that the table is a commutative group.

2 The regular pentagon has ten symmetries. Five are the rotations I, R_1, R_2, R_3, and R_4. The other five are reflections. Use a regular pentagon marked

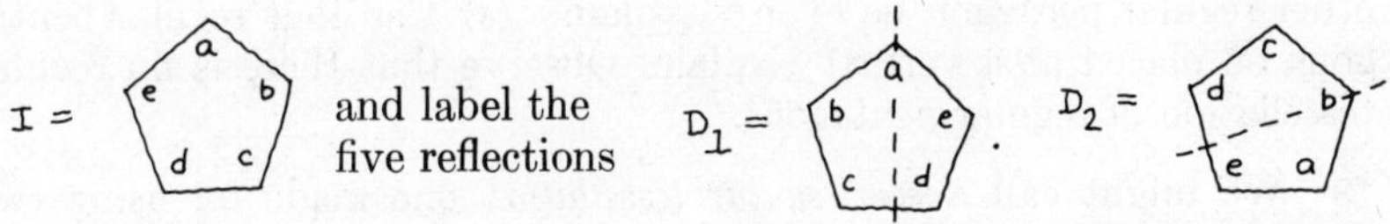

and so forth. (*a*) Make the table for the symmetries of the regular pentagon with the operation "followed by." (*b*) Show that the table is a group. (*c*) Is it a commutative group?

3 (*a*) What are the possible orders for subgroups of the symmetry group of the regular pentagon? (*b*) Find all the subgroups of each order.

***4** Consider any regular polygon of n sides where n is a prime number. The polygon has $2n$ symmetries, n rotations, and n reflections. (*a*) What are the possible orders for subgroups of the symmetry group of a regular polygon of n sides where n is a prime? (*b*) How many subgroups are there of each order?

5 (*a*) Construct the table for $(I_{11} - 0, \cdot)$. Since 11 is a prime number, the table is a group. (*b*) What are the possible orders for subgroups of $(I_{11} - 0, \cdot)$? (*c*) Find all the subgroups of each order.

6 The table for $(I_{12} - 0, \cdot)$ is not a group. However, $(I_{12} - 0, \cdot)$ has subsets which are groups. (These are not called subgroups since the original set is not a group.) (*a*) Find all the subsets of $(I_{12} - 0, \cdot)$ which are groups of order 2. (*b*) Find the one subset of $(I_{12} - 0, \cdot)$ which is a group of order 4, and make its table. (*c*) Is the table in part *b* a cyclic group or a Klein four group?

***7** Recall the group of eight matrices in Exercise 11, Looking Back, Chapter 4. (*a*) Find all the subgroups of order 2 of the group of eight matrices. (*b*) Which elements of the group of eight matrices are their own inverse? (*c*) Which elements of $(I_8,+)$ are their own inverse? (*d*) Which elements of $(S_4,*)$ are their own inverse? (*e*) Which elements of the quaternion group $(Q,\cdot)$ are their own inverse? (*f*) Are the group of eight matrices and $(I_8,+)$ isomorphic? (*Hint:* How many elements which are their own inverse has each?) (*g*) Are the group of eight matrices and $(S_4,*)$ isomorphic? (*Hint:* Is one commutative and one not?) (*h*) Are the group of eight matrices and $(Q,\cdot)$ isomorphic? (*Hint:* Use either of the hints above.)

8 (*a*) Find the number of degrees in the angle of the generator of a regular pentagon. (*b*) Find the number of degrees in the outside angle of a regular pentagon. (*c*) If two regular pentagons are placed at a vertex of a tessellation, can another regular pentagon be fit in? Explain. (*d*) If three regular pentagons are placed at a vertex of a tessellation, can another regular pentagon be fit in? Explain. (*e*) Can four regular pentagons be placed at a vertex? Explain. Observe that there is no regular tessellation of regular pentagons.

***9** We might call a *semiregular tessellation* one made by using two different regular polygons. This one uses the regular octagon and the square.

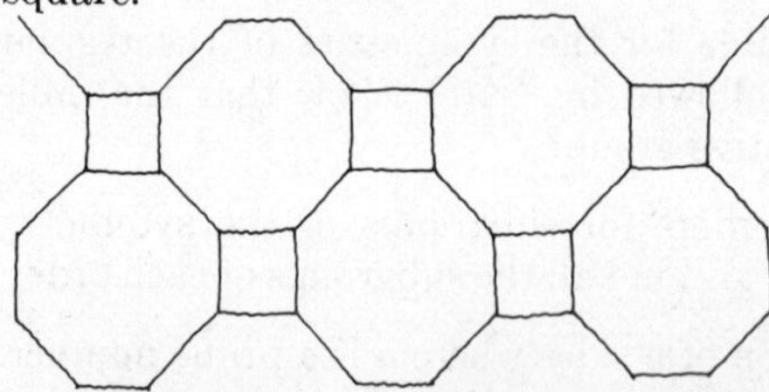

(*a*) Find the number of degrees in the angle of the generator of a regular octagon. (*b*) Find the number of degrees in the outside angle of a regular octagon. (*c*) Show that the semiregular tessellation above is possible by showing that the angles of one square and two octagons make 360 degrees. (*d*) If two regular pentagons are placed at a vertex of a tessellation, can another regular polygon be fit in? Explain. (*e*) If three regular pentagons are placed at a vertex of a tessellation, can another regular polygon be fit in? Explain. Observe that there is no semiregular tessellation which can be made using the regular pentagon with some other regular polygon.

Branching Out

1 Joseph-Louis Lagrange had two great contemporaries, Pierre-Simon de Laplace (1749–1827) and Adrien-Marie Legendre (1752–1833). Although all three have a great deal in common, Lagrange and Laplace are perhaps the better known, Lagrange in pure and Laplace in applied mathematics. Both Lagrange and Laplace (and especially Laplace) were affected by the French Revolution, particularly with respect to the later parts of their careers. All three had great influence on mathematicians who came after them and in particular on Cauchy. Although you will want to beware of parts of their work which are quite technical, the lives of these three French mathematicians make an interesting study in comparisons and contrasts.

2 A fourth great French mathematician of the revolutionary period belongs to the nineteenth century. Augustin-Louis Cauchy is usually considered to have outranked the three who came before him, having worked in both pure and applied mathematics. He was more deeply influenced by the Revolution, having grown up during it. In addition, he was a religious fanatic of sorts. His life is interesting on its own, and also in contrast with the fifth French mathematician of the Revolution, Galois.

3 The plans of many buildings are based on a principle of central symmetry, the same pattern as that of a kaleidoscope about a central point. Such symmetry, in both the building and the kaleidoscope, is generated by rotations and reflections and described by a regular polygon. The Pentagon in Washington is named for its central symmetry. Another famous building of another age is the Pantheon in ancient Rome, built in eight-sided, or octagonal, symmetry. You will find that symmetry in the building plan is itself intriguing. In addition, you can go on to symmetry in decoration such as friezes and columns. Although Hermann Weyl's *Symmetry* has some technical mathematics, it also has examples of symmetry in everything from moldings to molecules.

4 The rotation and reflection are two types of *transformations*. Two others are the *translation*, which moves an object such as this letter "**L**" in a given direction: **L** → **L**
and the *glide reflection*, which moves it and turns it over: **L** → **Γ**

The glide reflection combines a translation and a reflection. The late Dutch artist M. C. Escher has drawn many types of unusual pictures. As you look through his work, look for a type of design which fills the plane with identical objects. In one of his most famous of this type, the object is a knight on horseback. The patterns of these designs are generated entirely by reflections and glide reflections. There are two of them, with descriptions about how they are generated, in Coxeter's *Introduction to Geometry*. Coxeter uses symbols which may not be familiar to you, but the descriptions of the transformations and how they are used in the drawings should be understandable.

RACING
FORM

6

Calculated Chances

1

Probability as a Branch of Mathematics

Mathematicians like to say that the branch of mathematics called probability began in the back alleys of sixteenth-century Italy. At the same time, most mathematicians agree that the fathers of probability as a mathematical theory were two Frenchmen of the seventeenth century, Blaise Pascal (1623–1662) and Pierre de Fermat (1601–1665). The contrast between these two beginnings could not be greater.

The Italian mathematicians of the sixteenth century were the great equation-solvers, given to challenging one another to contests in solving cubic and quartic equations, and also given occasionally to stealing one another's methods of solution. Their association with the back alleys was through gambling.

Certainly much was known about things that were "likely," "unlikely," or "equally likely" well before the time of Pascal and Fermat. It would have been observed, for example, that if one tosses a coin, it is equally likely to land heads up or tails up. It would have been known that a total of 7 is more likely to be thrown on a pair of dice than 11. To prove

David Eugene Smith Collection, Columbia University Library

Among the most notable of the Italian equation-solvers and gamblers was Girolamo Cardano, who wrote the Ars magna, the most extensive work on algebra of its time (Chapter 1, Section 5). Cardano also wrote a gambler's manual. Here is Howard Eves'[1] description of the colorful Cardano.

"Girolamo Cardano is one of the most extraordinary characters in the history of mathematics. He was born in Pavia in 1501 as the illegitimate son of a jurist and developed into a man of passionate contrasts. He commenced his turbulent professional life as a doctor, studying, teaching and writing mathematics while practicing his profession. He once traveled as far as Scotland and upon his return to Italy successively held important chairs at the Universities of Pavia and Bologna. He was imprisoned for a time for heresy because he published a horoscope of Christ's life, showing that whatever Christ did he had to by the dictation of his stars. Resigning his chair in Bologna, he moved to Rome and became a distinguished astrologer, receiving, oddly enough, a pension as astrologer to the papal court. He died in Rome in 1576, by drinking poison, one story says, so as to fulfill his earlier astrological prediction of the date of his death. Many stories are told of his wickedness, as when in a fit of rage he cut off the ears of his younger son. At about the same time, his older son was executed for murder. An inveterate gambler, Cardano wrote a gambler's manual in which are considered some interesting questions on probability. It may be that history has somewhat maligned Cardano. Cardano's autobiography, of course, supports this view."

[1] *In Mathematical Circles.* (Boston: Prindle, Weber & Schmidt, Inc., 1969), 132°.

this fact, however, we must count all the possible combinations of numbers which can be thrown on a pair of dice. What Pascal and Fermat accomplished was the analysis of such counting problems. The difference between the interests of Pascal and Fermat and the interests of the gamblers who preceded them was that Pascal's and Fermat's interest was in the mathematical theory of the problem while their predecessors merely wanted to know what would win, not why.

It is said that a gambler, Antoine Gombaud, Chevalier de Méré, came to Pascal with a problem which had arisen from a gamblers' dispute. The problem, now a famous one called the *problem of points*, is concerned not with the outcome of the game but rather with the division of the stakes if the game is left unfinished. The counting problem within this problem is to determine all the ways the game can end, assuming the players are of equal skill. Pascal sent the problem to Fermat. While their analyses of the problem and their ways of counting the cases are different, their ultimate solutions are the same.

The correspondence between Pascal and Fermat continued, and many counting problems were discussed in their letters. It is said that one time Pascal made a mistake when he became impatient with the many cases to be counted in a problem. Fermat caught the mistake and wrote Pascal about it. In the following passage, Amy King and Cecil Read give an example of the type of reasoning each used to solve the problem of points.[1] Fermat's solution uses the theory of permutations, which we will consider in a later section. We will develop methods to understand his solution, however, in the next section.

We shall now turn to a consideration of the correspondence between Fermat and Pascal which forms the beginning of probability as a science, even though earlier work has been done in the field by Cardan, Galileo, and others. The correspondence began about 1654 with the famous "Problem of Points," a problem with which Cardan had been familiar. Pascal corresponded with Fermat concerning its solution, and both came up with the correct answer, but with different approaches.

Problems of this type deal with two players of equal skill who wish to stop playing a game before it is finished. If the number of points required to win, as

[1] Excerpted from *Pathways to Probability:* History of Mathematics of Certainty and Chance by Amy C. King/Cecil B. Read. Copyright © 1963 by Holt, Rinehart and Winston, Inc. Reprinted by special permission of the publishers, Holt, Rinehart and Winston, Inc.

BLAISE PASCAL

Blaise Pascal and Pierre de Fermat met through weekly meetings of a group which eventually became the Académie des Sciences. Their correspondence is considered to have laid the foundations for the theory of probability as a branch of mathematics.

Pascal was also important in geometry (Chapter 8, Section 2). Furthermore, he invented the first calculating machine, forerunner of the electrically driven wheel machines (now considered to be old-fashioned), giant electronic computers, and the now popular hand-held calculators.

Pascal is known in literature as well. His Lettres provinciales and Pensées are still read today. These works were written toward the end of his life as his attention was turning from mathematics to religion. Pascal had always been something of a religious fanatic. A story is told that one day in 1654, he was

Photo: Library of Congress

PIERRE DE FERMAT

in a carriage on the bridge at Neuilly. The horses ran out of control over the parapet, but the traces broke leaving Pascal safely on the bridge. From that day until his death in 1662, he devoted himself to religion, excepting eight days during which he developed a piece of geometry about a beautiful curve called a cycloid. Always of frail health, Pascal was only thirty-nine when he died.

Pierre de Fermat was a part-time mathematician in a different way. By profession he was a lawyer; mathematics was his hobby. He published few papers, most of his work being known through his correspondence with mathematicians, such as that with Pascal on probability. Even so, his contributions to mathematics are sufficient for him to be considered the greatest French mathematician of the seventeenth century by many historians.

Photo: David Eugene Smith Collection, Columbia University Library

well as the respective scores when they stop, are known, in what proportion should the stakes be divided? Basically the problem asks what probability each player has, at any stage of the game, of winning.

The following analysis of one such problem illustrates Pascal's method of reasoning: Assume two players play a game which requires three points to win and that each player has put up thirty-two pistoles (a former gold coin of Spain, worth about $4.00). Suppose that when they wish to cease play the first player has two points while the other has one. If they continue to play for one more point, the second player could either lose or tie the game. In either case, the first player is assured of at least thirty-two pistoles so the remaining thirty-two should be split evenly, the first player taking forty-eight and the second sixteen.

Next, suppose the first player has two points and the second player none. On the next trial the first player could either win or they could be in the same position as described in the previous paragraph. Therefore, the first player should have the forty-eight pistoles plus half of what is left, or a total of fifty-six pistoles; the second player then has eight pistoles.

Finally, suppose the first player has one point and the second none. If they play for an additional point, they would either be tied or in the position just described. If tied, the first payer is assured of thirty-two pistoles. In addition, he is entitled to half the difference between thirty-two and the fifty-six pistoles he would receive if he had two points and the second player none. This gives an additional twelve pistoles. Thus the first player should have a total of forty-four pistoles and the second twenty pistoles. The division is seen to be in the ratio of eleven to five.

Fermat's solution to the "Problem of Points" depends on the theory of permutations as follows: Assume A has one point and B has none, that is, A needs two points to win and B needs three. Certainly the game will be decided in not more than four trials. Let a and b represent a win by A and B respectively. There are a total of sixteen permutations of a and b possible (aaaa, aaab, aaba, aabb, abaa, abab, abba, abbb, baaa, baab, baba, babb, bbaa, bbab, bbba, bbbb). As will be noted, there are eleven cases favorable to A, i.e., where a occurs at least twice; but only five cases favorable to B, that is, where b occurs three times or more. Since these are assumed to be equally likely, A's chance is to B's as eleven is to five. This agrees with Pascal's solution.

2
Counting Problems

One of the most basic ways of counting all the possible outcomes of an experiment such as tossing coins or throwing dice is by means of a *tree diagram.* (The following discussion of tree diagrams is comparable to that of Chapter 2, Section 2.) Suppose we toss one coin. There are two possible outcomes, heads or tails. We use H to indicate heads and T for tails, and write

H
T

Now suppose we toss two coins, or toss one coin twice, which amounts to the same thing. In the case that the first coin, or the first toss, comes up heads, the second coin or toss could be either heads or tails. Similarly, in the case where the first coin or the first toss is tails, the second coin or toss could still be heads or tails. Thus, for each case for one coin, we have two cases for the second coin, or for the second toss of the coin. We illustrate the outcomes by showing the two new cases as branches of the original two cases.

If we toss three coins, or one coin three times, each of the four preceding cases has two new branches, one heads and one tails.

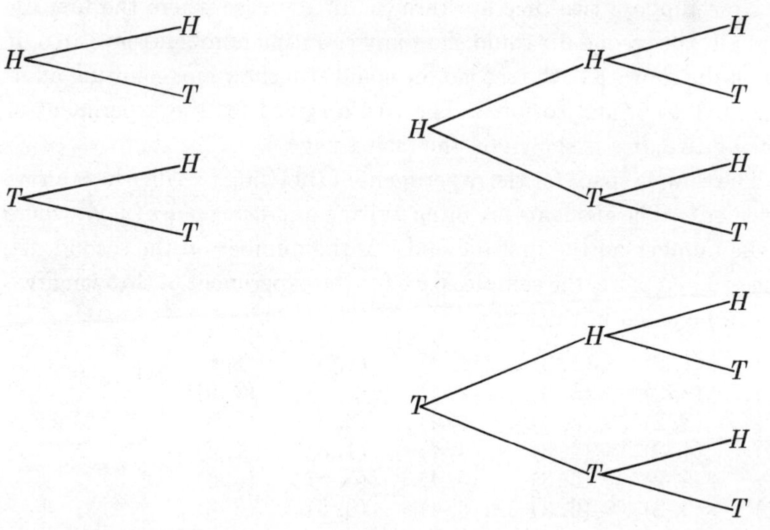

DEFINITION The *sample space* of an experiment is the set of all possible outcomes of the experiment. An *event* of an experiment is a subset of the sample space of the experiment.

The sample space for the experiment of tossing one coin is $\{H, T\}$. The event of getting heads is $\{H\}$ and the event of getting tails is $\{T\}$. The sample space for tossing two coins, or one coin twice, is $\{HH, HT, TH, TT\}$. The event of getting two heads is $\{HH\}$. The event of getting one head is $\{HT, TH\}$. The event of getting at least one head is $\{HH, HT, TH\}$. The event of getting no heads and the event of getting two tails are both $\{TT\}$.

If a single die is thrown, there are six possible outcomes. The number facing upward on the die could be any of the counting numbers from 1 to 6. Thus the sample space for throwing a die is $\{1, 2, 3, 4, 5, 6\}$. The event of throwing a 1 is $\{1\}$. The event of throwing a 1 or a 2 is $\{1, 2\}$. The event of throwing an even number is $\{2, 4, 6\}$.

An event can consist of the entire sample space. For example, suppose one die is thrown. The event that a counting number from 1 to 6 inclusive is thrown is $\{1, 2, 3, 4, 5, 6\}$, which is the same as the sample space for the experiment. At the opposite extreme, an event can have no elements at all. In this case the event is called the *empty set* and we write $\emptyset$, which is the symbol for the empty set. For example, if one die is thrown, the event of throwing a 7 is $\emptyset$.

Now suppose two dice are thrown. In the case where the first die shows a 1, the second die could show any counting number from 1 to 6. If the first die shows a 2, the second die could still show any counting number from 1 to 6, and so forth. The tree diagram for the experiment of throwing two dice is shown on the facing page.

The sample space for the experiment of throwing two dice has thirty-six elements. The elements are often written as *ordered pairs* (x, y), where x is the number on the first die and y is the number on the second die. Using ordered pairs, the sample space for the experiment of throwing two dice can be listed in the array

(1, 1)	(1, 2)	(1, 3)	(1, 4)	(1, 5)	(1, 6)
(2, 1)	(2, 2)	(2, 3)	(2, 4)	(2, 5)	(2, 6)
(3, 1)	(3, 2)	(3, 3)	(3, 4)	(3, 5)	(3, 6)
(4, 1)	(4, 2)	(4, 3)	(4, 4)	(4, 5)	(4, 6)
(5, 1)	(5, 2)	(5, 3)	(5, 4)	(5, 5)	(5, 6)
(6, 1)	(6, 2)	(6, 3)	(6, 4)	(6, 5)	(6, 6)

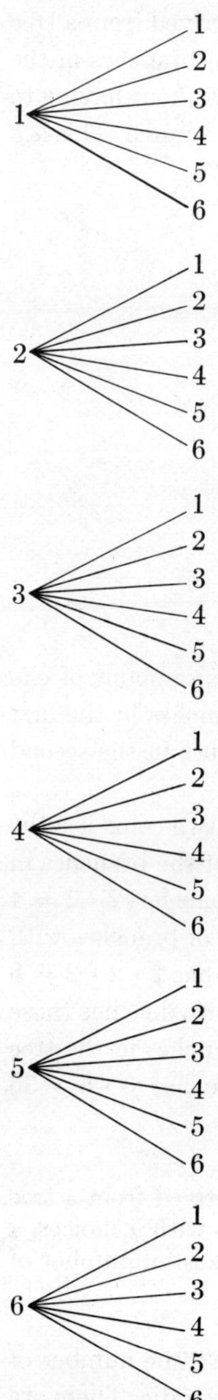
1
1
2
3
4
5
6
2
1
2
3
4
5
6
3
1
2
3
4
5
6
4
1
2
3
4
5
6
5
1
2
3
4
5
6
6
1
2
3
4
5
6

Finally, we observe that when a sample space is derived from a tree diagram, its number of elements is the total number of branches in the final column of the diagram. To calculate this number without having to draw the diagram, we use the number of choices of each branch. This is a branch with one choice,

1————1

a branch with two choices,

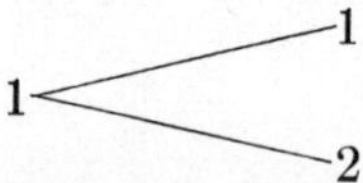

a branch with three choices,

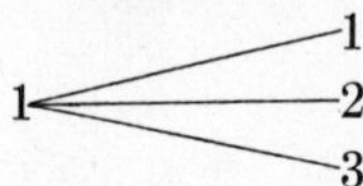

and so on. The final number of branches is the original amount of elements, multiplied by the number of choices of the branches in the first column, multiplied by the number of choices of branches in the second column, and so on.

Tossing a coin has two choices, H and T. Tossing two coins has the two original elements times 2, the number of choices of the branches in the tree diagram, so the sample space for tossing two coins has $2 \cdot 2 = 4$ elements. Tossing three coins joins on another column of branches with two choices, so the sample space for tossing three coins has $2 \cdot 2 \cdot 2 = 8$ elements. Throwing one die has six choices. Throwing two dice has those six elements times 6, the number of choices of the branches in the tree diagram, so the sample space for throwing two dice has $6 \cdot 6 = 36$ elements.

RULE FOR CHOICES If a sample space can be derived from a tree diagram with x original elements, a column of branches with y choices, a column of branches with z choices, and so on, then the total number of elements in the sample space is $x \cdot y \cdot z \cdot \ldots$.

Each branch of a tree diagram need not have the same number of choices. For example, suppose we toss a coin and throw a die. There are two choices for the coin and six for the die. Thus, the sample space for the experiment has $2 \cdot 6 = 12$ elements. The tree diagram is

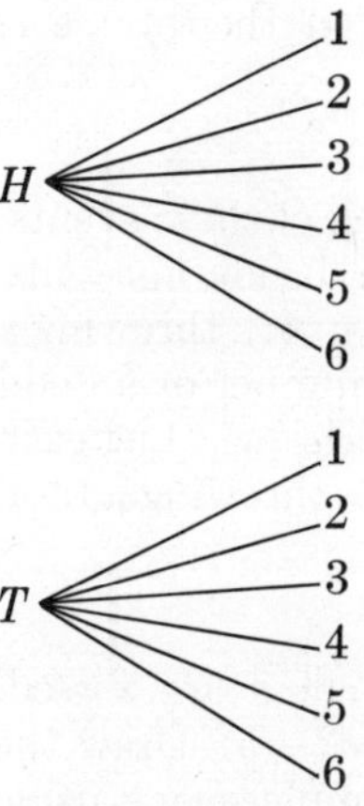

Using ordered pairs, the sample space for the experiment can be listed in the array

$(H, 1) \quad (H, 2) \quad (H, 3) \quad (H, 4) \quad (H, 5) \quad (H, 6)$
$(T, 1) \quad (T, 2) \quad (T, 3) \quad (T, 4) \quad (T, 5) \quad (T, 6)$

EXERCISE 1 (*a*) List the sample space for tossing three coins, or tossing one coin three times. (*b*) List each of these events of the experiment of tossing three coins: (*i*) getting three heads; (*ii*) getting two heads; (*iii*) getting at least two heads; (*iv*) getting one head; (*v*) getting at least one head; (*vi*) getting no heads; (*vii*) getting three tails.

EXERCISE 2 (*a*) Construct the tree diagram for the experiment of tossing four coins, or tossing one coin four times. (*b*) List the sample space for the experiment of tossing four coins. (*c*) List each of these events of the experiment of tossing four coins: (*i*) getting four heads; (*ii*) getting three heads; (*iii*) getting at least three heads; (*iv*) getting two heads; (*v*) getting one head.

EXERCISE 3 (*a*) For the experiment of throwing two dice, list the event that the total on the dice is 7. (*b*) For the experiment of throwing two dice, list the event that the total on the dice is 11. Observe that it is more likely that 7 is thrown than 11 in the sense that the event has more elements.

EXERCISE 4 (*a*) How many elements are there in the sample space for the experiment of throwing three dice? (*b*) List each of these events of the experiment of throwing three dice, writing each element of the event as an *ordered triple* (x, y, z): (*i*) the total is 3; (*ii*) the total is 4; (*iii*) the total is 5; (*iv*) the total is 6.

EXERCISE 5 (*a*) Suppose one coin is tossed. List each of these events: (*i*) getting either heads or tails; (*ii*) getting neither heads nor tails. (*b*) Suppose one die is thrown. List each of these events: (*i*) throwing a counting number less than 7; (*ii*) throwing an even number or an odd number; (*iii*) throwing a 0. (*c*) Suppose two dice are thrown. List each of these events: (*i*) throwing a total of at least 12; (*ii*) throwing a total of more than 12; (*iii*) throwing a total of 1.

EXERCISE 6 Suppose a die is thrown and two coins are tossed. (*a*) How many elements are there in the sample space? (*b*) Draw the tree diagram for the sample space. (*c*) List the sample space using ordered triples.

EXERCISE 7 Refer to Fermat's solution to the problem of points at the end of Section 1. The sample space can be constructed from a tree diagram with four branches, where there are two choices for each branch, *a* or *b*. (*a*) How many elements are there in the sample space? (*b*) Draw a tree diagram for the sample space. (*c*) List the sample space and check your list against Fermat's.

3

Theoretical Probability

If a single coin is tossed, the sample space for the experiment is $\{H, T\}$. From experience, we expect a fifty-fifty chance of heads or tails if the coin is true; that is, if the coin is not loaded toward one side or the other. We express this expectation by saying that the probability of getting heads is $\frac{1}{2}$ and the probability of getting tails is $\frac{1}{2}$. The event of getting heads is the set $\{H\}$, which has one element. The probability of getting heads is therefore the number of elements in the event of getting heads divided by the number of elements in the sample space. Similarly, the event of getting tails is the set $\{T\}$, and the probability of getting tails is the number of elements in the event divided by the number of elements in the sample space.

Now suppose two true coins are tossed. The sample space is $\{HH, HT, TH, TT\}$, which has four elements. The event of both coins coming up heads is $\{HH\}$. The probability of the two coins both being heads is $\frac{1}{4}$, the number of elements in the event divided by the number of elements in the sample space. Similarly, the event of no heads, or of two tails, is $\{TT\}$, so the probability of each of these events is also $\frac{1}{4}$. The event of getting exactly one head is $\{HT, TH\}$, so the probability of getting exactly one head is $\frac{2}{4} = \frac{1}{2}$. The event of getting at least one head is $\{HH, HT, TH\}$, so the probability of getting at least one head is $\frac{3}{4}$.

DEFINITION Suppose E is an event and S the sample space of an experiment. Then the *probability of* E is $N(E)/N(S)$, where $N(E)$ is the number of elements in E and $N(S)$ is the number of elements in S.

(In using this definition of probability, we assume all coins, dice, and so forth, involved are true.)

For any experiment, an event can at most be the entire sample space. The largest possible probability is the probability of an event E which is the same as the sample space S. In this case $N(E)/N(S) = N(S)/N(S) = 1$. An event with probability 1 is sure to occur. For example, if we neglect the possibility of a coin landing on its edge, or not landing at all, the probability of getting either heads or tails in the experiment of tossing one coin is 1. The least an event can be is the empty set, which has no elements. The smallest possible probability is $N(E)/N(S) = 0/N(S) = 0$. An event with probability 0 cannot occur. For example, under the assumptions above, the probability of the event that a coin lands neither heads up nor tails up is 0.

We may also observe that if two or more events are such that exactly one of the events must occur, then the sum of the probabilities of the events is 1. For example, under the assumptions above, if one coin is tossed exactly one of the events of getting heads or getting tails must occur. The probability of getting heads is $\frac{1}{2}$ and the probability of getting tails is $\frac{1}{2}$, and $\frac{1}{2} + \frac{1}{2} = 1$. If two coins are tossed, exactly one of the events of getting two heads, getting one head, or getting no heads must occur. The probability of getting two heads is $\frac{1}{4}$, the probability of getting one head is $\frac{1}{2}$, the probability of getting no heads is $\frac{1}{4}$, and $\frac{1}{4} + \frac{1}{2} + \frac{1}{4} = 1$.

Also if two coins are tossed, exactly one of the events of getting at least one head or getting no heads must occur. The probability of getting at least one head is $\frac{3}{4}$ and the probability of getting no heads is $\frac{1}{4}$, and $\frac{3}{4} + \frac{1}{4} = 1$. This observation provides a shortcut for computing certain probabilities. For example, if three coins are tossed, we know from the rule of choices that the number of elements in the sample space is eight. The event of getting no heads has one element, TTT. Thus, the probability of getting no heads is $\frac{1}{8}$. Since exactly one of the events, getting at least one head or getting no heads must occur, we can then compute its probability as $1 - \frac{1}{8} = \frac{7}{8}$ without listing the event of getting at least one head.

The theoretical definition of probability, when applied to experiments such as tossing coins or throwing dice, is based on the *assumption of equally likely outcomes.* The assumption of equally likely outcomes says that if an experiment has n equally likely outcomes, then the probability of each outcome is $1/n$. In tossing a true coin, heads or tails are equally likely outcomes. There are two outcomes, so each has probability $\frac{1}{2}$. If a true die is thrown, there are six possible outcomes. The assumption of equally likely outcomes then assigns a probability of $\frac{1}{6}$ to each outcome.

EXERCISE 8 Suppose three coins are tossed. Find the probability of each of these events: (*a*) getting three heads; (*b*) getting two heads; (*c*) getting at least two heads; (*d*) getting one head; (*e*) getting at least one head; (*f*) getting no heads; (*g*) getting three tails.

EXERCISE 9 (*a*) If four coins are tossed, how many elements are there in the sample space? (*b*) What is the probability of getting no heads? (*c*) What is the probability of getting at least one head?

EXERCISE 10 If one die is thrown, exactly one of the events of getting a 1, a 2, a 3, a 4, a 5, or a 6 must occur. (*a*) What is the probability of each of these events? (*b*) What is the sum of the probabilities of these events?

EXERCISE 11 If one die is thrown, find the probability of each of these events: (*a*) the number is 1; (*b*) the number is 1 or 2; (*c*) the number is even; (*d*) the number is more than 3; (*e*) the number is even or more than 3; (*f*) the number is more than 1; (*g*) the number is at least 1; (*h*) the number is less than 1.

EXERCISE 12 If two dice are thrown, find the probability of each of these events: (*a*) the total is 7; (*b*) the total is 11; (*c*) the total is 2; (*d*) the total is 1; (*e*) both dice show the same number; (*f*) the dice show different numbers.

4

Mutually Exclusive Events

Suppose we want to draw a spade from a deck of cards. An ordinary deck of playing cards has fifty-two cards, so there are fifty-two elements in the sample space. Since we assume any card is as likely to be drawn as any other, the assumption of equally likely outcomes gives probability $\frac{1}{52}$ of drawing any particular card. There are thirteen spades in the deck, so the probability of drawing a spade is $\frac{13}{52} = \frac{1}{4}$. There are four aces in the deck, one of each suit. The probability of drawing an ace is, therefore, $\frac{4}{52} = \frac{1}{13}$. There is only one ace of spades. The probability of drawing the ace of spades is $\frac{1}{52}$.

Now suppose we want to draw an ace or a king. Since there are four aces and also four kings, the event of drawing an ace or a king has eight elements. Thus, the probability of drawing an ace or a king is $\frac{8}{52} = \frac{2}{13}$. We must be more careful, however, if the event is that of drawing an ace or a spade. There are four aces and thirteen spades, but the event does not have $4 + 13 = 17$ elements. One of the elements of the event, the ace of spades, is both an ace and a spade, so it has been counted twice. The event has just sixteen elements, the four aces and the twelve remaining spades. The probability of drawing an ace or a spade is $\frac{16}{52} = \frac{4}{13}$.

EXERCISE 13 If the sample space is an ordinary deck of playing cards, find the probability of each of these events: (*a*) drawing a heart; (*b*) drawing a queen; (*c*) drawing the queen of hearts; (*d*) drawing a queen or a heart; (*e*) drawing a red card (hearts or diamonds); (*f*) drawing a face card (jack, queen, or king); (*g*) drawing a face card or a red card.

Let us return to the ordinary deck of playing cards and the event of drawing an ace or a king. We may think of this event as two events, that of drawing an ace and that of drawing a king. The event of drawing an ace has probability $\frac{4}{52} = \frac{1}{13}$, and the event of drawing a king also has probability $\frac{4}{52} = \frac{1}{13}$. The probability of drawing an ace or a king is $\frac{8}{52} = \frac{2}{13}$, which is the same as the sum of the separate probabilities of the two events, $\frac{1}{13} + \frac{1}{13} = \frac{2}{13}$. Further, we observe that if we draw an ace, we cannot have drawn a king; and if we draw a king, we cannot have drawn an ace.

DEFINITION Two events E_1 and E_2 are *mutually exclusive* if when E_1 occurs then E_2 cannot, and when E_2 occurs then E_1 cannot.

The events of drawing an ace and of drawing a king are mutually exclusive, and the probability of drawing an ace or a king is the sum of the probabilities of drawing an ace and of drawing a king.

On the other hand, if we consider the event of drawing an ace or a spade, the probability of the event is not the sum of the probabilities of drawing an ace and of drawing a spade. The probability of drawing an ace is $\frac{4}{52}$. The probability of drawing a spade is $\frac{13}{52}$. The probability of drawing an ace or a spade, however, is $\frac{16}{52} = \frac{4}{13}$.

The events of drawing an ace and of drawing a spade are not mutually exclusive. It is possible to draw an ace and also to draw a spade, if the ace of spades is drawn. Therefore, both events can occur at the same time. Since the ace of spades is both an ace and a spade, it is counted in both the probability of drawing an ace and the probability of drawing a spade. Were we simply to add these probabilities, therefore, we would have counted the probability of drawing the ace of spades twice. The probability of drawing the ace of spades is $\frac{1}{52}$. If we subtract this probability from the sum of the probabilities of drawing an ace and of drawing a spade, we will have counted the probability of drawing the ace of spades only once. Thus, we can compute the probability of drawing an ace or a spade as the sum of the probabilities of drawing an ace and of drawing a spade minus the probability of drawing the ace of spades, $\frac{4}{52} + \frac{13}{52} - \frac{1}{52} = \frac{16}{52} = \frac{4}{13}$.

RULE FOR MUTUALLY EXCLUSIVE EVENTS Let $P(E)$ mean the probability of an event E. If two events E_1 and E_2 are mutually exclusive, $P(E_1 \text{ or } E_2) = P(E_1) + P(E_2)$. If two events E_1 and E_2 are not mutually exclusive, $P(E_1 \text{ or } E_2) = P(E_1) + P(E_2) - P(E_1 \text{ and } E_2)$.

The two parts of the rule for mutually exclusive events could be written as only the second part because if E_1 and E_2 are mutually exclusive, only one of E_1 and E_2, but not both, can occur at the same time. Since both E_1 and E_2 cannot occur, $P(E_1 \text{ and } E_2) = 0$. Thus, when E_1 and E_2 are mutually exclusive, the second part of the rule for mutually exclusive events becomes $P(E_1 \text{ or } E_2) = (PE_1) + P(E_2) - P(E_1 \text{ and } E_2) = P(E_1) + P(E_2) - 0 = P(E_1) + P(E_2)$. This result is the first part of the rule for mutually exclusive events.

EXERCISE 14 If the sample space is an ordinary deck of playing cards, which of these events are mutually exclusive: (*a*) drawing a heart or a diamond; (*b*) drawing a jack or a queen; (*c*) drawing a queen or a heart; (*d*) drawing a jack, a queen, or a king; (*e*) drawing a face card or a red card; (*f*) drawing a black card or a red card; (*g*) drawing a heart or a red card.

EXERCISE 15 For the events of Exercise 14 which are mutually exclusive, use the first part of the rule for mutually exclusive events to find $P(E_1 \text{ or } E_2)$ (or E_1 or E_2 or E_3 when appropriate). For the events of Exercise 14 which are not mutually exclusive, use the second part of the rule for mutually exclusive events to find $P(E_1 \text{ or } E_2)$.

EXERCISE 16 Refer to the sample space for Fermat's solution to the problem of points at the end of Section 1, or in Exercise 7 at the end of Section 2. Find the probability of each of these events. (*a*) Player A has at least two points; that is, a occurs at least twice. (*b*) Player B has at least three points; that is, b occurs at least three times. (*c*) What is the probability that player A has at least two points or player B has at least three points?

5
Independent Events

We now consider the probability of one event followed by another. For example, suppose we want to get heads on a toss of a coin and then get heads on a second toss. We have seen that this event is $\{HH\}$ in the sample space for tossing two coins $\{HH, HT, TH, TT\}$, so the probability is $\frac{1}{4}$. Alternatively, we can approach the problem by considering that the probability of heads on the first toss is $\frac{1}{2}$ and the probability of heads on the second toss is also $\frac{1}{2}$. Using the rule for choices, we have one choice out of two for the first toss and one choice out of two for the second toss. The result is $\frac{1}{2} \cdot \frac{1}{2} = \frac{1}{4}$, which is the correct probability. Similarly, the probability of getting heads on three successive tosses is $\frac{1}{2} \cdot \frac{1}{2} \cdot \frac{1}{2} = \frac{1}{8}$, which also agrees with our previous results.

If we throw one die, the probability of getting a 1 on the die is $\frac{1}{6}$. The probability of getting a second 1 on another throw is also $\frac{1}{6}$. Thus, the probability of getting a 1 followed by another 1 is $\frac{1}{6} \cdot \frac{1}{6} = \frac{1}{36}$. This result agrees with the probability of getting double 1s on two dice.

Of course, we need not be looking for the same event twice. The probability of throwing a 2 on a die followed by throwing a 3 is also $\frac{1}{6} \cdot \frac{1}{6} = \frac{1}{36}$. The probability of tossing a coin twice and getting heads followed by tails is $\frac{1}{2} \cdot \frac{1}{2} = \frac{1}{4}$. The probability of getting tails followed by heads is also $\frac{1}{2} \cdot \frac{1}{2} \equiv \frac{1}{4}$. The probability of getting one head and one tail, which is the probability of the two mutually exclusive events, heads followed by tails or tails followed by heads, can be calculated as $\frac{1}{2} \cdot \frac{1}{2} + \frac{1}{2} \cdot \frac{1}{2} = \frac{1}{4} + \frac{1}{4} = \frac{1}{2}$. The situations described above can be defined by the rule for calculating their probabilities.

DEFINITION Two events E_1 and E_2 are *independent* if $P(E_1$ followed by $E_2) = P(E_1) \cdot P(E_2)$.

When events are independent, the probability of the second is not affected by the outcome of the first. So far, the events we have considered have been independent. One throw of a die does not affect the probability of the next throw; one toss of a coin does not affect the next toss. If we consider the experiment of drawing a card from a deck of playing cards, however, the events of drawing one card followed by another are not independent unless we replace the first card before the second is drawn. If the first card is replaced, the deck is the same as it was originally, and the probability of the second event is not affected by the first.

If, however, the first card is not replaced, the second event has a sample space of only fifty-one cards. For example, if an ace has been drawn as the first event and is not replaced, the probability of then drawing a king as the second event is $\frac{4}{51}$. Moreover, if an ace has been drawn and not replaced, and the second event is also that of drawing an ace, there are only three aces left. Therefore, if an ace has been drawn and not replaced, the probability of drawing a second ace is $\frac{3}{51} = \frac{1}{17}$.

So long as we know how one event affects another, we can find the probability of the first event followed by the second by multiplying the probability of the first and the adjusted probability of the second. If the events are not independent, however, the probability of the second event is not the same as it would have been had the first event not happened. For example, the probability of drawing a ace followed by a king, if we replace the ace, is $\frac{4}{52} \cdot \frac{4}{52} = \frac{1}{169}$. If we do not replace the ace, however, there are four kings in the remaining fifty-one cards, so the probability of drawing an ace followed by a king without replacement is $\frac{4}{52} \cdot \frac{4}{51} = \frac{4}{663}$.

The probability of drawing two aces in succession, if we replace the first ace, is $\frac{4}{52} \cdot \frac{4}{52} = \frac{1}{169}$. But if we do not replace the first ace, there are only three aces left in fifty-one cards, so the probability of drawing an ace followed by an ace without replacement is only $\frac{4}{52} \cdot \frac{3}{51} = \frac{1}{221}$.

Similarly, the probability of drawing a spade followed by a heart is $\frac{13}{52} \cdot \frac{13}{52} = \frac{1}{16}$ if the space is replaced. Without replacement, however, the probability is $\frac{13}{52} \cdot \frac{13}{51} = \frac{13}{204}$. The probability of drawing a spade followed by a spade is also $\frac{13}{52} \cdot \frac{13}{52} = \frac{1}{16}$ if the first spade is replaced. This time, if the spade is not replaced, there are only twelve spades left in fifty-one cards, so the probability of drawing a spade followed by a spade without replacement is $\frac{13}{52} \cdot \frac{12}{51} = \frac{1}{17}$.

In considering sequences of events, a common mistake is to confuse the probability of a single event in the sequence with the probability of the entire sequence. For example, suppose a coin is tossed four times and each time it comes up heads. You might think that on the fifth toss the probability of tails should be greater than the probability of heads. This is not the case, however, because each toss of the coin is an independent event, and the occurrence of heads on one toss does not affect the occurrence of heads on any other. The probability of heads on the fifth toss of the coin is $\frac{1}{2}$, and the probability of tails is also $\frac{1}{2}$. Now, if we consider the entire sequence of five tosses of the coin, the probability that each of the five tosses comes up heads is $\frac{1}{2} \cdot \frac{1}{2} \cdot \frac{1}{2} \cdot \frac{1}{2} \cdot \frac{1}{2} = \frac{1}{32}$. Consequently, the probability of heads on any individual toss is $\frac{1}{2}$, but the probability of heads on five out of five tosses is only $\frac{1}{32}$.

EXERCISE 17 Suppose a coin is tossed four times. (*a*) What is the probability of getting heads on each separate toss? (*b*) What is the probability of getting heads on all four tosses?

EXERCISE 18 Suppose two dice are thrown. (*a*) What is the probability of throwing a 7? (*b*) What is the probability of throwing a number other than 7? (*Hint:* Subtract the probability of throwing a 7 from 1.) (*c*) What is the probability of throwing a number other than 7 on three consecutive throws?

EXERCISE 19 Suppose the sample space is an ordinary deck of playing cards. Find $P(E_1$ followed by $E_2)$ (or E_1 followed by E_2 followed by E_3 when appropriate) if each card is replaced after it has been drawn, for each of these sequences of events: (*a*) Drawing a heart followed by a diamond. (*b*) Drawing a heart followed by a heart. (*c*) Drawing a red card followed by a black card. (*d*) Drawing a red card followed by a red card. (*e*) Drawing a queen followed by a king. (*f*) Drawing a king followed by a king. (*g*) Drawing a jack, followed by a queen, followed by a king. (*h*) Drawing a king, followed by a king, followed by a king. (*i*) Drawing the king of spades followed by the king of hearts. (*j*) Drawing the king of spades followed by the king of spades.

EXERCISE 20 Find the probability of each of the sequences of events in Exercise 19 if each card is not replaced after it has been drawn.

EXERCISE 21 Suppose a coin is tossed ten times. (*a*) If the first nine tosses are heads, what is the probability that the tenth toss will be heads? (*b*) What is the probability that all ten tosses will be heads? Suppose two dice are thrown. (*c*) If the first two throws are double 1s, what is the probability that the third throw will be double 1s? (*d*) What is the probability that all three throws will be double 1s?

6
Some Famous Errors

It was mentioned in Section 1 that Pascal is said to have made a mistake in counting cases in a problem on probability; and indeed it is easy to make mistakes in counting the elements of the sample space of an experiment or the elements of an event. Probably the most famous such mistake is D'Alembert's error, made by Jean-le-Rond D'Alembert (1717–1783). Perhaps his is the mistake always remembered because he had the misfortune to include it in an article in the French *Encyclopédie.* In fairness, we should point out that D'Alembert made many important contributions in several areas of mathematics, including suggestions which have since become important in contemporary mathematics. Perhaps another reason the mistake is remembered is because it was made by a mathematician who was otherwise of the first rank. King and Read describe D'Alembert's famous error.[2]

His most classic error concerns the following problem: Two players toss a coin which is to be thrown twice. If heads appear on either trial the first player wins, and if a head does not appear at all the second player wins. D'Alembert argues that if the first toss is a head then the first player wins and the game is over, or if the first toss is tails then the first player wins on the second toss if it is heads and the second player wins if it is tails. He therefore erroneously concludes that there are three cases, and the probability of the first player winning should be 2/3.

D'Alembert's argument appears to be very convincing. We know, at least, it convinced him. It is true that there are three cases as D'Alembert described them: the first player could win on one toss (heads), the first player could win on two tosses (tails followed by heads), or the second player could win on two tosses (tails followed by tails). But D'Alembert used a probability of $\frac{1}{3}$ for each case, applying the assumption of equally likely outcomes. What is wrong is that the three cases are not equally likely.

King and Read suggest two ways to solve the problem correctly. In the first, we simply list the sample space of the game, which is just the sample space for tossing two coins. The first player wins in the event that the first toss is heads or the first toss is heads but the second tails. This event has probability $\frac{3}{4}$. The second player wins only in the event that both tosses are tails, which has probability $\frac{1}{4}$.

Another way to solve the problem is to use the assumption of equally likely outcomes correctly; that is, applied to the toss of a true coin.

[2] Excerpted from *Pathways to Probability:* History of Mathematics of Certainty and Chance by Amy C. King/Cecil B. Read. Copyright © 1963 by Holt, Rinehart and Winston, Inc. Reprinted by special permission of the publishers, Holt, Rinehart and Winston, Inc.

Either heads or tails is equally likely, so each has a probability of $\frac{1}{2}$. The probability of the first case, that the first player gets heads on the first toss, is then $\frac{1}{2}$. The second case is that the first player gets tails on the first toss followed by heads on the second. These are independent events, so the probability of this case is $\frac{1}{2} \cdot \frac{1}{2} = \frac{1}{4}$. Since these two cases are mutually exclusive, the probability that the first player wins is $\frac{1}{2} + \frac{1}{4} = \frac{3}{4}$.

EXERCISE 22 Consider the first solution to D'Alembert's problem. (*a*) List the sample space. (*b*) List the event that the first player wins. (*c*) List the event that the second player wins. (*d*) What is the probability that the first player wins? (*e*) What is the probability that the second player wins?

EXERCISE 23 Consider the following approach to D'Alembert's problem (a variation of the second solution). (*a*) What is the probability of tails on the first toss? (*b*) What is the probability of tails on the second toss? (*c*) What is the probability that the second player wins; that is, tails on the first toss followed by tails on the second toss? (*d*) What is the probability that the first player wins? (*Hint:* Subtract the probability that the second player wins from 1.)

Most books ascribe the beginnings of probability theory to the problem of points sent to Pascal by Antoine Gombauld, the Chevalier de Méré. In *Flaws and Fallacies in Statistical Thinking*, Stephen Campbell describes a different problem. In Campbell's story, the Chevalier de Méré made an error in calculating probabilities concerned with throwing a die, and also with throwing two dice. The first error did not affect his fortunes, for the correct answer still let him win over the long run although not by as much as he had calculated. The story has a sad ending, however, for in the second case he loses. It is this problem, according to Campbell, about which de Méré consulted Pascal. Ignoring the question of which version is historically accurate (the consensus seems to lie with that of King and Read given in Section 1), we can enjoy Campbell's version as the story of another mistake in calculating probabilities.[3]

De Méré had prospered for a time by betting that he could get at least one six in four rolls of a single die. He reasoned that this bet would win more often than it loses. He happened to be right but for the wrong reason. He figured that,

[3] Stephen K. Campbell, *Flaws and Fallacies in Statistical Thinking* © 1974, pp. 116–117. Reprinted by permission of Prentice-Hall, Inc., Englewood Cliffs, New Jersey.

since he had an even chance of rolling any one of the six numbers on the die on the first roll, the liklihood [sic] of a six is the same as for any of the other numbers, namely 1/6. For four rolls, he reasoned—and here is where the error creeps in—the chance would be four times as good. The probability of getting at least one six on the four rolls, therefore, should be 4/6 or 2/3. This reasoning led de Méré to conclude that he would, in the long run, win two wagers for every one he would lose. Actually the reasoning and the conclusion it leads to are wrong, but de Méré's fortunes were becoming progressively greater so his error went undetected. (The correct probability is about 0.52 in favor of his getting at least one six in four rolls. This figure is obtained by multiplying 5/6—the probability of getting something other than a six in a single roll—together four times to determine the probability of his failing to get a six on all of the four rolls, and then subtracting the resulting product from 1.0.)

The wager that gave probability its start was a variation on the preceding one. For reasons unknown, de Méré switched to betting that within a sequence of 24 rolls of two dice, he could get at least one 12. Using the same line of reasoning as before, he decided that, because the probability of getting a 12 on one roll is 1/36 (there being 36 possible numbers that could be showing, only one of which is 12), in 24 rolls there must be a probability of 24/36 of getting at least one 12. In truth, however, this bet loses slightly more often than it wins as de Méré's dwindling fortunes attested.

At this point we might observe that a probability of 0.5, or $\frac{1}{2}$, is break-even. More than 0.5 wins and less than 0.5 loses. A probability of 0.52 is just over 0.5, so it is just barely winning. Thus, if the story above is true at all, we might suggest that this probability could account for de Méré's change of strategy. Even if he was not too good at calculating probabilities, de Méré was nobody's fool as a gambler. He would soon have observed that he was for some reason winning just barely over half the time, a far cry from two wins for every one loss, and would have changed his game.

EXERCISE 24 Consider de Méré's first wager. (*a*) What is the probability of throwing a 6 on one die? (*b*) What is the probability of throwing a number other than 6 on one die? (*c*) What is the probability of throwing a number other than 6 on four consecutive throws? (*d*) What is the probability of throwing a 6 on at least one of the four throws? (*Hint:* Subtract the probability of throwing a number other than 6 on four throws from 1.) Observe that the result of part *d* is more than 0.5, so de Méré would tend to win in the long run.

EXERCISE 25 Consider de Méré's second wager. (*a*) What is the probability of throwing a 12 on two dice? (*b*) What is the probability of throwing a number other than 12 on two dice? (*c*) What calculation must be made to find the probability of throwing a number other than 12 on twenty-four consecutive throws? (Do not try to calculate the number; working on an eight-place hand calculator, I get an approximation of 0.5086.) (*d*) What is the probability of throwing a 12 on at least one of the twenty-four throws? (*Hint:* Subtract the approximation above from 1.) Observe that the result of part *d* is slightly less than 0.5, so de Méré would tend to lose in the long run.

7
Permutations and Combinations

In counting sample spaces and events of experiments, we are often concerned with the different ways the elements of some set can be arranged. For example, let us consider the experiment of throwing three dice and the event of getting a total of 6. There are three combinations of numbers which give a total of 6 on three dice: 2, 2, and 2; 1, 1, and 4; and 1, 2, and 3. There are more than three ways to get a total of 6 on three dice, however, since 1, 1, and 4 can be thrown in any of the arrangements (1, 1, 4), (1, 4, 1), or (4, 1, 1); and there are six arrangements for 1, 2, and 3.

To consider the six ways of throwing 1, 2, and 3, we are concerned with the possible arrangements of the set of numbers 1, 2, and 3. The arrangements can be derived from a tree diagram.

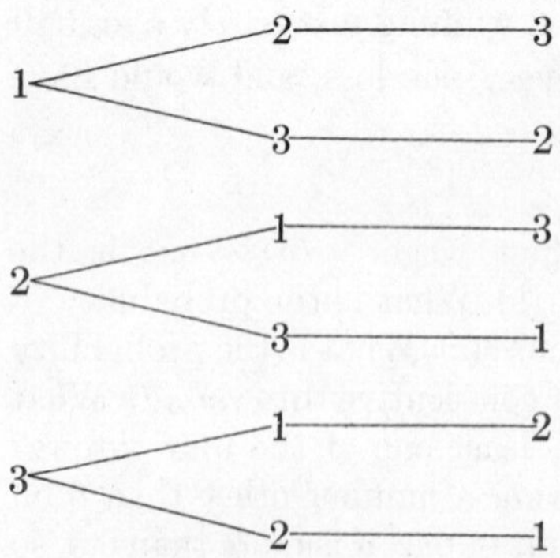

The six arrangements are (1, 2, 3), (1, 3, 2), (2, 1, 3), (2, 3, 1), (3, 1, 2), and (3, 2, 1). The tree diagram we have used is different from our previous tree diagrams in that once an element has been used, it is not repeated in the next branch. Thus, there are three original choices, 1, 2, and 3. For the first branch there are only two choices: where 1 is the original choice only 2 or 3 can be chosen in the first branch, where 2 is the original choice only 1 or 3 can be chosen in the first branch, and where 3 is the original choice only 1 or 2 can be chosen in the first branch. For the second branch there is only one choice: whichever of the three numbers is left. Using the rule for choices, we then have $3 \cdot 2 \cdot 1 = 6$ arrangements of the set of numbers 1, 2, and 3.

DEFINITION A *permutation* of a set is an arrangement of the elements of the set in any specific order.

A set with one element has only one permutation, since a single element can be written in only one order. For example, the element 1 can be written only in the order (1). A set with two elements has two permutations. There are two choices for the first element in an arrangement, and then the second element must be the one that is left; so the rule of choices gives $2 \cdot 1 = 2$. For example, the set of 1 and 2 can be written in the arrangements (1, 2) and (2, 1). We have seen the $3 \cdot 2 \cdot 1 = 6$ permutations of a set with three elements.

In general, if a set has n elements, it has $n \cdot (n - 1) \cdot (n - 2) \cdot \cdots \cdot 3 \cdot 2 \cdot 1$ permutations. There are n choices, any of the n elements of the set, for the first element in any specific arrangement. With one element used as the first element of the arrangement, there are $n - 1$ choices for the second element. When two elements are used, there are $n - 2$ elements left as choices for the third element in the arrangement, and so on, until there is only one element left for the last place. A set with ten elements has $10 \cdot 9 \cdot 8 \cdot 7 \cdot 6 \cdot 5 \cdot 4 \cdot 3 \cdot 2 \cdot 1 = 3{,}628{,}800$ permutations. The symbol 10! ("ten factorial") is used to represent 10 multiplied by each counting number less than 10; that is, $10! = 10 \cdot 9 \cdot 8 \cdot 7 \cdot 6 \cdot 5 \cdot 4 \cdot 3 \cdot 2 \cdot 1$. Similarly, $3! = 3 \cdot 2 \cdot 1 = 6$, $2! = 2 \cdot 1 = 2$, and $1! = 1$. For reasons we shall see later, we also define zero factorial as $0! = 1$.

DEFINITION If n is any counting number, *n factorial* is $n! = n \cdot (n-1) \cdot (n-2) \cdot \cdots \cdot 3 \cdot 2 \cdot 1$. If $n = 0$, $0! = 1$.

For some types of problems we need to know the number of permutations of a subset of a set. Suppose a club has ten members and is planing to elect three officers: a president, a vice-president, and a secretary-treasurer. There are ten choices, any of the ten members, for president. Once the president is chosen, there are nine choices for vice-president. Finally, there are eight choices for secretary-treasurer. Thus, there are $10 \cdot 9 \cdot 8 = 720$ permutations, or 720 ways to choose three officers in a club of ten members. We call this number the number of permutations of ten elements taken three at a time.

The number of permutations of ten elements taken three at a time is simply the first three factors of 10 factorial. We can write this number as

$$\frac{10!}{7!} = \frac{10 \cdot 9 \cdot 8 \cdot 7 \cdot 6 \cdot 5 \cdot 4 \cdot 3 \cdot 2 \cdot 1}{7 \cdot 6 \cdot 5 \cdot 4 \cdot 3 \cdot 2 \cdot 1} = 10 \cdot 9 \cdot 8 = 720$$

Finally, we observe that $7 = 10 - 3$, so the number of permutations of ten elements taken three at a time can be written as $10!/(10-3)!$.

In general, we use the symbol ${}_nP_r$ to mean *the number of permutations of n elements taken r at a time.* Then

$$\begin{aligned} {}_nP_r &= \frac{n!}{(n-r)!} \\ &= \frac{n \cdot (n-1) \cdot (n-2) \cdot \cdots \cdot (n-r+1) \cdot (n-r) \cdot (n-r-1) \cdot \cdots \cdot 3 \cdot 2 \cdot 1}{(n-r) \cdot (n-r-1) \cdot \cdots \cdot 3 \cdot 2 \cdot 1} \\ &= n \cdot (n-1) \cdot (n-2) \cdot \cdots \cdot (n-r+1) \end{aligned}$$

For example, ${}_{10}P_3 = 10!/(10-3)! = 10 \cdot 9 \cdot 8 = 720$, as above.

EXERCISE 26 Compute each of these factorials: (*a*) $3!$ (*b*) $4!$ (*c*) $5!$ (*d*) $6!$

EXERCISE 27 Compute each of these numbers of permutations of n elements taken r at a time: (*a*) ${}_{10}P_4$. (*b*) ${}_6P_2$. (*c*) ${}_6P_6$. (*d*) ${}_{12}P_6$.

EXERCISE 28 Suppose a club has eight members and is planning to elect two officers, a president and a secretary. (*a*) In how many ways can the officers be chosen? (*b*) If the same club wants to elect three officers, in how many ways can the officers be chosen?

EXERCISE 29 There are nine players on a baseball team. (*a*) Assuming the pitcher bats last, in how many ways can a batting order of the eight other plays be arranged? (*b*) Suppose a baseball team has fifteen players other than the pitchers. In how many ways can a batting order of eight players be chosen from among the fifteen players available?

In many types of problems, the arrangement of the elements in a set or subset is irrelevant. For example, suppose a club with ten members is planning to select a committee of three people. Which person is the first committee member, which the second, and which the third does not matter. If we were to count the number of permutations of ten elements taken three at a time, we would have counted, say, (Jane, Jim, John), (Jane, John, Jim), (Jim, Jane, John), (Jim, John, Jane), (John, Jane, Jim), and (John, Jim, Jane), when any one of these arrangements would suffice to represent the committee. Thus we would have six arrangements for each one that we wished to count. We may represent these six arrangements as

$$ {}_3P_3 = \frac{3!}{(3-3)!} = \frac{3!}{0!} = \frac{6}{1} = 6 $$

If we divide the number of permutations of ten elements taken three at a time, ${}_{10}P_3$, by the number of permutations of three elements, ${}_3P_3$, we will have the possible committees of three people without regard to the order in which they are arranged. The number of possible committees is then

$$ \begin{aligned} \frac{{}_{10}P_3}{{}_3P_3} &= \frac{10!/(10-3)!}{3!/(3-3)!} = \frac{10!/7!}{3!/1} = \frac{10!}{7!} \cdot \frac{1}{3!} = \frac{10!}{7!\,3!} \\ &= \frac{10 \cdot 9 \cdot 8 \cdot 7 \cdot 6 \cdot 5 \cdot 4 \cdot 3 \cdot 2 \cdot 1}{(7 \cdot 6 \cdot 5 \cdot 4 \cdot 3 \cdot 2 \cdot 1)(3 \cdot 2 \cdot 1)} = \frac{10 \cdot 9 \cdot 8}{3 \cdot 2 \cdot 1} \\ &= 120 \end{aligned} $$

DEFINITION A *combination* is a subset in which the order of the elements is not specified.

In general, *the number of combinations of n elements taken r at a time* can be computed as the number of permutations of n elements taken r at a time divided by the number of permutations of the r elements. We use the symbol ${}_nC_r$ for the number of combinations of n elements taken r at a time. Then

$$_nC_r = \frac{_nP_r}{_rP_r}$$

$$= \frac{n!/(n-r)!}{r!/(r-r)!} = \frac{n!/(n-r)!}{r!/1} = \frac{n!}{(n-r)!} \cdot \frac{1}{r!} = \frac{n!}{(n-r)!\,r!}$$

For example,

$$_{10}C_3 = \frac{10!}{(10-3)!\,3!} = 120$$

Two common examples which use combinations of n elements taken r at a time are the numbers of possible poker hands and bridge hands. A poker hand contains five cards from a fifty-two-card deck, assuming that the game being played does not use a joker. Since the order in which the cards are dealt into the hand is irrelevant, the number of possible poker hands is

$$_{52}C_5 = \frac{52!}{(52-5)!\,5!} = 2{,}598{,}960$$

The highest winning hand is the royal flush, consisting of the ace, king, queen, jack, and ten of any one suit. There are four such hands, one for each suit, so the probability of a player's being dealt a royal flush is

$$\frac{4}{2{,}598{,}960} = \frac{1}{649{,}740}$$

A bridge hand contains thirteen cards dealt from a fifty-two-card deck, so there are

$$_{52}C_{13} = \frac{52!}{(52-13)!\,13!} = 635{,}013{,}559{,}600$$

possible bridge hands. Since there are four suits of thirteen cards each, the probability of a player's being dealt a hand in which all thirteen cards are of the same suit is

$$\frac{4}{635{,}013{,}559{,}600} = \frac{1}{158{,}753{,}389{,}900}$$

EXERCISE 30 Compute each of these numbers of combinations of n elements taken r at a time: (*a*) $_{10}C_4$. (*b*) $_6C_2$. (*c*) $_6C_6$. (*d*) $_{12}C_6$.

EXERCISE 31 Suppose a club with eight members is planning to select a committee of two people. (*a*) In how many ways can the committee be chosen? (*b*) If the same club wants to select a committee of three people, in how many ways can the committee be chosen?

EXERCISE 32 Suppose a baseball team has fifteen players other than the pitchers. In how many ways can a team of eight players be selected, if the order in which they bat is not to be considered?

EXERCISE 33 A rummy hand consists of seven cards dealt from a fifty-two-card deck. (*a*) How many possible rummy hands are there? (*b*) What is the probability of a player's being dealt a hand which consists of a run of the seven through the king of the same suit?

EXERCISE 34 In poker, the royal flush is a special case of a straight flush. A straight flush consists of five consecutive cards of the same suit. What is the probability of a player's being dealt a straight flush? (*Hint:* Aces may count both high and low in poker; therefore, the hand can contain ace through five, two through six, and so forth up to ten through ace of any suit.)

8

The Normal Curve

The symbols $_nP_r$ for the number of permutations of n elements taken r at a time and $_nC_r$ for the number of combinations of n elements taken r at a time are useful because of their similarity. In other contexts, however, when we wish to concentrate on combinations only, we often use the symbol $\binom{n}{r}$ instead of $_nC_r$. We will call the symbol $\binom{n}{r}$ *the n-r symbol.* The n-r symbol has the same meaning as $_nC_r$; that is,

$$\binom{n}{r} = \frac{n!}{(n-r)!\,r!}$$

To derive a new interpretation for the n-r symbol, we return to the experiments of tossing coins. Suppose the experiment is tossing one coin. There are two possible outcomes, heads or tails or, in other words, heads or no heads. We may represent these two possibilities by the

n-r symbol, where n is the number of coins and r is the number of heads. For the event of getting one head we have

$$\binom{1}{1} = \frac{1!}{(1-1)!\,1!} = 1.$$

For the event of getting no heads we have

$$\binom{1}{0} = \frac{1!}{(1-0)!\,0!} = 1.$$

Since the events of getting one head and no heads are $\{H\}$ and $\{T\}$, respectively, the results of the n-r symbol agree with the numbers of elements in the corresponding events.

If two coins are tossed, there are three possible outcomes, two heads, one head, or no heads. Again using n for the number of coins and r for the number of heads, for the event of getting two heads we have

$$\binom{2}{2} = \frac{2!}{(2-2)!\,2!} = 1.$$

For the event of getting one head we have

$$\binom{2}{1} = \frac{2!}{(2-1)!\,1!} = 2.$$

For the event of getting no heads we have

$$\binom{2}{0} = \frac{2!}{(2-0)!\,0!} = 1.$$

Since these events are $\{HH\}$, $\{HT, TH\}$, and $\{TT\}$, the results of the n-r symbols again agree with the numbers of elements in the events.

If three coins are tossed, the possibilities are three heads, two heads, one head, or no heads. Then n-r symbols for these possibilities are

$$\binom{3}{3} = \frac{3!}{(3-3)!\,3!} = 1,$$

$$\binom{3}{2} = \frac{3!}{(3-2)!\,2!} = 3,$$

$$\binom{3}{1} = \frac{3!}{(3-1)!\,1!} = 3,$$

$$\binom{3}{0} = \frac{3!}{(3-0)!\,0!} = 1.$$

The events are $\{HHH\}$, $\{HHT, HTH, THH\}$, $\{HTT, THT, TTH\}$, and $\{TTT\}$, and the results of the *n-r* symbols agree with the numbers of elements in the events.

EXERCISE 35 If four coins are tossed, compute the *n-r* symbol for each of these possibilities. (*a*) The event of getting four heads. (*b*) The event of getting three heads. (*c*) The event of getting two heads. (*d*) The event of getting one head. (*e*) The event of getting no heads. (*f*) List each of the events above. Observe that the results of the *n-r* symbols agree with the numbers of elements in the events.

Pascal made great use of a simple diagram by which the results of the *n-r* symbols for a given value of n can be derived from the results of the *n-r* symbols for the preceding value of n. Since the diagram takes the shape of a triangle, it is called *Pascal's triangle.* From the values we have computed for the *n-r* symbols for $n = 1, 2, 3$, and 4. we write

$$
\begin{array}{ccccccccc}
 & & & 1 & & 1 & & & \\
 & & 1 & & 2 & & 1 & & \\
 & 1 & & 3 & & 3 & & 1 & \\
1 & & 4 & & 6 & & 4 & & 1
\end{array}
$$

Each new line of Pascal's triangle is computed by pacing a 1 at each end and then adding the number to the upper left and the number to the upper right to get the numbers between. Thus, from the last line above, for $n = 4$, we get the line for $n = 5$ by adding $1 + 4 = 5$ and $4 + 6 = 10$:

$$
\begin{array}{ccccccccccc}
 & 1 & & 4 & & 6 & & 4 & & 1 & \\
1 & & 5 & & 10 & & 10 & & 5 & & 1
\end{array}
$$

The line for $n = 5$ represents the results of the *n-r* symbols

$$\binom{5}{5} = \frac{5!}{(5-5)!\,5!} = 1 \qquad \binom{5}{4} = \frac{5!}{(5-4)!\,4!} = 5$$

$$\binom{5}{3} = \frac{5!}{(5-3)!\,3!} = 10 \qquad \binom{5}{2} = \frac{5!}{(5-2)!\,2!} = 10$$

$$\binom{5}{1} = \frac{5!}{(5-1)!\,1!} = 5 \qquad \binom{5}{0} = \frac{5!}{(5-0)!\,0!} = 1$$

We obtain an interpretation of the *n-r* symbol by graphing the results of the n-r symbols for increasing values of n. For example, for $n = 4$ the graph is

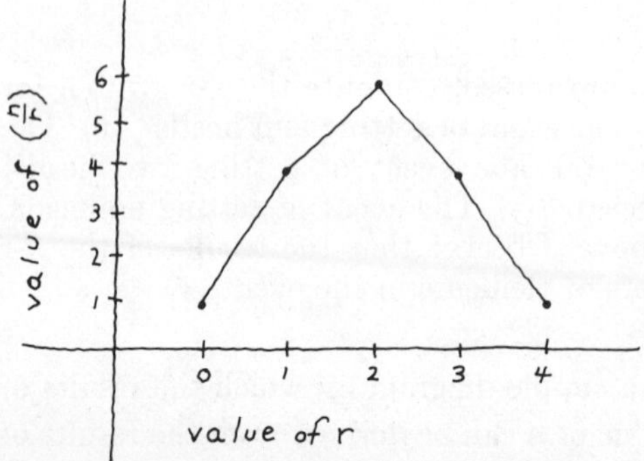

For $n = 5$, the graph becomes

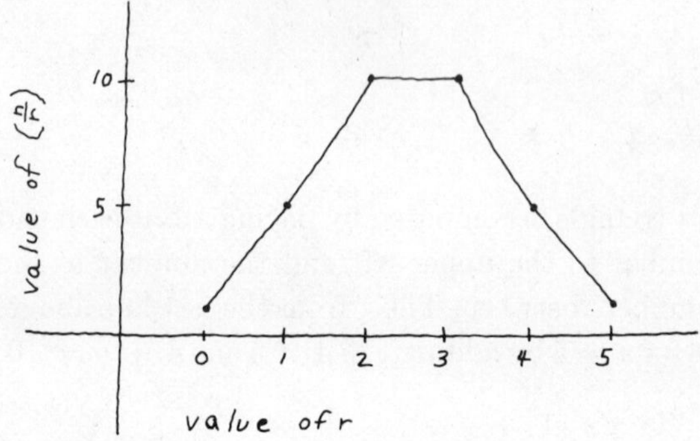

As n increases, the graph of the values of the n-r symbols for n approaches the general shape of a bell. This bell-shaped curve is a graph basic to statistics called *the normal curve.* The normal curve was discovered by Abraham De Moivre (1667–1754), an English mathematician of French descent. It is usually drawn with the vertical axis at its center, and in this position has the fancy equation

$$y = \frac{1}{\sqrt{\pi}} e^{-\frac{1}{2}x^2}$$

where e, like π, is a special constant of mathematics.

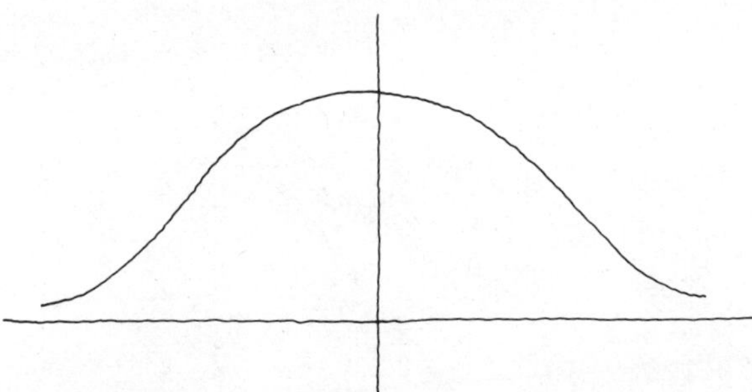

EXERCISE 36 (*a*) Write the line of Pascal's triangle for $n = 6$. (*b*) Write the *n-r* symbol represented by each number in the line of Pascal's triangle for $n = 6$. (*c*) Draw De Moivre's graph for the *n-r* symbols for $n = 6$.

EXERCISE 37 (*a*) Extend Pascal's triangle to the line for $n = 10$. (*b*) Draw De Moivre's graph for the *n-r* symbols for $n = 10$. (On the vertical axis, use a scale of 50, 100, 150, etc.)

9

Mean, Median, and Mode

In this section, we digress briefly into the field of statistics, a field so closely related to probability that the two are often termed together "probability and statistics."

Suppose we give a standard IQ test to a large number of people. The average score on an IQ test is 100, so we expect the highest point of the curve representing the scores to be at 100. The curve tapers off sharply at scores of, say, 150 on the high end and 50 on the low end, so we expect it to have roughly the shape of the normal curve.

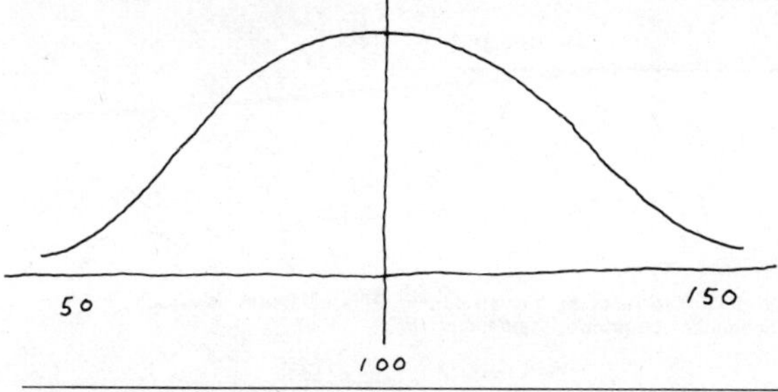

Photograph by Sol Mednick. Reprinted by permission of Mrs. Miriam Mednick, Scientific American, *September 1964.*

This device is a probability demonstrator. It is also sometimes called a Galton Board after Sir Francis Galton, who built the first one. It contains a large number of tiny balls which are started from the funnel at the top. Each ball falls through the array of hexagonal barriers and comes to rest in one of the troughs at the bottom. When all the balls are in the troughs, their distribution is an approximation of the normal curve. The approximation is constructed in exactly the same way as De Moivre's graphs.

At the first hexagon, each ball may fall to the left or to the right. At the second row, with two hexagons, a ball may fall to the left or to the right of either hexagon, and so on. The hexagons thus form a physical model of a tree diagram branching downward. Each hexagon offers two choices, left or right, just as the branches of tree diagrams for tossing coins have two choices, H or T.

If we apply the assumption of equally likely outcomes, the probability is that one-half of the balls will go to the left and one-half to the right at the first hexagon. At the second two hexagons, one-fourth of the balls will go to the far left and one-fourth will go to the far right. Two-fourths will go between the two hexagons, one-fourth from each hexagon. At the third row, one-eighth will go to the far left and far right, and three-eighths between each pair of hexagons (one-eighth from the outside hexagon and two-eighths—half of the two-fourths that came down the center—from the center hexagon), and so on.

The approximate distribution of balls in the passages through the first three rows of hexagons is

$$\frac{1}{2}, \frac{1}{2}$$
$$\frac{1}{4}, \frac{2}{4}, \frac{1}{4}$$
$$\frac{1}{8}, \frac{3}{8}, \frac{3}{8}, \frac{1}{8}$$

The numerators of these fractions are the first three rows of Pascal's triangle. Since each of the numerators is the sum of the contributions from the hexagon on each side in the row above, we see how the numbers in each row of Pascal's triangle are derived by adding the numbers in the row above.

The final distribution of the balls in the troughs is actually only an approximation of a De Moivre graph because, even with a great many balls, not exactly half will go each way at each hexagon.

However, the normal curve does not even approximately fit cases where we do not have a large number of results or where the distribution of the results cannot be expected to form what is called a *normal distribution*. In such cases we must be careful how we use the word "average," for it can be used in three different ways.

The first type of average is called the *mean*. The mean is the computation most people think of as an average. To compute the mean we add all the results and divide by the total number of results. For example, suppose that during a term a class is given five tests with possible scores on each test of 0 to 100. If a student has scores of 68, 86, 90, 88, and 79, then his mean score is $(68 + 86 + 90 + 88 + 79)/5 = 82.2$.

The *median* is the middle result of all the available results. We write the results in descending order, so we have 90, 88, 86, 79, 68 for the test scores above. Since the middle score is 86, the student's median score is 86. If there is an even number of results, then there are two middle results. In this case we find the median by adding the two middle results and dividing by 2. For example, suppose there are six tests, and a student has scores of 90, 88, 86, 79, 68, and 67. Then his median score is $(86 + 79)/2 = 82.5$.

The mean and the median are called *measures of central tendency*, since they are two different ways of indicating a center of a set of results. The third measure of central tendency is the *mode*. The modes of a set of results are the results which occur most often. For small amounts of results, such as the test-score examples given above, the mode has no meaning. For very large amounts of scores, there can be one or more modes. For example, when an IQ test is given to a large number of people, the scores that occur most often should be near 100. However, 101 could actually occur most often, and then 101 would be the mode. Moreover, 101 and 102 could occur the same number of times, and more often than any other score, in which case 101 and 102 would both be modes.

If there are a great many results, and the distribution is a normal distribution, the mean, median, and modes should all be close together. On the theoretical normal curve, which represents an infinite amount of results, the mean, median, and mode are identical and all at the highest point.

EXERCISE 38 Suppose student A has test scores of 93, 60, 88, and 92. (*a*) What is his mean score? (*b*) What is his median score? Suppose student B has test scores of 91, 88, 89, and 97. (*c*) What is his mean score? (*d*) What is his median score? (*e*) Do you consider the mean or the median to be the fairer measure of central tendency for the two students' scores?

EXERCISE 39 Suppose a class of twenty-five students makes the following scores on a test:

95
90, 90
85, 85, 85
80, 80, 80, 80
75, 75, 75
70, 70
65
60, 60
40, 40, 40, 40
25, 25
20

(*a*) What is the mean score of the class? (*b*) What is the median score of the class? (*c*) What are the modes of the class? (*d*) Which do you consider to be the best measure of central tendency of the class scores?

Of course, almost nothing you encounter in real life even vaguely resembles the theoretical normal curve. Any teacher will tell you that the set of test scores for a class will be almost any distribution except a normal distribution. In real-life situations, an average will depend on two things: first, how you get the figures you use; and second, which of the three types of averages you use. That is why people say you can prove anything with statistics. And so you can, if you are slightly unscrupulous about it. Statistics properly used is an excellent tool. In the wrong hands, like many other types of tools, statistics is often misused. Darrell Huff, in his popular book *How to Lie with Statistics*,[4] gives this example of abuse of the average.

That's why when you read an announcement by a corporation executive or a business proprietor that the average pay of the people who work in his

[4] Reprinted from *How to Lie with Statistics* by Darrell Huff, Pictures by Irving Geis. By permission of W. W. Norton & Company, Inc. Copyright 1954 by Darrell Huff and Irving Geis.

establishment is so much, the figure may mean something and it may not. If the average is a median, you can learn something significant from it: Half the employees make more than that; half make less. But if it is a mean (and believe me it may be if its nature is unspecified) you may be getting nothing more revealing than the average of one $45,000 income—the proprietor's—and the salaries of a crew of underpaid workers. "Average annual pay of $5,700" may conceal both the $2,000 salaries and the owner's profits taken in the form of a whopping salary.

Let's take a longer look at that one. . . . The boss might like to express the situation as "average wage $5,700"—using the deceptive mean. The mode, however, is more revealing: most common rate of pay in this business is $2,000 a year. As usual, the median tells more about the situation than any other single figure does; half the people get more than $3,000 and half get less.

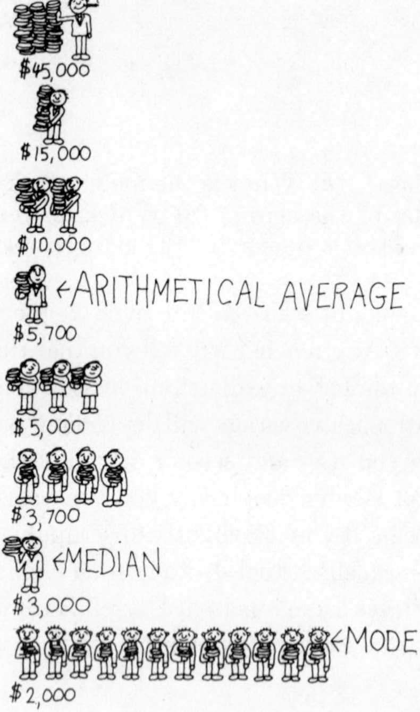

Looking Back

1 (*a*) Write the definition of a sample space. (*b*) Write the definition of an event. (*c*) Construct the tree diagram for tossing five coins, or tossing one coin five times. (*d*) List the sample space for tossing five coins. (*e*) List each of these events for the experiment of tossing five coins: (*i*) getting five heads; (*ii*) getting four heads; (*iii*) getting at least four heads; (*iv*) getting three heads; (*v*) getting at least three heads; (*vi*) getting two heads; (*vii*) getting one head; (*viii*) getting no heads.

2 (*a*) Write the definition of the probability of an event E. (*b*) List the sample space for throwing two dice. (*c*) Find the probability of each of these events for the experiment of throwing two dice: (*i*) the total is 2; (*ii*) the total is 3; (*iii*) the total is 4 and so on up to the total is 12; (*iv*) the total is 1; (*v*) both dice show an even number; (*vi*) both dice show an odd number; (*vii*) one die shows an even and the other an odd number; (*viii*) both dice show a number less than 4; (*ix*) at least one die shows a number less than 4.

***3** In the fifteenth and sixteenth centuries, it was observed that if three dice are thrown, a total of 10 occurs slightly more often than a total of 9. Before the famous correspondence between Pascal and Fermat, Galileo Galilei, perhaps best known today for his work in physics and astronomy, had shown that indeed a total of 10 occurs more often than a total of 9. Following Galileo's method, (*a*) List all the ways a total of 10 can be thrown on three dice. (*Hint:* Be sure to include all permutations of each way; there is a total of twenty-seven ways.) (*b*) List all the ways a total of 9 can be thrown on three dice. (*Hint:* Be sure to include all permutations of each way; there is a total of twenty-five ways.) (*c*) Similarly, a total of 11 occurs slightly more often on three dice than a total of 12. Explain why.

4 Suppose the sample space is an ordinary deck of playing cards. (*a*) Find the probability of each of these events: (*i*) Drawing an ace. (*ii*) Drawing a face card (jack, queen, or king). (*iii*) Drawing an ace or a face card. (*iv*) Drawing a spade. (*v*) Drawing a queen or a spade. (*vi*) Drawing a queen and a spade. (*b*) Find the probability of each of these events, assuming that the first card is replaced before the second is drawn. (*i*) Drawing a queen followed by a king. (*ii*) Drawing a queen followed by a spade. (*iii*) Drawing a queen followed by a queen. (*iv*) Drawing a spade followed by a club. (*v*) Drawing a spade followed by a spade. (*vi*) Drawing a queen of spades followed by a queen of spades. (*c*) Find the probability of each of the events in part (*b*), assuming that the first card is not replaced before the second is drawn.

5 A pinochle deck consists of forty-eight cards: two aces, two kings, two queens, two jacks, two tens, and two nines, of each suit. (*a*) Find the probability of each of these events where the sample space is a pinochle deck: (*i*) drawing an ace; (*ii*) drawing a face card (jack, queen, or king); (*iii*) drawing a red card (heart or diamond); (*iv*) drawing an ace or a red card; (*v*) drawing an ace or a heart; (*vi*) drawing an ace of hearts. (*b*) Find the probability of each of these events, assuming that the first card is replaced before the second is drawn: (*i*) drawing an ace followed by a king; (*ii*) drawing a heart followed by a diamond; (*iii*) drawing an ace followed by a heart; (*iv*) drawing an ace of hearts followed by an ace of hearts; (*c*) Find the probability of each of the events in part (*b*), assuming that the first card is not replaced before the second is drawn.

***6** A classical problem of probability concerns balls of different colors in a box, or traditionally, an urn. The sample space consists of all the balls in the urn. An event consists of balls of a given color. Suppose an urn contains four red balls, eight white balls, and twelve blue balls. (*a*) Find the probability of each of these events: (*i*) drawing a red ball; (*ii*) drawing a white ball; (*iii*) drawing a blue ball. (*b*) Find the probability of each of these events, assuming that the first ball is replaced before the second is drawn: (*i*) drawing a red ball followed by a white ball; (*ii*) drawing a red ball followed by a blue ball; (*iii*) drawing a white ball followed by a blue ball; (*iv*) drawing a red ball followed by a red ball; (*v*) drawing a white ball followed by a white ball; (*vi*) drawing a blue ball followed by a blue ball; (*c*) Find the probability of each of the events in part (*b*), assuming that the first ball is not replaced before the second is drawn.

7 (*a*) Suppose there are fifteen horses entered in a race. There are three winning positions: win, place, and show. In how many ways can the horses be arranged in the winning positions? (*b*) Suppose fourth place also counts, in case one of the winners is disqualified. In how many ways can the horses be arranged in the winning positions? (*c*) Suppose there are three winning positions but no distinction made among them. In how many ways can the three winners be chosen? (*d*) Suppose there are four winning positions with again no distinction made among them. In how many ways can the four winners be chosen? (*e*) Suppose there are three winning positions with no distinction made among them, but also a fourth place, distinguished from the other three, in case one of the winners is disqualified. In how many ways can three winners and one runner-up be chosen?

8 A pinochle hand consists of twelve cards dealt from the forty-eight-card pinochle deck. (*a*) How many possible pinochle hands are there? (*b*) What is the probability of getting the two nines through two aces of hearts? (*c*) What is the probability of getting the two nines through the two aces all of any one suit?

9 (*a*) Write the first eight lines of Pascal's triangle. (*b*) Write the n-r symbol represented by each number in the line of Pascal's triangle for $n = 7$. (*c*) Draw De Moivre's graph for $n = 7$. (*d*) Draw De Moivre's graph for $n = 8$. (*e*) Sketch the normal curve approximated by De Moivre's graphs.

10 Suppose membership in a club is open to people aged eighteen to twenty-five, and that in a given year the membership has this distribution:

Two members are eighteen
Five members are nineteen
Fifteen members are twenty
Fifteen members are twenty-one
Eight members are twenty-two
Ten members are twenty-three
No members are twenty-four
Five members are twenty-five

(*a*) What is the mean age of the club? (*b*) What is the median age of the club? (*c*) What are the modes of the ages of the club? (*d*) In the theoretical normal curve, where are the mean, median, and mode? (*e*) Is the distribution above a good approximation of the normal curve? Explain why or why not.

Branching Out

1 Blaise Pascal was perhaps one of the oddest characters of mathematics. Always of frail health, his father did not allow him to study mathematics for fear he would strain himself. This prohibition naturally aroused his curiosity about mathematics and he began to study geometry on his own. His last mathematical work came when, having given himself up to a life of contemplation, he began to think about a graceful curve called the cycloid to take his mind off a toothache and spent the next eight days in its study. He is known for two literary works, the *Pensées* and the *Lettres provinciales*, the first of which, according to E. T. Bell, reveals some unnatural tendencies. Depending on your interests, you might pursue his mathematical or his nonmathematical endeavors, or both.

2 Number theory is a branch of mathematics which deals with special properties of numbers such as evens, odds, primes, squares, numbers called triangular numbers, perfect numbers, and so on. The Pythagoreans were fascinated with such numbers. You will find these types of numbers and others described in Eves' *An Introduction to the History of Mathematics.* Pierre de Fermat is famous for his work in number theory as well as in probability and many other branches of mathematics. Of particular fame is a conjecture known as Fermat's last theorem. In a margin of a book, Fermat wrote that he had a proof of the theorem but no room to write it. To this day, no one has been able to prove Fermat's last theorem. The type of number theory connected with Fermat and his famous conjecture is more difficult than that of the Pythagoreans, but many people find it one of the most absorbing topics in mathematics.

3 After that of Pascal and Fermat, the next great work in probability was done by a Swiss mathematician Jacob Bernoulli (1654–1705), also referred to as Jakob, Jacques, and James, in different books. There were eight outstanding mathematicians in three generations of the Bernoulli family, of which Jacob and his brother Johannes were the first. Like Pascal, Jacob Bernoulli was forbidden by his father to study mathematics, but he studied calculus on his own and continued into many branches of mathematics, writing the *Ars Conjectandi* on the theory of probability and also teaching mathematics to Johannes. According to Kasner and Newman, Jacob Bernoulli originated the principle of insufficient reason, which would be the forerunner of our assumption of equally likely outcomes. Nicholas and Daniel Bernoulli, sons of Johannes, are known for a famous problem of probability called the Petersburg paradox, which you will find described in many books, including King and Read's *Pathways to Probability* and Eves' *An Introduction to the History of Mathematics.*

4 Jean-le-Rond D'Alembert's accomplishments are sufficiently impressive that most histories of mathematics make no mention at all of D'Alembert's error nor connect him at all with probability. His most famous work is in mathematical physics, where there is a principle named for him. Morris Kline, in *Mathematical Thought from Ancient to Modern Times,* credits him with predicting such modern concepts as time as a fourth dimension. Before you start on D'Alembert's work, you might find out where the name Jean-le-Rond came from, a story found in several books.

5 There are two important concepts of probability we did not discuss in this chapter. One, which you no doubt have heard of, is the concept of odds. Whereas the probability of an event is the ratio of the number of favorable cases to the total number of cases, the odds in favor of an event is the ratio of the number of favorable cases to the number of unfavorable cases. The other is "mathematical expectation," introduced by the Dutch physicist Christian Huygens (1629–1695). Huygens, whose work in probability preceded that of the Bernoullis, became interested in the subject directly from the correspondence between Pascal and Fermat. Although Huygens' life was not dramatic, and his other work is of a technical nature, you can find out about mathematical expectation in almost any elementary book which includes a study of probability.

6 Statistics, which King and Read call "a Child of Probability," is one of the most used and abused fields of contemporary mathematics. Almost every edition of a newspaper or magazine contains statistics of some type. *How to Lie with Statistics* by Darrell Huff is a charming little book on the abuses of statistics. In support of the well-known claim that you can prove anything with statistics, Huff shows you how to choose your results so that they give the result you want, and how to choose an average from the mean, median, and mode to ensure that you get that result. He also discusses different kinds of graphs and how to draw them so that they show only what you want to show. You will not find De Moivre's graphs or the normal curve in Huff's book, but you will find tricks used every day by people from politicians to publishers. If you prefer the more serious side, in elementary books which include concepts of statistics you can study more about measures of central tendency and continue on to measures of deviation. In particular, you will want to find out about standard deviation and its relationship to the normal curve.

7 There is a close connection between probability and logic. If you have studied logic, either in this book or elsewhere, you might like to pursue that connection. The 1 and 0 which are sometimes used for T and F on truth tables correspond to the 1 which means an event must happen and the 0 which means it cannot in probability. This idea can even be extended to many-valued logics, using 1 for true, 0 for false, and the fractions between, which correspond to measures of probabilities, for the other truth values. There is a brief discussion of this extension in Eves and Newsom's *An Introduction to the Foundations and Fundamental Concepts of Mathematics*. Kasner and Newman have a brief discussion of the connection between probability and two-valued logic in *Mathematics and the Imagination*.

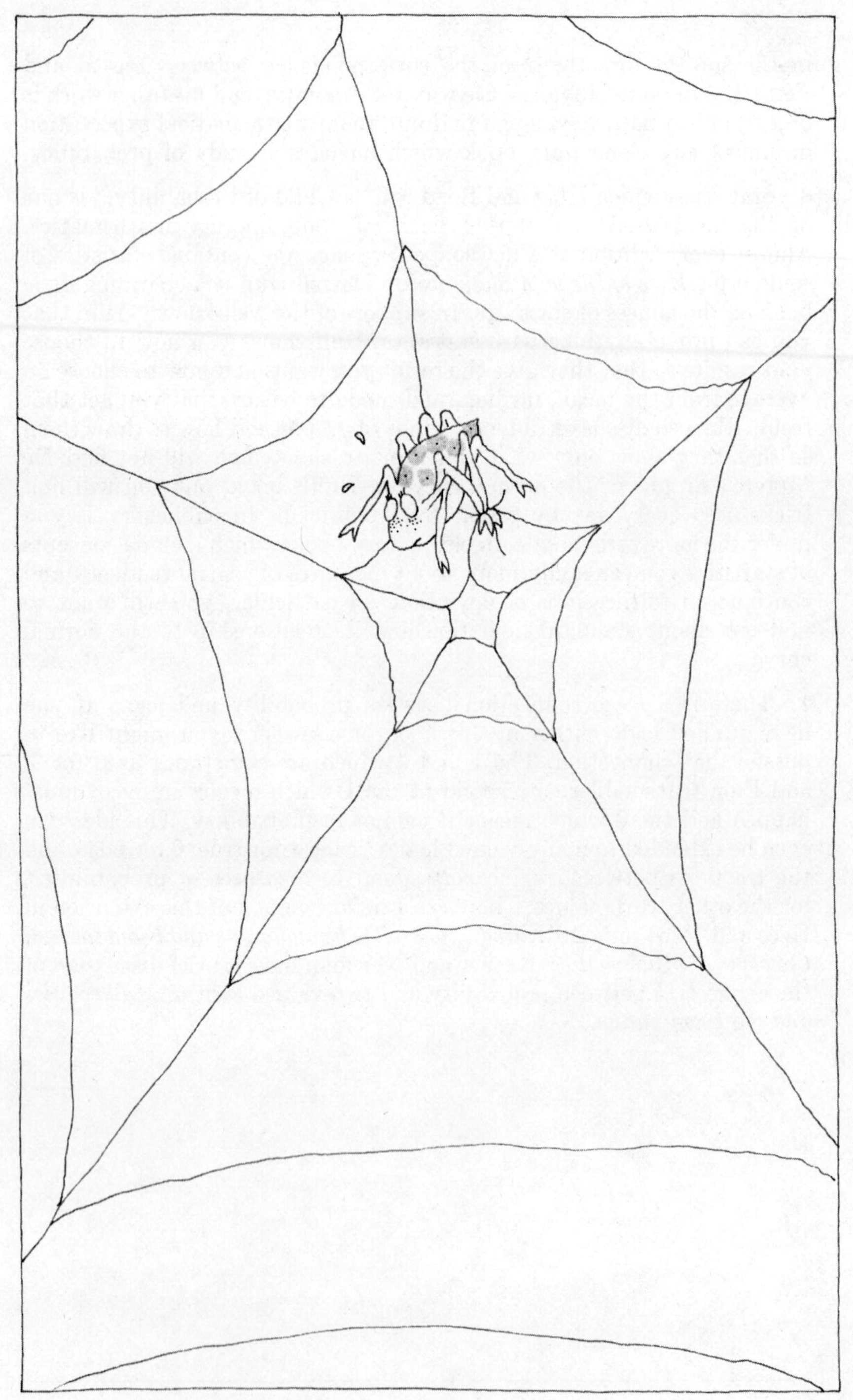

7

Finite Geometries

1

Graphs and Networks

Literally, the word "geometry" means "to measure the earth." In actuality, geometry has gone far beyond the surveying of land and measuring of figures that it was, for instance, to the ancient Egyptians. In contemporary mathematics there are many different kinds of geometries.

The basic object of geometries is generally the *point.* A point has no *dimensions;* that is, no extent in any direction. We use a dot as a drawing of a point:

Although the dot represents a point, the dot is not actually a point because it has length and breadth, and even to a small extent, depth.

We may think of a *line* as one type of a set of points. All lines have *one dimension.* The drawing on the left is of a straight line, and the drawing on the right is of a line which is not straight:

We realize that the drawings are not actually lines, since each has breadth and depth as well as length. Two distinct points of a line, together with the set of points of the line between them, form a *line segment.* The drawing on the left is of a straight line segment, and the drawing on the right is of one which is not straight:

Finite geometries are kinds of geometries constructed from a finite set of points called *vertices.* One type of finite geometry important in contemporary mathematics is the *graph* (but not the type of graph you may have drawn in an algebra course). Any two vertices of the finite set of points of a graph may be connected by a line segment, not necessarily straight, called an *arc.* Every arc of a graph must lie on two distinct vertices. No arc may have a vertex at only one end, and no arc may have both its ends on the same vertex. These are examples of finite geometries which are graphs:

This finite geometry is not a graph because it has a line with a vertex at only one end:

This finite geometry is not a graph because it has a line which has both its ends at the same vertex:

If you can get from any vertex of a graph to any other vertex along a series of one or more arcs, the graph is *connected.* These graphs are connected:

This graph is not connected:

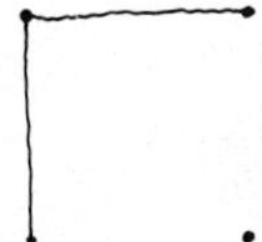

This graph is totally disconnected:

We will assume that all our graphs are connected, so that each vertex must have at least one arc on it. Thus, a connected graph has exactly two vertices on each arc; but the number of arcs on any vertex may be any counting number. This graph has a total of three vertices and two arcs. The middle vertex has two arcs on it; the others each have one.

This graph has five vertices and six arcs. The middle vertex has four arcs on it; the others each have two.

Finally, we recall that an arc need not be a straight line segment. This graph has two vertices and two arcs, with two arcs on each vertex.

A graph may have arcs which appear to cross but with the apparent crossing not considered to be a vertex. This is a connected graph with four vertices and four arcs. The apparent crossing at the center is not indicated as a vertex.

If all the crossings of arcs of a graph are vertices, the graph is a *plane graph*. The graph above is not a plane graph, but the following are:

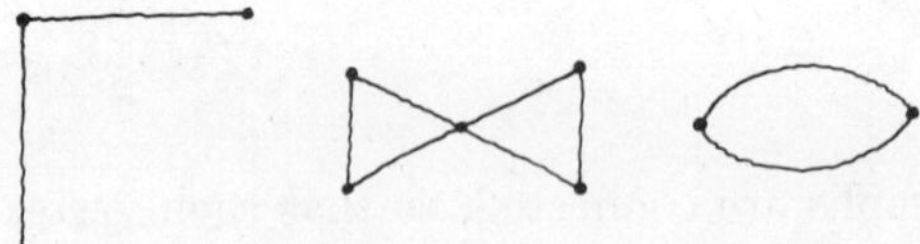

We will call a connected plane graph a *network*. Since each of the plane graphs above is connected, each is a network.

A part of a surface completely surrounded by arcs is called a *region*. The part outside all the arcs is also a region. Every network has at least one region, the surface it is drawn on, but the number of regions of a network may be any counting number. The first network above has just one region, the surface it is drawn on. The last network has two regions, one inside and one outside its arcs; and the middle network has three regions.

This network has four vertices, four arcs, and two regions:

If we draw one diagonal, the new network has four vertices, five arcs, and three regions.

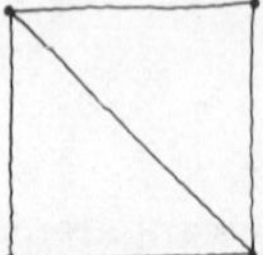

If we draw the other diagonal and want the graph to be a network, we must consider the crossing of arcs at the center to be a vertex. Then each of the diagonals must be considered to be two arcs separated by the vertex at the center. The resulting network has five vertices, eight arcs, and five regions.

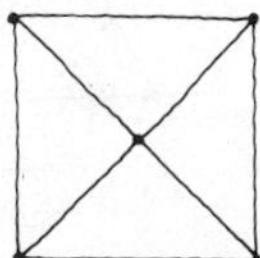

EXERCISE 1 Which of these finite geometries are graphs? For those which are not, say why not.

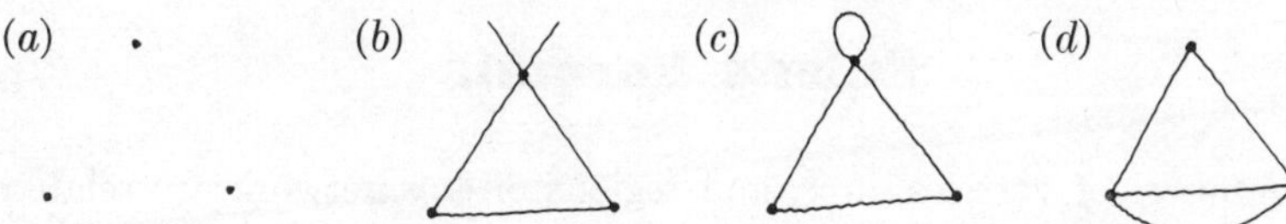

EXERCISE 2 Which of these graphs are connected?

EXERCISE 3 Which of these connected graphs are networks?

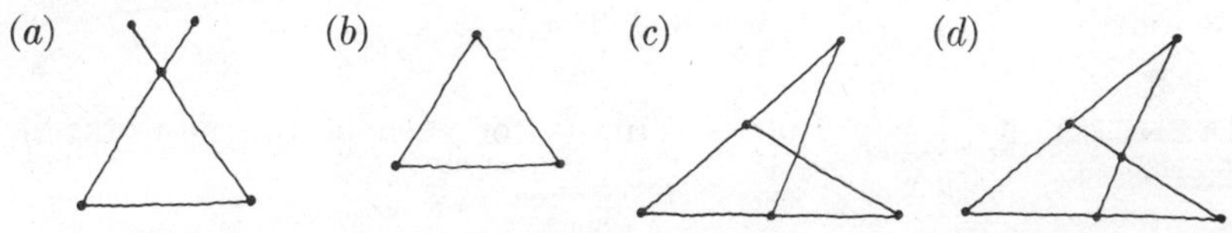

EXERCISE 4 List the numbers of vertices, arcs, and regions of each of these networks.

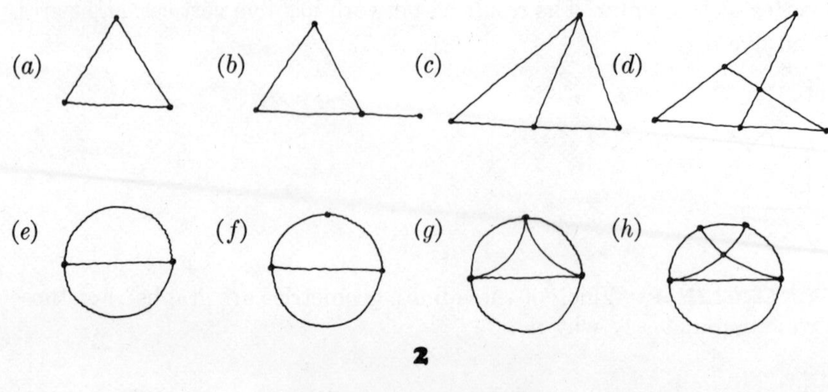

2

Euler's Formula

The numbers of vertices, arcs, and regions of any network are related by a famous formula. Although it was known before his time, the formula is named for Leonhard Euler.

EULER'S FORMULA If V is the number of vertices, A is the number of arcs, and R is the number of regions of a network, then $V - A + R = 2$.

For example, consider these networks:

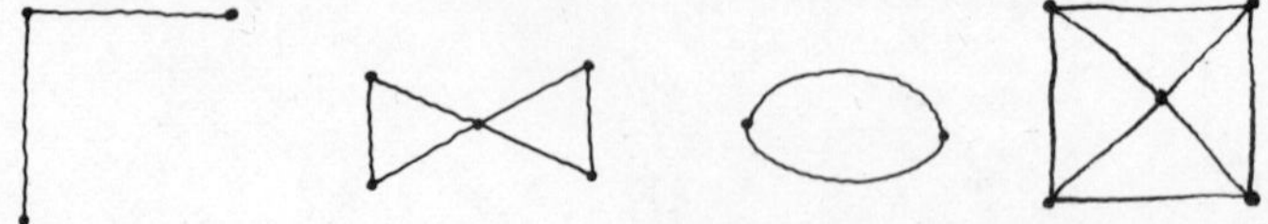

The first network has three vertices, two arcs, and one region, so $V - A + R = 3 - 2 + 1 = 2$. For the second network, $V - A + R = 5 - 6 + 3 = 2$. For the third network, $V - A + R = 2 - 2 + 2 = 2$, and for the fourth, $V - A + R = 5 - 8 + 5 = 2$.

EXERCISE 5 Verify Euler's formula for each of the networks in Exercise 4.

Library of Congress

Without contest, Leonhard Euler wrote more books and papers than anyone else in the history of mathematics. He supposedly was the first to use circular regions to diagram syllogisms (Chapter 3). The formula of the current section is named for him, although he was not the first to use it. His name is found throughout mathematics in such fields as calculus and differential equations, and also in physics and other mathematics-related disciplines. He wrote books on calculus, mechanics, and even navigation.

Euler was Swiss, but Kasner and Newman refer to Germany as his fatherland (Section 4). When a young man, he spent a few years in Russia, then twenty-five years in Germany, and then his last seventeen years again in Russia. He was blind during those last years but still produced mathematics at as great a rate as ever, or perhaps greater: some books and more than 400 papers, according to Kline. Euler's total works will fill some seventy-four volumes.

To derive Euler's formula, we may start with the simplest possible network, a single vertex:

This network has one vertex, no arcs, and one region, so $V - A + R = 1 - 0 + 1 = 2$. The only way to extend this network is to draw an arc from the original vertex to a new vertex:

We cannot have a new vertex with no arc, for then the graph would not be connected; and we cannot have a line with no arc at its other end or with both ends on the original vertex, for then the finite geometry would not be a graph. The new network has two vertices, one arc, and one region, so $V - A + R = 2 - 1 + 1 = 2$.

We can extend this network in any of three ways. One way is again to draw an arc from an original vertex to a new vertex:

In this case we have added a new vertex and also a new arc, but no new regions. Since vertices are added in Euler's formula, the new vertex adds 1 to the original total. Since arcs are subtracted, the new arc subtracts 1 from the original total. Thus, we have both added and subtracted 1, and the total of $V - A + R$ remains 2.

A second way to extend the original network is to introduce a new arc connecting the same two vertices:

In this case, we have a new arc and also a new region. Since arcs are subtracted in Euler's formula, the new arc subtracts 1 from the original total. Since regions are added, the new region adds 1 to the original total. Therefore, we have both subtracted and added 1, and the total of $V - A + R$ remains 2.

Finally, we can put a new vertex on the existing arc, creating two arcs from the original:

Again we have added a new vertex but we also have a new arc. We have added 1 and subtracted 1, so the total of $V - A + R$ remains 2.

Any network, no matter how complicated, can be constructed by a series of these three steps, starting from the network consisting of two vertices and one arc. In each step we draw a new arc to a new vertex, connect two old vertices by a new arc, or put a new vertex on an existing arc, until the desired network is constructed. In this way, the total of $V - A + R$ always remains 2.

EXERCISE 6 The steps below give a construction of the following network.

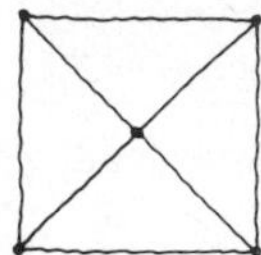

For each step of the construction, say which of the three extensions of a network is used; list the numbers of vertices, arcs, and regions; verify Euler's formula.

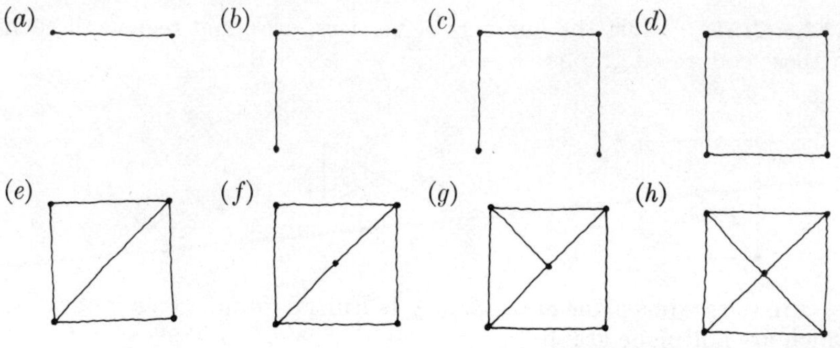

EXERCISE 7 Use the three extensions of a network to construct this network.

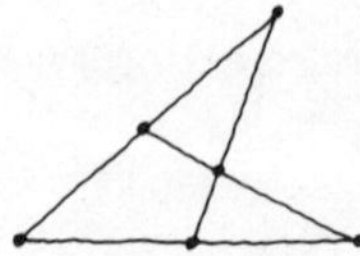

Start with this basic network as a first step.

For each step of the construction, say which of the three extensions of a network is used; list the numbers of vertices, arcs, and regions; verify Euler's formula.

EXERCISE 8 List the numbers of vertices, arcs, and regions of each of these graphs:

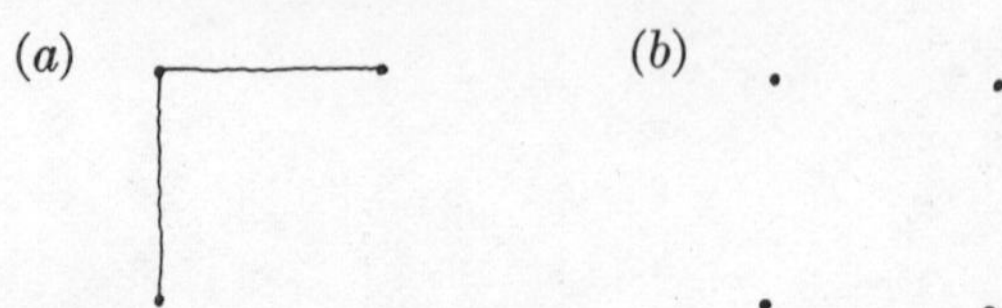

(*c*) Are the graphs connected? (*d*) Is Euler's formula true for graphs which are not connected?

EXERCISE 9 List the numbers of vertices, arcs, and regions of each of these connected graphs.

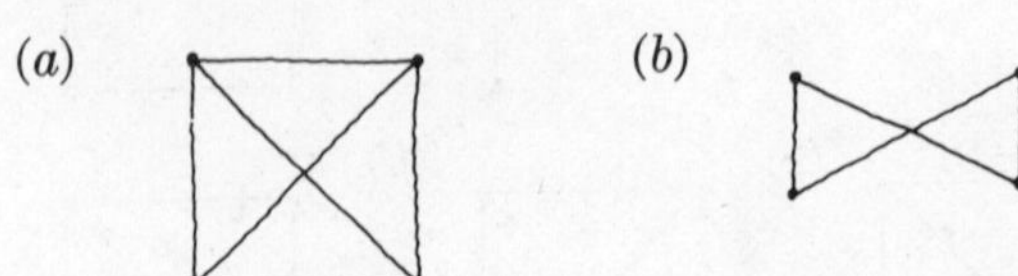

(*c*) Are the graphs plane graphs? (*d*) Is Euler's formula true for graphs which are not plane graphs?

Suppose after the first diagonal is drawn, the other two vertices are connected by an arc which does not cross the first.

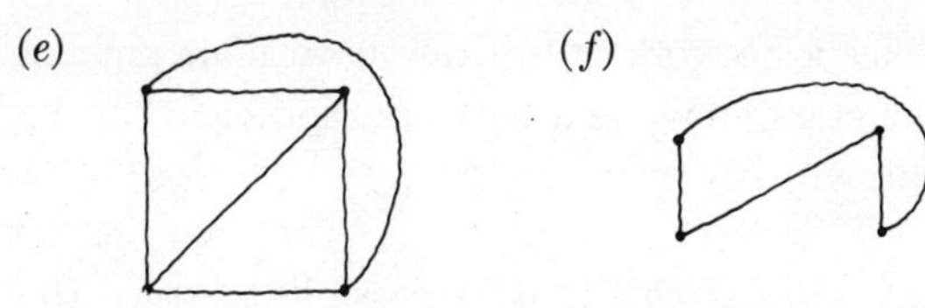

(*g*) Is each of the graphs a network? (*h*) Verify Euler's formula for each of the graphs.

3

Traversable Networks

One of the oldest mathematical pastimes, and also a type of problem enjoyed today, is that of trying to draw a geometric figure without picking up your pencil. The ancient mystic sign called the *pentagram*, a symbol of the Pythagoreans, is usually drawn without picking up the pencil.

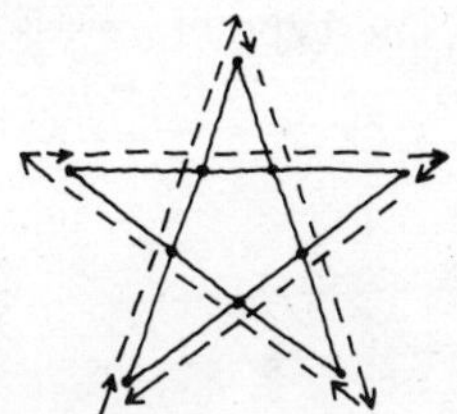

To draw the *hexagram*, symbol of Israel, most people draw two separate triangles, one right side up and one upside down, picking up the pencil between. It is possible, however, to draw the hexagram without picking up the pencil.

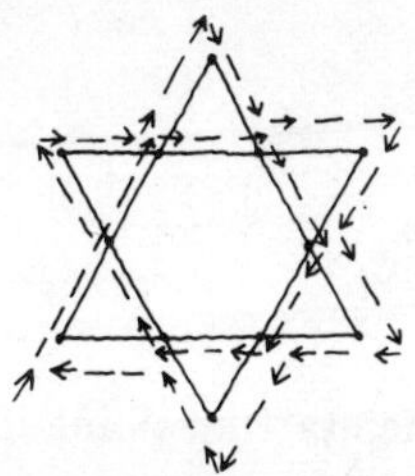

We will say that to *trace* a network is to draw it without picking up your pencil. Each of the stars above is a network, and each can be traced so that each arc is drawn only once.

DEFINITION A network is *traversable* if it is possible to trace the network covering each arc once and only once.

Some networks are easily traversable. Some are traversable if you start at the right vertex. And some networks are not traversable, no matter where you start.

EXERCISE 10 (*a*) Show how to traverse this network starting at any vertex.

(*b*) Show how to traverse this network starting at one type of vertex, but that it is not traversable starting at the other.

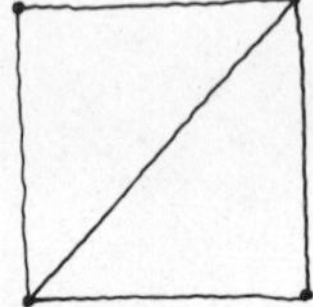

(*c*) Try to traverse this network until you are convinced it cannot be traversed regardless of where you start.

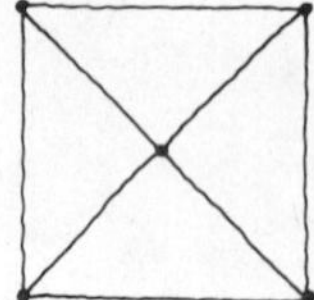

Whether or not a network is traversable, and the starting point if it is, depend on the nature of the vertices.

DEFINITION A vertex is *even* if there is an even number of arcs on it. A vertex is *odd* if there is an odd number of arcs on it.

Even vertices are desirable if a network is to be traversable. When we approach an even vertex along one arc, since the vertex is even we are sure to have another arc at that vertex to leave on.

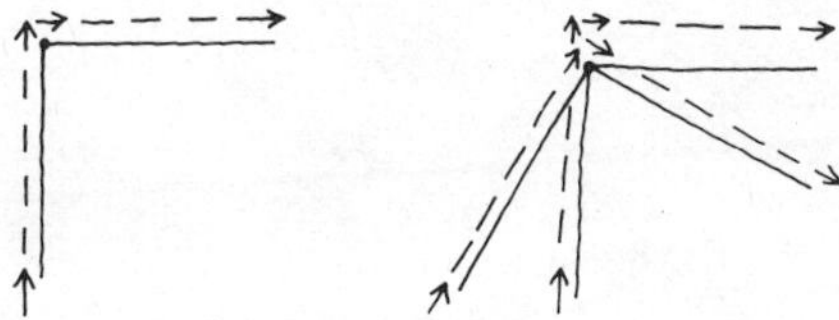

Odd vertices, on the other hand, are not desirable, for eventually we will come to an odd vertex along one arc and have no arc left to leave on.

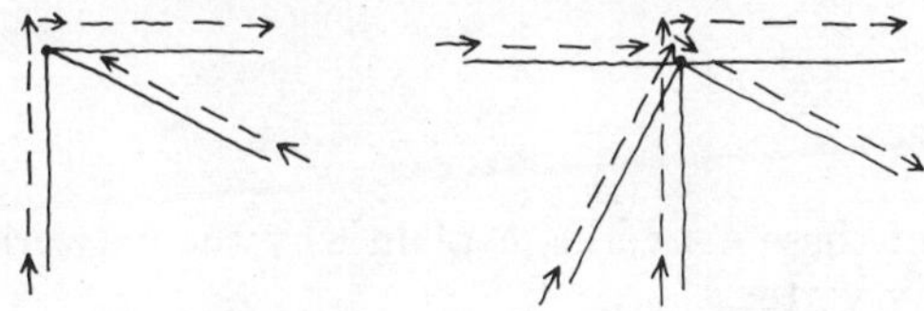

Of course, if the arc we approach an odd vertex on is the last arc to be covered, we can end the tracing of the network at that vertex. Similarly, we can begin tracing at an odd vertex. Thus, a network is traversable if it has no more than two odd vertices.

RULE FOR TRAVERSABLE NETWORKS A network is traversable if it has no more than two odd vertices. If the network has two odd vertices, the tracing must begin at one odd vertex and end at the other. If the network has more than two odd vertices, it is not traversable.

This network has four even vertices, so it is traversable starting at any vertex.

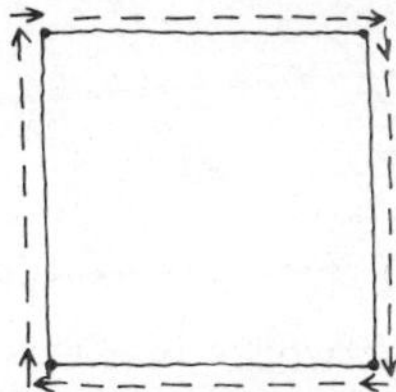

This network has two even vertices and two odd vertices. It is traversable, but we must start the tracing at one odd vertex and end at the other.

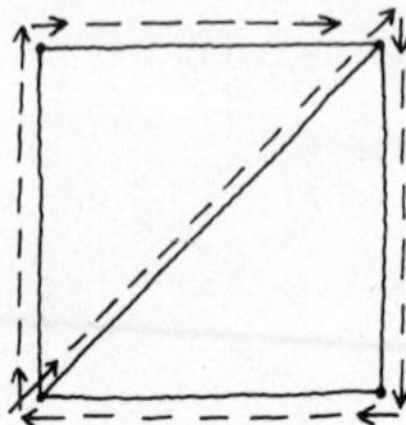

EXERCISE 11 Explain why this network is not traversable.

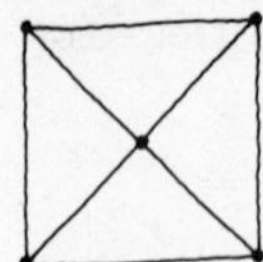

EXERCISE 12 For each of these networks, explain why the network is traversable, starting at any vertex.

(*a*)

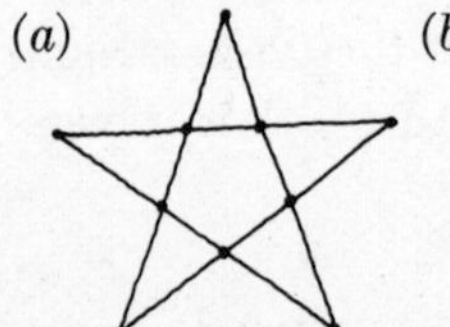

(*b*)

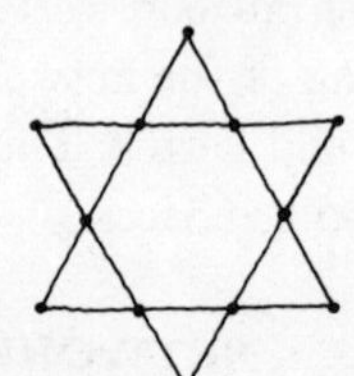

EXERCISE 13 For each of these networks, list the number of even vertices; list the number of odd vertices; say whether or not the network is traversable.

(*a*)

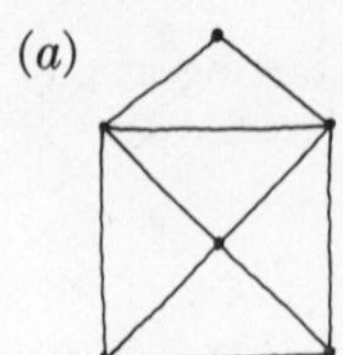

(*b*)

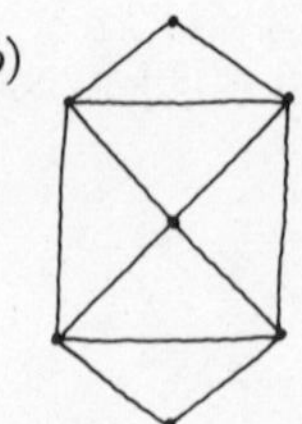

(*c*)

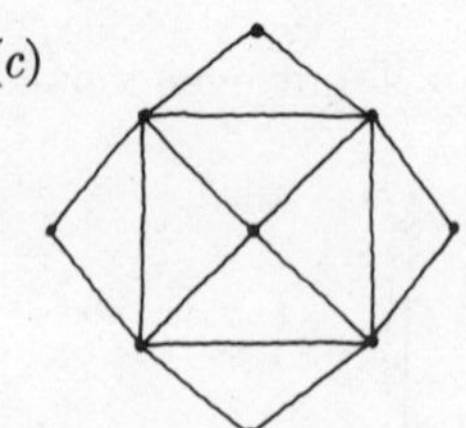

EXERCISE 14 Decide whether or not each of the networks in Exercise 4 is traversable.

EXERCISE 15 (*a*) Decide whether or not each of graphs *a* and *b* in Exercise 8 is traversable. (*b*) Can a graph which is not connected ever be traversable?

EXERCISE 16 (*a*) Decide whether or not each of graphs *a* and *b* in Exercise 9 is traversable. (*b*) Can a graph which is not a plane graph ever be traversable? (*c*) Decide whether or not this graph is traversable:

(*d*) Is this graph a plane graph? (*e*) Under what condition is a graph which is not a plane graph traversable?

EXERCISE 17 Claudia Zaslavsky says (*Africa Counts*, p. 106), "Shongo children draw these networks in the sand in a continuous line, without lifting the finger." Thus the networks are traversable. Indicate by the letters A and B the only two vertices of each of these networks at which you can begin and end.

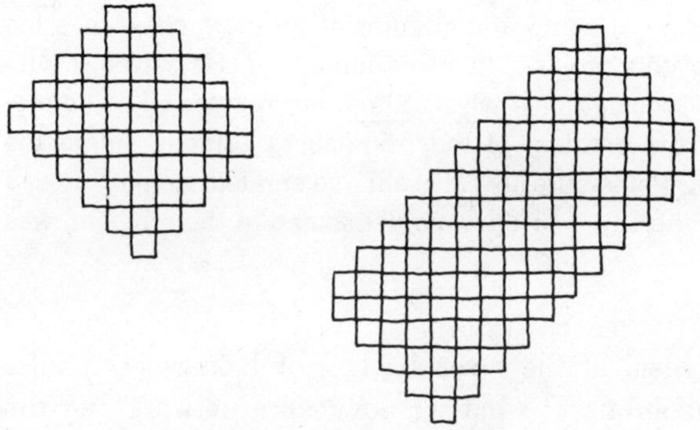

After Emil Torday, *On the Trail of the Bushongo* (Philadelphia: J. B. Lippincott Co., 1925)

4
Topological Equivalence

Kasner and Newman discuss several examples of networks, both traversable and not traversable. In the following passage, they describe one of the most famous of all problems about traversable networks and the role Leonhard Euler played in this aspect of the study of networks.[1]

Once upon a time seven bridges crossed the river Pregel as it twisted through the little German university town of Königsberg. Four of them led from opposite banks to the small island, Kneiphof. One bridge connected Kneiphof with another island, the other two joined this with the mainland. These seven bridges of the eighteenth century furnished the material for one of the celebrated problems of mathematics.

Seemingly trivial problems have given rise to the development of several mathematical theories. Probability rattled out of the dice cups of the young noblemen of France; Rubber-Sheet Geometry was brewed in the gemütliche air of the taverns of Königsberg. The simple German folk were not gamblers, but they did enjoy their walks. Over their beer steins they inquired: "How can a Sunday afternoon stroller plan his walk so as to cross each of our seven bridges without recrossing any of them?"

Repeated trials led to the belief that this was impossible, but a mathematical proof is based neither on beliefs nor trials.

Far away in St. Petersburg, the great Euler shivered in the midst of honors and emoluments, as mathematician at the court of Catherine the Great. To Euler, homesick and weary of pomp and circumstance, there came in some strange fashion news of this problem from his fatherland. He solved it with his customary acumen. Topology, or Analysis Situs was founded when he presented his solution to the problem of the Königsberg bridges before the Russian Academy at St. Petersburg in 1735. This celebrated memoir proved that the journey across the seven bridges, as demanded in the problem, was impossible.

To solve the problem of the seven bridges of Königsberg, Euler treated each bridge as an arc of what we now call a network and the islands and banks they connected as vertices. First, each island is shrunk to a single point. Then, treating the bridges as, say, pieces of wire so that they can be bent, the banks are also each reduced to a single point.

[1] Edward Kasner and James R. Newman, *Mathematics and the Imagination*. Copyright © 1940, by Edward Kasner and James R. Newman. Reprinted by permission of Simon and Schuster, Inc.

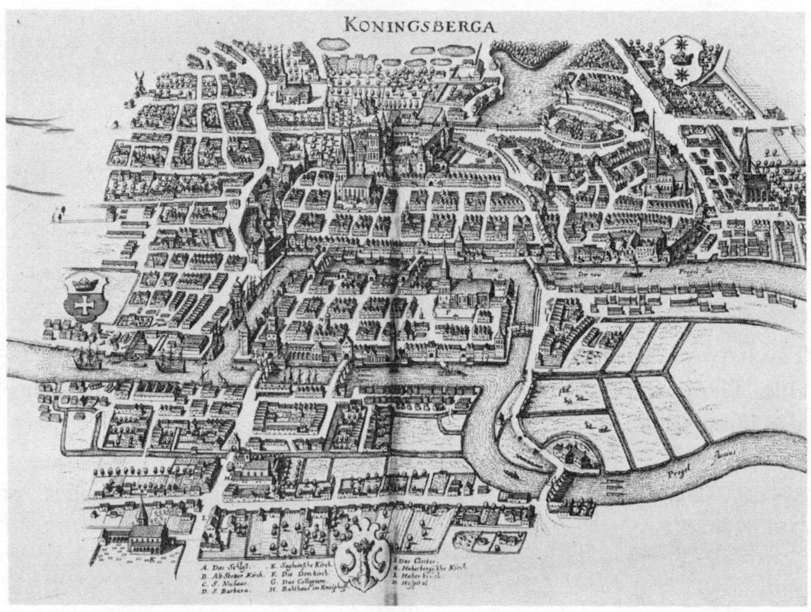

This drawing of the town of Königsberg shows how it might have looked when Euler solved the problem of its bridges. The map is reproduced from a work called Topographiae by Martin Zeiller (1589–1661), with illustrations by several contributors, first published in Frankfurt from 1642 to 1661, about one hundred years before the problem was sent to Euler. The work consists of twenty-nine volumes, illustrated by maps and charts such as the one here which is from the fourteenth volume. Germany (then Prussia), Switzerland, and France are included, with a general index published in 1672. Two more volumes covering Rome and Italy were added in 1681 and 1688, and a new edition of the index in 1726. The last date is of interest because it falls within Euler's life span, making the final work contemporary with Euler.

Photo: Rare Book Division, The New York Public Library
Astor, Lenox and Tilden Foundations

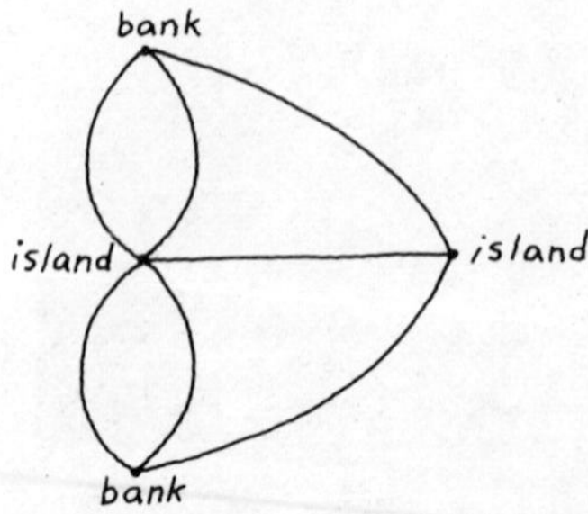

The network so constructed has four odd vertices, so it is not traversable. The journey across the bridges as required in the problem is then also impossible.

EXERCISE 18 Consider this diagram of the seven bridges of Königsberg.

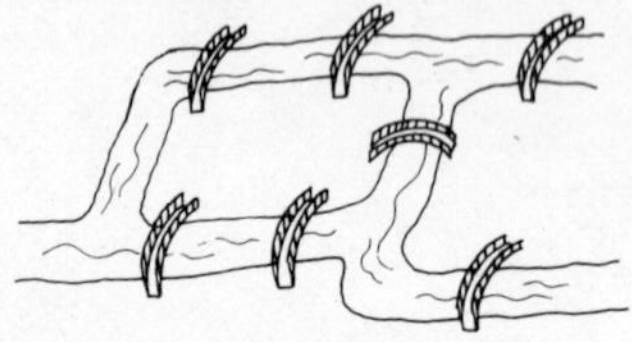

Try to trace the diagram so that you cross each bridge once without recrossing any of them until you are convinced it cannot be done.

EXERCISE 19 Consider this diagram with eight bridges, where a second bridge connects the two islands.

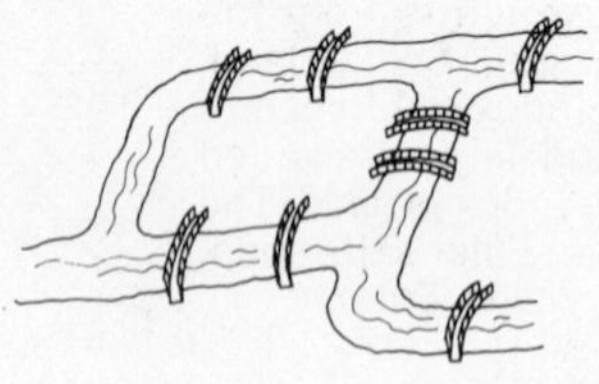

(*a*) Shrink the islands and bridges to vertices and arcs of a network.
(*b*) Is the network traversable? If so, where must the tracing begin and end?

EXERCISE 20 Consider this diagram with eight bridges, suggested by Kasner and Newman, where the eighth bridge connects the two shores:

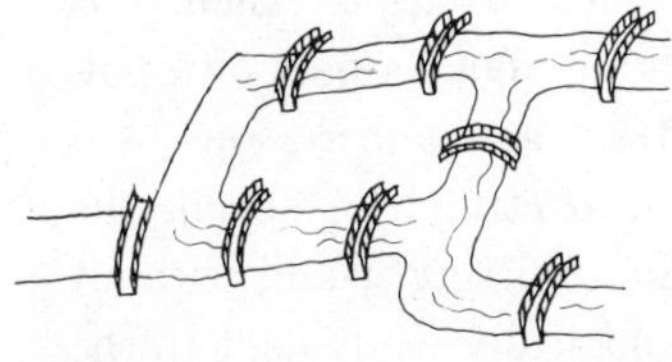

(*a*) Shrink the islands and bridges to vertices and arcs of a network. (*b*) Is the network traversable? If so, where must the tracing begin and end?

EXERCISE 21 (*a*) Draw a diagram with nine bridges, where both the bridges of Exercises 19 and 20 are included. (*b*) Shrink the islands and bridges to vertices and arcs of a network. (*c*) Is the network traversable? If so, where must the tracing begin and end?

EXERCISE 22 (*a*) Draw a diagram with ten bridges, with the eighth bridge connecting the two shores as in Exercise 20, and the ninth and tenth bridges connecting the small island once again with each shore. (*b*) Shrink the islands and bridges to vertices and arcs of a network. (*c*) Is the network traversable? If so, where must the tracing begin and end?

The diagrams above are traversable when the networks to which they are reduced are traversable, because even though land areas have been shrunk to vertices and bridge areas to arcs, no new vertices, arcs, or regions are formed in the process. Bridges which meet on the same land areas become arcs which meet at the same vertices, and bridges which do not meet are arcs which still do not meet. Arcs and regions can be stretched or shrunk as long as no new ones are created and none are pasted together or cut apart.

DEFINITION Two networks are *topologically equivalent* if one can be transformed into the other by stretching or shrinking, but without cutting or pasting.

A *topological property* is a property shared by topologically equivalent networks. For example, traversability is a topological property. If a network is traversable, then any network topologically equivalent to it is also traversable.

There is a large field of mathematics called *topology.* When it is based on stretching and shrinking, we are talking about a part of topology often called *Rubber Sheet Geometry.* We treat arcs and regions as if they were made of elastic or rubber. If we can stretch or shrink the rubber into the shape of another figure, without cutting it and without pasting pieces together, the new figure is topologically equivalent to the old one.

A *torus* is a figure shaped like a doughnut. If we think of a torus made of very stretchy rubber, we can pull out one side and stretch it into a cup shape without cutting or pasting. The rest of the torus becomes the handle of the cup. The two figures are topologically equivalent from the point of view of Rubber Sheet Geometry—hence the mathematicians' joke that a topologist can't tell his doughnut from his cup of coffee!

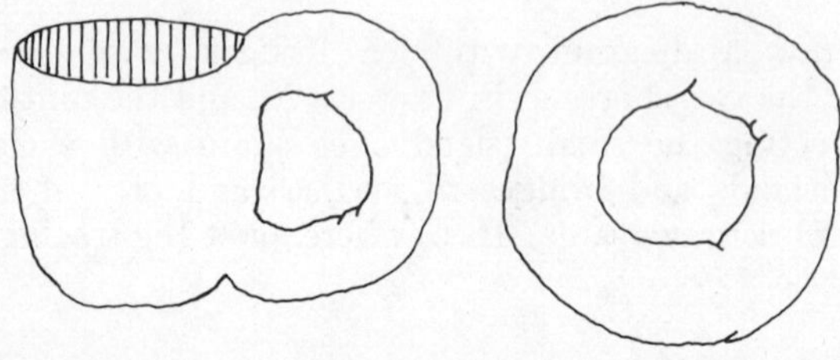

5

Polygons and Polyhedra

In the terminology of the preceding sections, a *polygon* can be described as a network which has straight line segments as its arcs, with exactly two arcs at each vertex, and exactly two regions, one inside and one outside. The following networks are polygons.

Suppose we connect two nonadjacent vertices of a polygon by an arc, where the arc is again a straight line segment. If the arc falls entirely within the inside region of the polygon, we say the polygon is *convex*. The following are convex polygons, where the dotted lines represent straight line segments joining nonadjacent vertices.

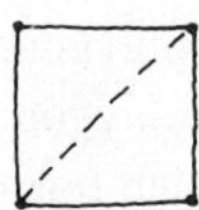 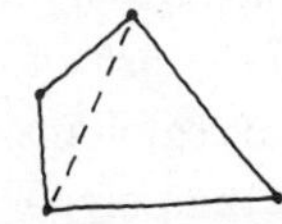 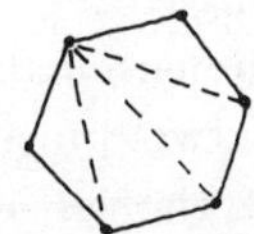

If two nonadjacent vertices of a polygon can be connected by a straight line segment which does not fall entirely within its inside region, then the polygon is *concave*. This is a concave polygon, with the dotted straight line arc falling outside the polygon.

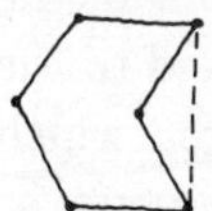

EXERCISE 23 Explain why each of these networks is not a polygon.

(*a*)  (*b*)  (*c*) 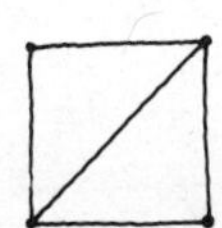 (*d*) 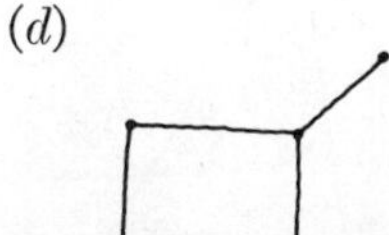

EXERCISE 24 Which ones of these polygons are convex polygons?

(*a*) 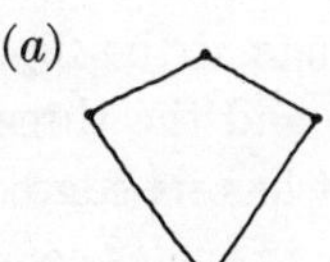 (*b*) 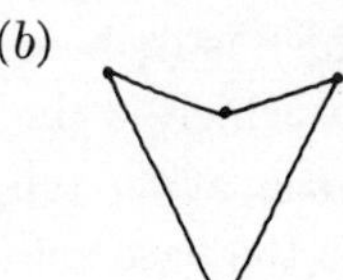 (*c*) 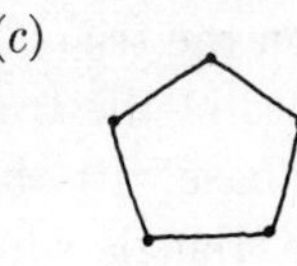

(*d*) 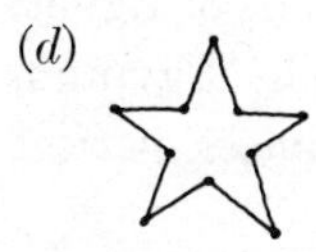 (*e*)  (*f*) 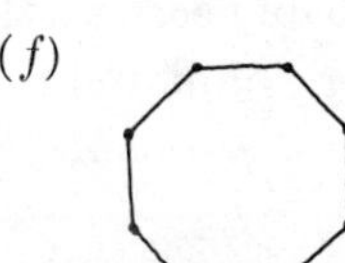

A polygon along with its inside and outside regions forms a *plane.* A plane, like a line, may be thought of as a set of points. However, a line has one dimension while a plane has *two dimensions*, length and breadth but not depth. Local *space*, such as the space around you in a room, has *three dimensions*, length, breadth, and depth.

Polygon means "many-angled." The *polyhedron*, which means "many-based," is the three-dimensional counterpart of the polygon. Each *face* of a polyhedron is the two-dimensional region inside a polygon. The *vertices* of a polyhedron are the vertices of the polygons which form its faces. The line segments which are the sides of the polygons are called the *edges* of the polyhedron. Any polyhedron has exactly two faces on each edge, exactly two vertices on each edge, and exactly two three-dimensional regions, one inside and one outside.

One of the simplest types of polyhedra is the *pyramid.* A pyramid consists of a convex polygon, which is its *base*, and a point not in the plane of that polygon with each vertex of the polygon connected to the point to form the other edges and faces. Although the polygon which is the base of the pyramid has points that are vertices, the point at the top is called the *vertex* of the pyramid.

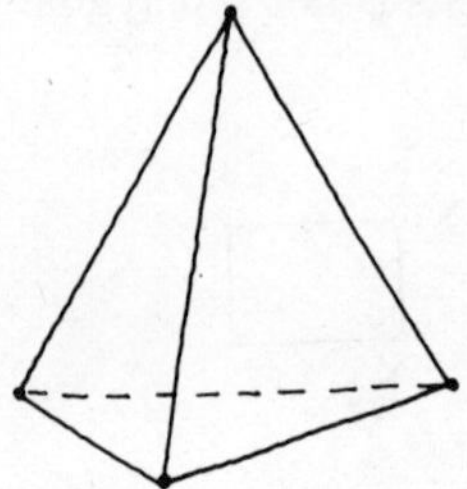

This is a *triangular pyramid* because its base is a triangle. It has four vertices: three on the triangle which is its base and one at the top. Its edges are the sides of the triangle which is the base and the three edges connecting the base with the vertex at the top; so it has six edges. It has four faces: the triangle which is the base and three triangles connecting the edges of the base with the vertex at the top.

You have probably seen pictures of the ancient pyramids of Egypt, some of which were built more than 3,000 years ago. The Egyptain pyramids are *square pyramids*, since their bases are almost perfect squares.

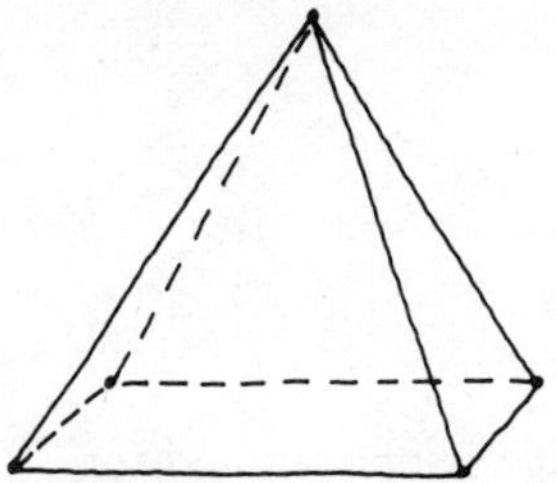

The square pyramid is a special case of a *quadrilateral pyramid*, where the base is a quadrilateral, or four-sided polygon. The quadrilateral pyramid has five vertices: four around the base and one at the top. It has eight edges: four around the base and four connecting the base with the vertex at the top. It has five faces: the quadrilateral base and four triangles connecting the edges of the base with the vertex at the top.

Triangular and quadrilateral pyramids are the types most commonly seen. However, a pyramid can be constructed on any polygonal base. The base of a *pentagonal pyramid* is a pentagon. The pentagonal pyramid has six vertices, ten edges, and six faces.

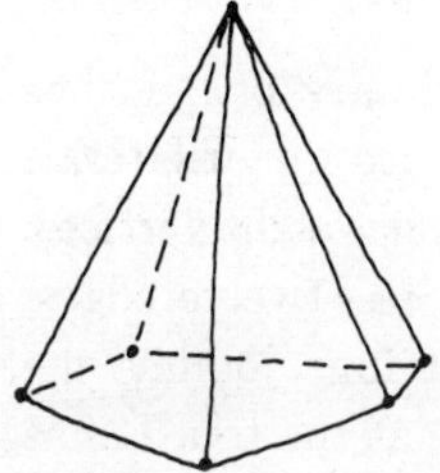

An *n-gon* is an n-sided polygon, where n is a counting number at least 3. In general, if a pyramid is constructed on an n-gonal base, the pyramid has $n + 1$ vertices, the n vertices of the n-gonal base and one at the top of the pyramid. The n-gonal base has n edges, and also there are n edges connecting the vertices of the base with the vertex at the top of the pyramid; so it has $n + n = 2n$ edges. There are $n + 1$ faces, the n-gonal base and n triangular faces connecting the edges of the base with the vertex at the top of the pyramid.

A *prism* is similar to a pyramid in that it has a polygonal base. Its top, however, is also a polygon, identical to the polygon at the bottom.

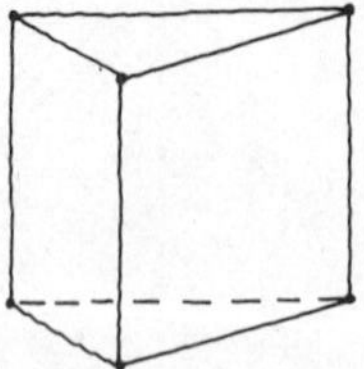

The *triangular prism* has triangular bases at the bottom and top. A triangular prism has six vertices: three on the bottom triangle and three on the top triangle. It has nine edges: three on each triangle and three joining the vertices of the two triangles. It has five faces: the two triangles and the three quadrilaterals joining the edges of the bottom triangle with the edges of the top triangle.

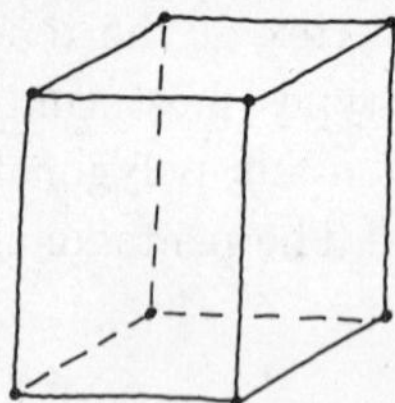

The *quadrilateral prism* is somewhat like an ordinary box. It has identical quadrilaterals at the bottom and top, and also quadrilaterals joining them around the sides. The quadrilateral prism has eight vertices: four around the bottom and four around the top. It has twelve edges: four around the bottom, four around the top, and four joining the vertices of the quadrilateral at the bottom with those at the top. It has six faces: the bottom and top quadrilaterals and those connecting their edges.

The *pentagonal prism* has pentagons at the bottom and top. It has ten vertices, fifteen edges, and seven faces.

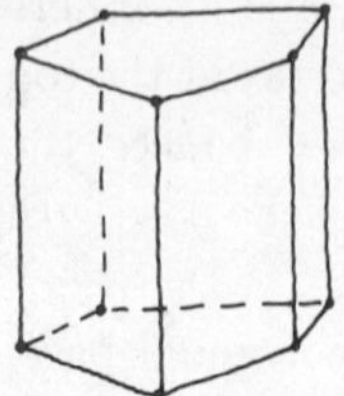

In general, a prism with n-gons at the bottom and top has n vertices at the bottom and n vertices at the top, so it has $n + n = 2n$ vertices. It has n edges at the bottom and n edges at the top, and also n edges connecting the vertices of the bottom n-gon with the top n-gon, so it has $n + n + n = 3n$ edges. It has two n-gonal faces, one at the bottom and one at the top, and n quadrilateral faces connecting the edges of the n-gon at the bottom with the edges of the n-gon at the top, so it has $n + 2$ faces.

EXERCISE 25 (*a*) Draw a hexagonal pyramid. (*b*) How many vertices does it have? (*c*) How many edges does it have? (*d*) How many faces does it have?

EXERCISE 26 (*a*) Draw a hexagonal prism. (*b*) How many vertices does it have? (*c*) How many edges does it have? (*d*) How many faces does it have?

EXERCISE 27 (*a*) Verify the formulas $n + 1$ for the number of vertices, $2n$ for the number of edges, and $n + 1$ for the number of faces of a pyramid, by filling in the following chart:

	n	***Vertices*** $n + 1$	***Edges*** $2n$	***Faces*** $n + 1$
Triangular				
Quadrilateral				
Pentagonal				
Hexagonal				

(*b*) Can you find a formula which connects the numbers of vertices, edges, and faces of any pyramid?

EXERCISE 28 (*a*) Verify the formulas $2n$ for the number of vertices, $3n$ for the number of edges, and $n + 2$ for the number of faces of a prism, by filling in the following chart:

	n	*Vertices* $2n$	*Edges* $3n$	*Faces* $n + 2$
Triangular				
Quadrilateral				
Pentagonal				
Hexagonal				

(*b*) Can you find a formula which connects the numbers of vertices, edges, and faces of any prism?

6

Three-dimensional Networks

Pyramids, prisms, and other polyhedra can be thought of as types of *three-dimensional networks.* Three-dimensional networks are made up of vertices, edges, and faces. Euler's formula for two-dimensional networks can be adapted to the vertices, edges, and faces of polyhedra.

EULER'S FORMULA FOR POLYHEDRA If V is the number of vertices, E the number of edges, and F the number of faces of a polyhedron, then $V - E + F = 2$.

For example, the triangular pyramid has four vertices, six edges, and four faces. Euler's formula for the triangular pyramid is $V - E + F = 4 - 6 + 4 = 2$. The triangular prism has six vertices, nine edges, and five faces. Euler's formula for the triangular prism is $V - E + F = 6 - 9 + 5 = 2$.

Suppose we pick up the triangular pyramid and look through its base triangle toward its vertex. What we would see is this diagram, where the outside triangle is the base of the pyramid, the point inside is its vertex, and the lines inside are the edges connecting the base with the vertex.

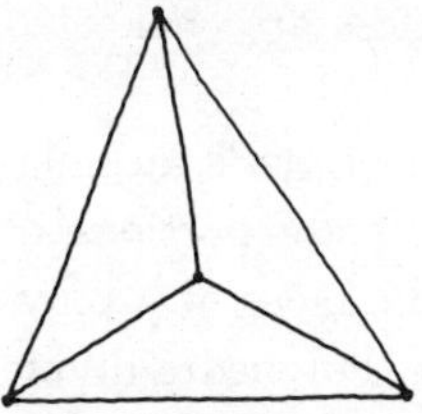

The diagram is called a *Schlegel diagram*. The Schlegel diagram is apparently named for Victor Schlegel (1843–1905), a German teacher of science and mathematics but relatively undistinguished in the history of mathematics.

A Schlegel diagram represents the polyhedron as a two-dimensional network. All the vertices and edges of the polyhedron are present as vertices and arcs of the network. The faces of the polyhedron are the two-dimensional regions of the network, except the face we are looking through since the region that would represent it is filled by the other regions. However, we recall that for two-dimensional networks the region outside the network is counted in Euler's formula. The outside region can count in place of the face we are looking through, and Euler's formula for the triangular pyramid is the same as Euler's formula for the network in the Schlegel diagram that represents it.

There are two Schlegel diagrams for the triangular prism, one looking through one triangular base toward the other, and one looking through a quadrilateral face that connects the bases.

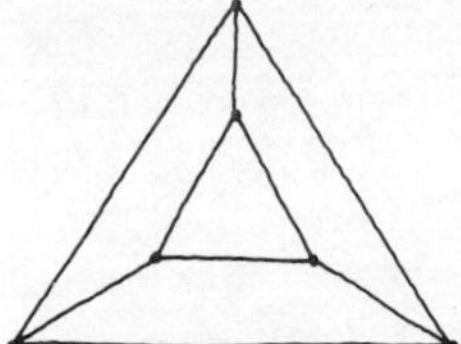

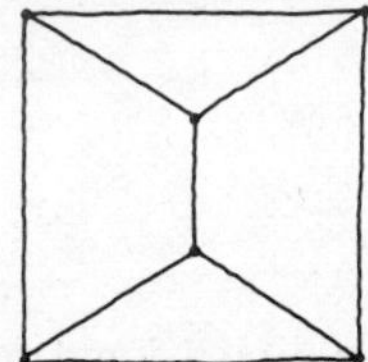

For the network in each diagram, counting the outside region in place of the face we are looking through, Euler's formula is the same as for the triangular prism.

Since the two Schlegel diagrams for the triangular prism represent the same polyhedron, you might feel that there should be some connection between them. They are called *isomorphic* graphs. Two graphs are isomorphic if there is a one-to-one correspondence between their vertices and their arcs such that corresponding vertices are connected by corresponding arcs.

In space, two-dimensional regions become faces in Euler's formula. There are three-dimensional regions, however, which are portions of three-dimensional space completely surrounded by the faces of a polyhedron and the portion of space outside a polyhedron. Polyhedra divide space into two three-dimensional regions, one inside and one outside the polyhedron. We can extend Euler's formula to include three-dimensional regions. If we use Euler's formula for polyhedra and subtract the two three-dimensional regions of the polyhedron, the result is zero.

EXTENSION OF EULER'S FORMULA If V is the number of vertices, E the number of edges, F the number of faces, and R the number of three-dimensional regions in a three-dimensional network, then $V - E + F - R = 0$.

The extension of Euler's formula holds for three-dimensional networks which are not polyhedra. For example, suppose we pass a plane vertically through the top vertex and one base vertex of a triangular pyramid. The plane cuts the inside region of the pyramid into two regions.

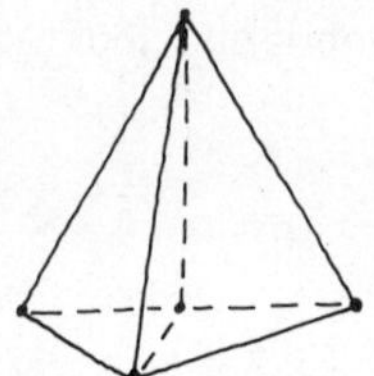

There are now three three-dimensional regions in the network, two inside and one outside. Counting carefully, we find five vertices, nine edges, and seven faces. The back edge of the base is now two edges separated by a new vertex. Also, the back triangle and base triangle are each now two triangles, each counting as two faces with a new edge separating them. The triangle through the center of the pyramid also counts as a new face, separating the inside of the pyramid into two regions. Thus $V - E + F - R = 5 - 9 + 7 - 3 = 0$.

EXERCISE 29 Verify Euler's formula for polyhedra for each of these polyhedra: (*a*) The quadrilateral pyramid. (*b*) The pentagonal pyramid. (*c*) The hexagonal pyramid. (*d*) The quadrilateral prism. (*e*) The pentagonal prism. (*f*) The hexagonal prism.

EXERCISE 30 (*a*) Draw the Schlegel diagram for the quadrilateral pyramid looking through its quadrilateral base toward its vertex. Verify that Euler's formula for the Schlegel diagram is the same as for the quadrilateral pyramid. (*b*) Draw the Schlegel diagram for the quadrilateral prism looking through any of its quadrilateral faces. Verify that Euler's formula for the Schlegel diagram is the same as for the quadrilateral prism. (*c*) Draw the Schlegel diagram for the quadrilateral pyramid looking through any of the triangular faces which connect its base with its vertex. Verify that Euler's formula for this Schlegel diagram is the same as in part *a*. (This is not, however, sufficient to show that the graphs are isomorphic.)

EXERCISE 31 Pass a plane diagonally through two opposite edges of a quadrilateral prism, as shown in this figure:

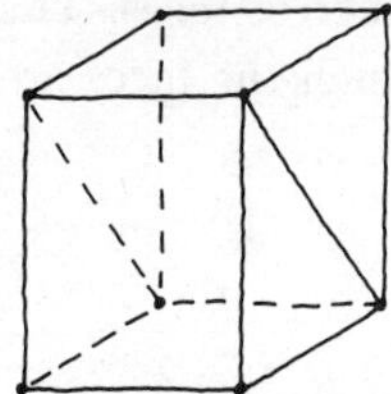

(*a*) How many vertices does it have? (*b*) How many edges does it have? (*c*) How many faces does it have? (*d*) How many regions does it have? (*e*) Verify the extension of Euler's formula.

EXERCISE 32 Pass another plane through two opposite edges not adjacent to those of the figure in Exercise 31 to construct this figure:

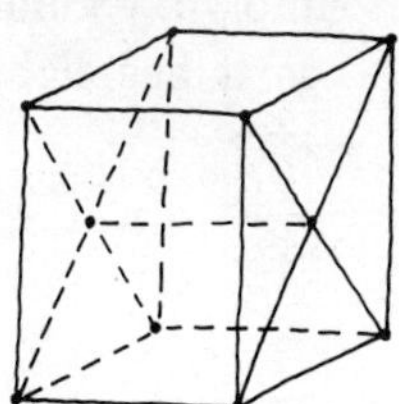

(*a*) How many vertices does it have? (*b*) How many edges does it have? (*c*) How many faces does it have? (*d*) How many regions does it have? (*e*) Verify the extension of Euler's formula.

7
The Platonic Solids

A *regular polygon* is a polygon with all its sides equal and all its angles equal. A *regular polyhedron* is a polyhedron with all its faces identical regular polygons and the same number of those polygons at each vertex. The regular polyhedra are called *Platonic solids* after the Greek philosopher Plato (late fifth–early fourth centuries B.C.). There are exactly five Platonic solids. All five were probably known to Greek mathematicians before Plato, but his is the earliest description of all five to come down to us.

Each of the Platonic solids is named for its number of faces. The first is the *tetrahedron*, where *tetra* means "four," in which the faces are four equilateral triangles.

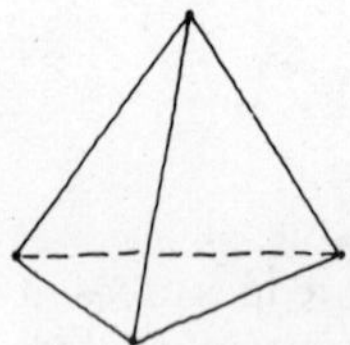

The tetrahedron is simply a regular triangular pyramid, so it has four vertices, six edges, and four faces.

The formal name of the next Platonic solid is the *hexahedron*, where *hex* means "six," but it is commonly called a *cube*. It is simply a regular quadrilateral prism in which the faces are six squares, so it has eight vertices, twelve edges, and six faces.

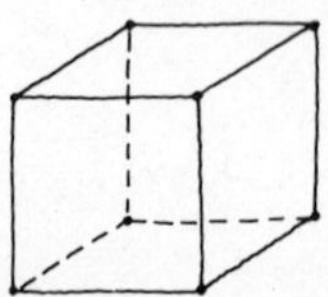

The *octahedron* is made up of eight equilateral triangles. One way to visualize the octahedron is as two square pyramids, base to base, one right side up and one upside down. The common square base of the square pyramids does not count as a face of the octahedron because it is a polyhedron with only one region inside.

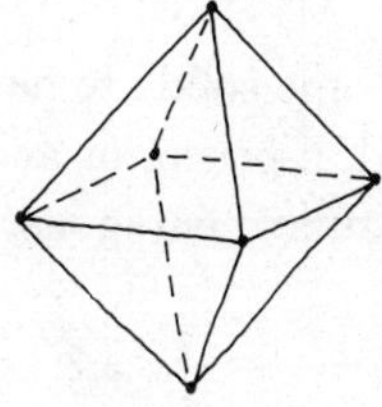

The octahedron can also be constructed as an *antiprism*. An antiprism is a variation of the prism, whe̍re the bottom and top polygons are oppositely oriented.

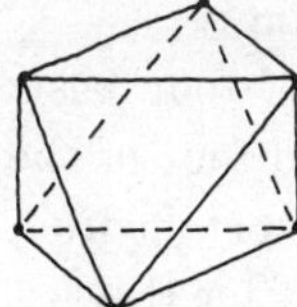

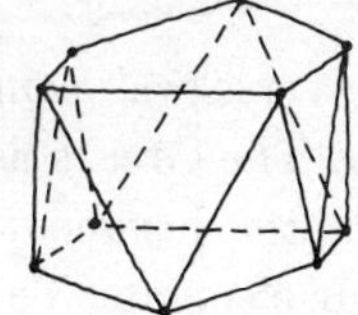

In the *triangular antiprism*, on the left, the bottom triangle points toward you and the top triangle points away from you. Observe that the figure is a different view of the octahedron, tilted so that its bottom and top vertices are the vertices of the triangles that point toward and away from you. The figure on the right is a *pentagonal antiprism*. The pentagon at the bottom points toward you and the pentagon at the top points away from you.

The pentagonal antiprism is not a regular polyhedron because even if the triangular faces are all equilateral triangles, its bottom and top are pentagons. It can be made into a Platonic solid by constructing five equilateral triangles around the bottom pentagon, one on each edge, and also five equilateral triangles around the top pentagon. The resulting polyhedron consists of twenty equilateral triangles. It is called the *icosahedron*.

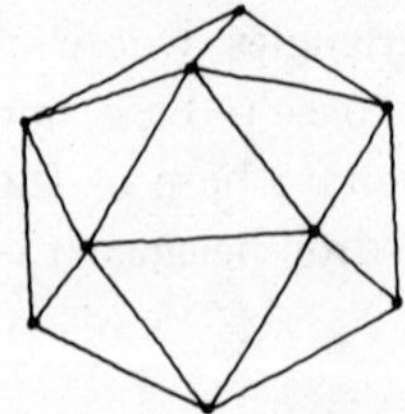

The *dodecahedron* was probably the last of the Platonic solids to be discovered, for it cannot be constructed from a pyramid, a prism, or an antiprism. The dodecahedron is made up of twelve regular pentagons.

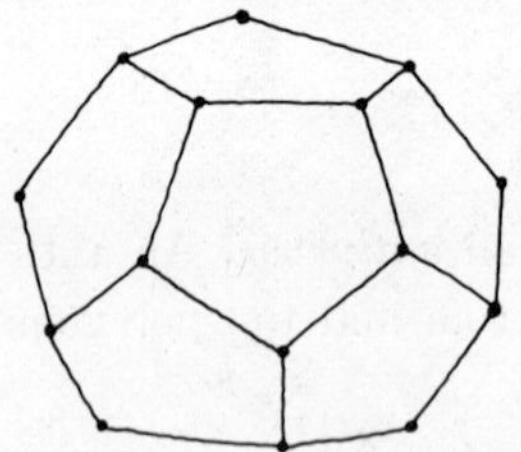

To the ancient Greeks, the universe was composed of four basic elements—fire, earth, air, and water. The Greeks associated four of the regular polyhedra—the tetrahedron, the hexahedron, the octahedron, and the icosahedron—with these four elements. To account for the discovery of the fifth regular polyhedron, Plato associated the dodecahedron with a fifth and highest element, the quintessence, where *quint* means "five" and *essence* means "element" in the medieval Latin of the alchemists, those early chemists who tried to find a way to turn lead to gold. It is a curious bit of etymology that we still use quintessence, "the fifth essence," to mean purest or ultimate.

The proof that there are exactly five regular polyhedra is a bit demanding algebraically but worth the effort because of the surprising nature of the fact. We use Euler's formula and also two new letters, p and q. The letter p stands for the number of edges on each face of a regular polyhedron, and q for the number of edges on each vertex.

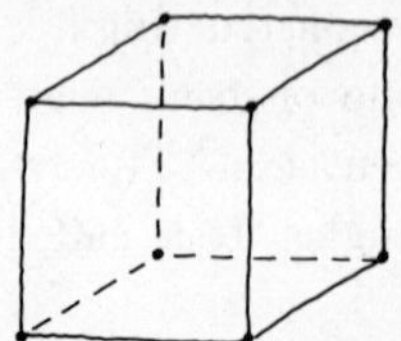

Escher describes this lithograph as "the life cycle of a little alligator." Reptiles were one of Escher's favorite subjects for tessellations of the plane (Chapter 5, Section 6). But this tessellation of reptiles takes on a new dimension, as one of the alligators comes to life. He climbs up various objects to what Escher calls "the highest point of his existence," where we see his snort of triumph. Then he descends to become part of the plane again.

Escher uses the phrases quoted above in his own description of the picture in The Graphic Work of M. C. Escher. Nowhere in this description does he mention the name of the object on which the alligator signals his high point, nor does he give that object any symbolic significance. The object is, of course, a dodecahedron. The choice of one of the Platonic solids was probably deliberate, since Escher was apparently interested in them and made many pictures involving different types of polyhedra.

A question remains unanswered: why, of the Platonic solids, did Escher choose the dodecahedron. The actual answer probably is that it has a flat top for the alligator to rest on but is more interesting than, for instance, the cube. One would like to think, however, that Plato's association of the dodecahedron with the fifth essence, the ultimate, had some significance in the choice of it to be the high point of the alligator's life cycle.

Photo: Escher Foundation, Haags Gemeentemuseum, The Hague

In the cube, for example, $p = 4$, because the cube is made up of squares which have four edges; and $q = 3$, because three edges meet on each vertex.

In a regular polyhedron with q edges on each vertex and V vertices in all, we might think that there are qV edges in all. Each edge, however, connects two vertices, so we have counted each edge twice. Therefore, the number of edges is $E = qV/2$ and $2E = qV$. Similarly, there are p edges on each face and F faces, so we might think there are pF edges. But each edge is also on two faces, so again we have counted each edge twice. Thus $E = pF/2$ and $2E = pF$. Starting with Euler's formula, we have

$$\begin{aligned} V - E + F &= 2 \\ pqV - pqE + pqF &= 2pq \\ p(2E) - pqE + q(2E) &= 2pq \\ (2p - pq + 2q)E &= 2pq \end{aligned}$$

Since E is positive and $2pq$ is positive, $2p - pq + 2q$ must also be positive. Then $-(2p - pq + 2q) = pq - 2p - 2q$ is negative. Therefore,

$$\begin{aligned} pq - 2p - 2q &< 0 \\ pq - 2p - 2q + 4 &< 4 \\ (p - 2)(q - 2) &< 4 \end{aligned}$$

The quantities $p - 2$ and $q - 2$ are counting numbers since p, the number of edges on a polygon, and q, the number of edges on a vertex of a polyhedron, are each at least 3. Their product is less than 4. There are exactly five such products: $1 \cdot 1$, $1 \cdot 2$, $2 \cdot 1$, $1 \cdot 3$, and $3 \cdot 1$. Each product results in one of the Platonic solids.

$(p-2)(q-2)$	$p-2$	$q-2$	p	q	
$1 \cdot 1$	1	1	3	3	Tetrahedron
$2 \cdot 1$	2	1	4	3	Hexahedron
$1 \cdot 2$	1	2	3	4	Octahedron
$3 \cdot 1$	3	1	5	3	Dodecahedron
$1 \cdot 3$	1	3	3	5	Icosahedron

As a by-product of this proof, we have formulas to check our counts of the numbers of vertices, edges, and faces of each Platonic solid. Since $(2p - pq + 2q)E = 2pq$, we have $E = 2pq/(2p - pq + 2q)$. Also, since $2E = qV$, we have $V = 2E/q$; and since $2E = pF$, we have $F = 2E/p$.

Observe that we must compute E first from the values of p and q, and then V and F from the value of E.

EXERCISE 33 Use the drawings of the Platonic solids to count the numbers of vertices and edges of each, and complete the following chart:

	Vertices	***Edges***	***Faces***
Tetrahedron			
Hexahedron			
Octahedron			
Dodecahedron			
Icosahedron			

EXERCISE 34 Verify Euler's formula for polyhedra for each of the Platonic solids.

EXERCISE 35 Use the formulas for E, V, and F above to complete the following chart, where the results for V, E, and F should agree with Exercise 33:

	p	q	V	E	F
Tetrahedron					
Hexahedron					
Octahedron					
Dodecahedron					
Icosahedron					

EXERCISE 36 The figure next to each of these Platonic solids is a *net* which when folded along the lines and taped together makes a model of the solid.

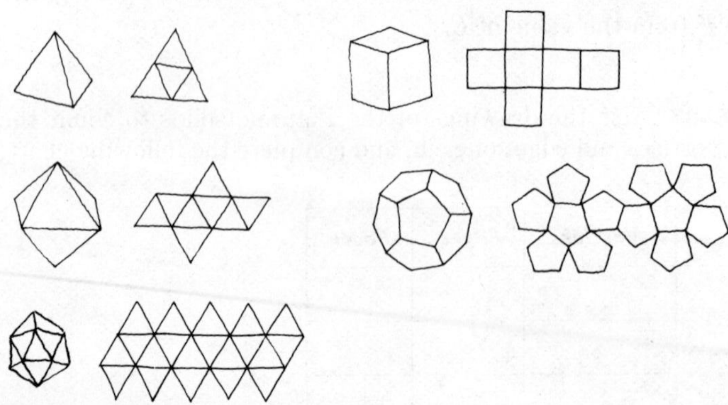

Coxeter says: "Apparantly Plato himself used nets to make models of polyhedra" (*Introduction to Geometry*, p. 152). Make a larger drawing of each net on heavy paper or cardboard and construct a model of each Platonic solid.

8
The Fourth Dimension

You might find it difficult to imagine four-dimensional figures or even to believe them possible. This difficulty is not surprising, for we live in three-dimensional space. However, it is intriguing to speculate on the possibility that this three-dimensional space of ours is somehow enveloped in four-dimensional space, just as the two-dimensional page on which these words are printed is surrounded by our three-dimensional space. Almost a century ago a Shakespearean scholar, Edwin A. Abbott, wrote an allegorical tale of a Square to whom the whole universe is a two-dimensional space. A Sphere comes into the Square's two-dimensional space to teach him the concept of three-dimensional space. In the following exchange,[2] the Sphere attempts to explain the concept of a cube to a Square, who can imagine only a square. The

[2] Edwin A. Abbott, *Flatland, A Romance of Many Dimensions* (New York: Barnes & Noble, Inc., 1963), pp. 78–81.

Square, a Flatland businessman with a deep interest in mathematics, Flatland variety, is the narrator.

Sphere. Tell me, Mr. Mathematician; if a point moves Northward, and leaves a luminous wake, what name would you give to the wake?

I. A straight Line.

Sphere. And a straight Line has how many extremities?

I. Two.

Sphere. Now conceive the Northward straight Line moving parallel to itself, East and West, so that every point in it leaves behind it the wake of a straight Line. What name will you give to the Figure thereby formed? We will suppose that it moves through a distance equal to the original straight Line.—What name, I say?

I. A Square.

Sphere. And how many sides has a Square? How many angles?

I. Four sides and four angles.

Sphere. Now stretch your imagination a little, and conceive a Square in Flatland, moving parallel to itself upward.

I. What? Northward?

Sphere. No, not Northward; upward, out of Flatland altogether.

If it moved Northward, the Southern points of the Square would have to move through the positions previously occupied by the Northern points. But that is not my meaning.

I mean that every Point in you—for you are a Square and will serve the purpose of my illustration—every Point in you, that is to say in what you call your inside, is to pass upwards through Space in such a way that no Point shall pass through the position previously occupied by any other Point; but each Point shall describe a straight Line of its own. This is all in accordance with Analogy; surely it must be clear to you.

Restraining my impatience—for I was now under a strong temptation to rush blindly at my Visitor and to precipitate him into Space, or out of Flatland, anywhere so that I could get rid of him—I replied:—

"And what may be the nature of the Figure which I am to shape out by this motion which you are pleased to denote by the word 'upward'? I presume it is describable in the language of Flatland."

Sphere. Oh, certainly. It is all plain and simple, and in strict accordance with Analogy—only, by the way, you must not speak of the result as being a Figure, but as a Solid. But I will describe it to you. Or rather not I, but Analogy.

We begin with a single Point, which of course—being itself a Point—has only one terminal Point.

One Point produces a Line with two terminal Points.

One Line produces a Square with four terminal Points.

Now you can give yourself the answer to your own question: 1, 2, 4, are evidentally in Geometrical Progression. What is the next number?

I. Eight.

Sphere. Exactly. The one Square produces a Something-which-you-do-not-as-yet-know-a-name-for-but-which-we-call-a-Cube with eight terminal Points. Now are you convinced?

I. And has this Creature sides, as well as angles or what you call "terminal Points"?

Sphere. Of course; and all according to Analogy. But, by the way, not what you call sides, but what we call sides. You would call them solids.

I. And how many solids or sides will appertain to this Being whom I am to generate by the motion of my inside in an "upward" direction, and whom you call a Cube?

Sphere. How can you ask? And you a mathematician! The side of anything is always, if I may say so, one Dimension behind the thing. Consequently, as there is no Dimension behind a Point, a Point has 0 sides; a Line, if I may say, has 2 sides (for the Points of a Line may be called by courtesy, its sides); a Square has 4 sides; 0, 2, 4, what Progression do you call that?

I. Arithmetical.

Sphere. And what is the next number?

I. Six.

Sphere. Exactly. Then you see that you have answered your own question. The Cube which you will generate will be bounded by six sides, that is to say, six of your insides. You see it all now, eh?

"Monster," I shrieked, "be thou juggler, enchanter, dream, or devil, no more will I endure thy mockeries. Either thou or I must perish." And saying these words I precipitated myself upon him.

It should be told, perhaps, that the Sphere does eventually succeed in convincing the Square of a third dimension; but to find out how this is accomplished, you will have to read the book. The Square tries to teach the concept to his fellow Flatlanders and is eventually imprisoned as a dangerous heretic.

To us, four-dimensional figures are as three-dimensional figures are to the Square. When we draw on a page, the only figures we actually can draw are two-dimensional. For example, we can draw a square.

If we wish to draw a cube, however, we can draw only a *projection* of the three-dimensional cube onto the two-dimensional page.

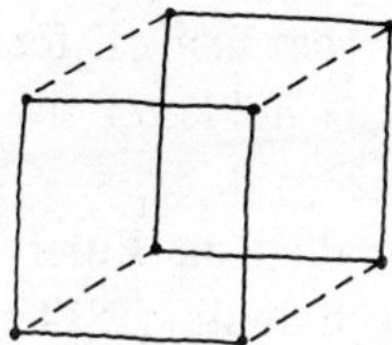

We do this by "moving" the square diagonally across the page to represent the direction upward from the page. We call the result a cube, with six faces each a square, although only the front and back are actually squares in the drawing. The other four faces are projections of squares.

Similarly, we cannot draw an actual four-dimensional analog of a cube, but we can draw its projection onto the page. We simply move our projection of the cube in a different diagonal direction across the page to represent the direction outward from three-dimensional space. The resulting figure is called a *hypercube.* Only two of the cubes in the hypercube look like the original cube. The others are projections of it.

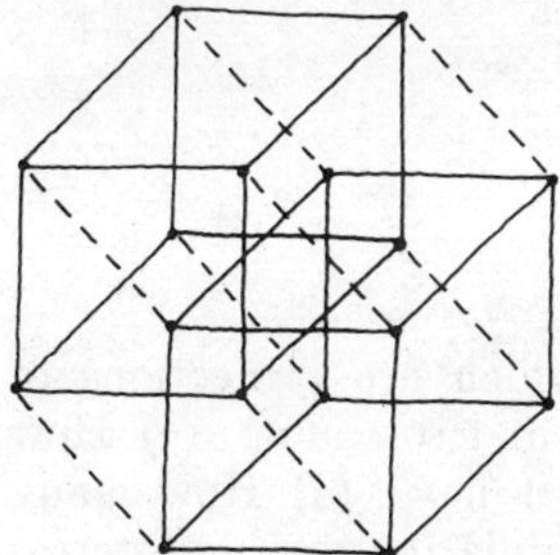

The four-dimensional analogs of polyhedra are made up of vertices, edges, faces, and three-dimensional sides which are polyhedra. The polyhedra which are the sides are called *cells.* The hypercube has sixteen vertices: eight on the original cube and eight on the new cube which results from moving the cube in a fourth direction. There are twelve edges on each of the two cubes, and eight more are formed by the movement of the vertices of the original cube, so the hypercube has $12 + 12 + 8 = 32$ edges. Similarly, each of the two cubes has six faces, and twelve more are formed by the edges of the original cube as it is moved, so the hypercube has $6 + 6 + 12 = 24$ faces. Finally, each of the six faces of the original cube makes a new cube as it moves, which with the original and new cubes make eight cells. We observe that using C for cells in place of regions, our extension of Euler's formula holds for the hypercube, since $V - E + F - C = 16 - 32 + 24 - 8 = 0$.

The formula $V - E + F - C = 0$ might be called an Euler-Schläfli formula for the Swiss mathematician Ludwig Schläfli (1814–1895), who discovered it around the mid-nineteenth century. At that time, according to Coxeter, only Schläfli, Cayley, Grassmann (Branching Out, Chapter 4), and Möbius (Chapter 8), "had ever conceived the possibility of geometry in more than three dimensions" (*Regular Polytopes*, p. 141).

EXERCISE 37 The *regular simplex*, shown here, can be interpreted as a four-dimensional analog of the tetrahedron. (The *hypertetrahedron* cannot be drawn by moving a tetrahedron in a fourth direction, for that would create faces which are squares rather than equilateral triangles.)

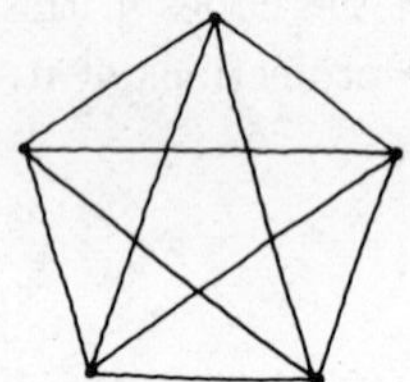

(*a*) Find the hypertetrahedron's five cells, which are projections of tetrahedra. Use the figure and the cells you found to count. (*b*) How many vertices it has. (*c*) How many edges it has. (*d*) How many faces it has. (*e*) Verify the Euler-Schläfli formula for the hypertetrahedron.

Looking Back

1 (*a*) Describe what is meant by a graph. (*b*) Describe what is meant by a connected graph. (*c*) Describe what is meant by a plane graph. (*d*) Describe what is meant by a network.

2 Which of these graphs are connected? Which of these graphs are plane graphs? Which of these graphs are networks?

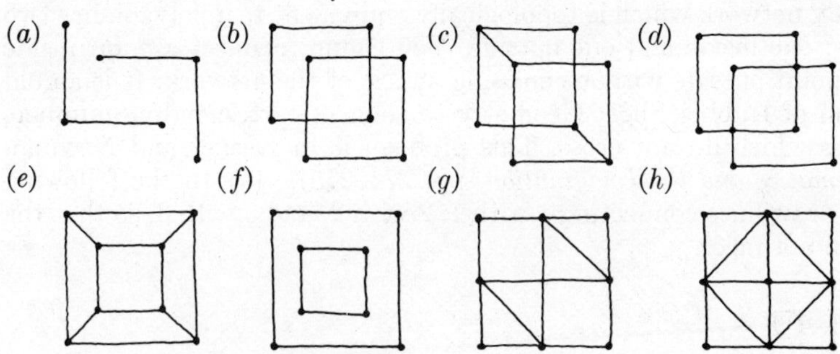

3 Write Euler's formula for two-dimensional networks. How many vertices, arcs, and regions has each of these networks? Verify Euler's formula for each of these networks:

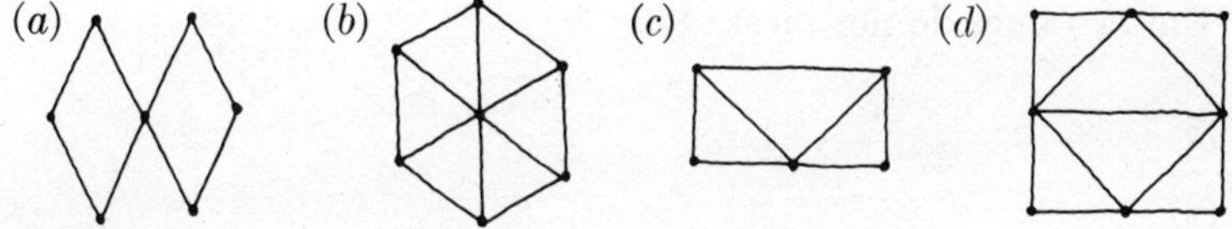

4 (*a*) Write the definition of traversability for networks. (*b*) Write the definition of an even vertex. (*c*) Write the definition of an odd vertex. (*d*) Explain the conditions under which a network is traversable. (*e*) How many even vertices and how many odd vertices has each of the networks in Exercise 3? (*f*) Which of the networks in Exercise 3 are traversable? (*g*) Which of the traversable networks in Exercise 3 have special conditions, and why? (*h*) Which of the networks in Exercise 3 is not traversable, and why not?

5 (*a*) Draw a map of the seven bridges of Königsberg. (*b*) Reduce the map of the seven bridges of Königsberg to a topologically equivalent network. (*c*) Explain why the network is not traversable. (*d*) Reduce the following map to a topologically equivalent network.

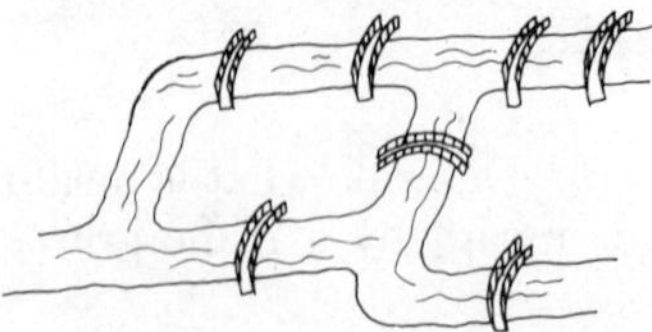

(*e*) Decide whether or not the network is traversable. Are there any special conditions?

***6** Any network which is topologically equivalent to a polygon has two regions, one inside and one outside. You cannot connect a point inside and a point outside without crossing an arc of the network. It is an old problem of Rubber Sheet Geometry to connect corresponding numbers by lines which do not cross. This problem is in Kasner and Newman (*Mathematics and the Imagination*, pp. 274–275). (*a*) In the following figure, draw lines connecting 1 with 1, 2 with 2, and 3 with 3, so that the lines do not meet.

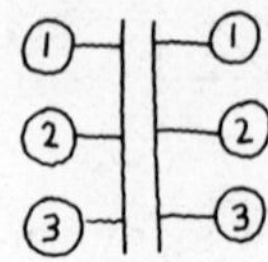

(*b*) In the following figure draw lines connecting 1 with 1 and 3 with 3 so that the two lines do not meet.

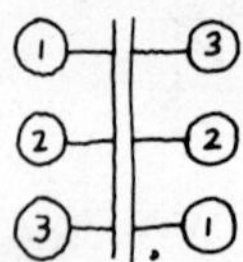

(*c*) Use the concept of topological equivalence to explain why any line connecting 2 with 2 must cross one of the other two lines.

***7** A traversable network where you must begin at one vertex and end at another is said to have an *Euler path.* A network has an *Euler circuit* if it is not only traversable but also is such that you can begin and end the tracing at the same vertex. Refer to the networks of Exercises 18 to 22 of Section 4, and Exercise 6 above. (*a*) Which ones of the networks in these exercises have Euler circuits? (*b*) What condition must the vertices of a network fulfill in order to have an Euler circuit? Some books state the bridges of Königsberg problem in terms of an Euler circuit; that is, the walk must begin and end at the same place. (*c*) If a network does not have an Euler path, explain why it cannot have an Euler circuit.

***8** A *Hamilton path* is a journey to each vertex of a network exactly once, rather than a journey across each arc exactly once. For a *Hamilton circuit*, the journey must begin and end at the same vertex. For example, this network has a Hamilton circuit:

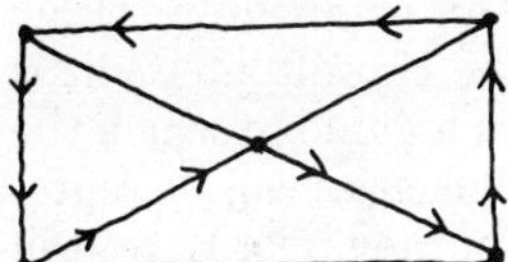

Decide whether or not each of these networks has a Hamilton circuit:

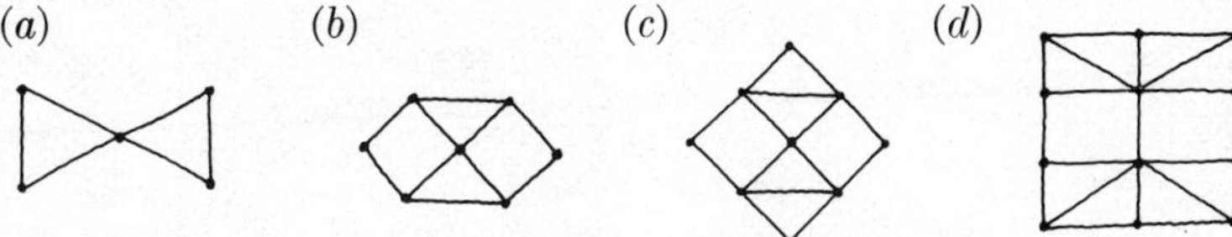

(Unlike Euler paths and circuits, no general method has been discovered to tell, for any network, whether or not the network has a Hamilton circuit.)

9 (*a*) Draw a triangular pyramid. (*b*) Draw a triangular prism. (*c*) Decide whether or not the triangular pyramid is traversable. (*d*) Decide whether or not the triangular prism is traversable. (*e*) Is any pyramid or prism traversable? Explain.

***10** (*a*) Consider the following graphs (the two Schlegel diagrams for the quadrilateral pyramid):

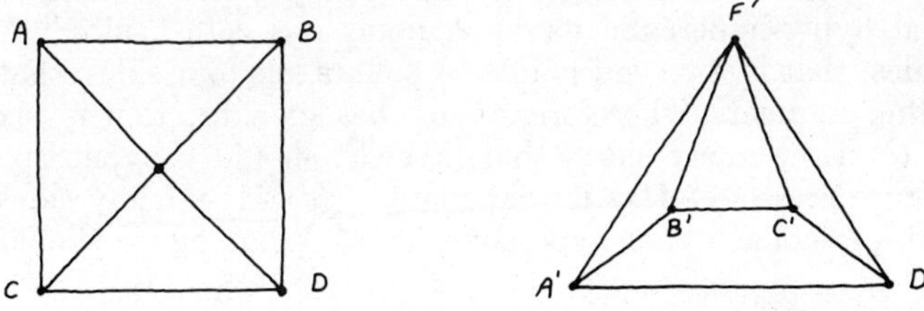

The second graph is isomorphic to the first if the arc connecting A' and B' can be labeled a', the arc connecting B' and C' can be labeled b', and so on; thus, corresponding vertices are connected by corresponding arcs.

Label the arcs of the second graph as suggested, thus showing that the graphs are isomorphic. (*b*) Draw the two Schlegel diagrams for the triangular prism. Lable the vertices and arcs in a way similar to that described in part *a*, thus showing that the graphs are isomorphic.

11 (*a*) Name the five Platonic solids. (*b*) Which one is the only Platonic solid that is traversable? Explain. (*c*) For each of the other four Platonic solids, explain why it is not traversable.

***12** (*a*) Draw a tetrahedron, a hexahedron, and an octahedron. Draw a Hamilton circuit on each of these solids. (*b*) The Hamilton circuit is named for William Rowan Hamilton, who invented a game in which the goal was to make a trip to each of twenty cities. The cities can be represented by the vertices of a dodecahedron, and the journey by a Hamilton circuit on the dodecahedron. Starting with vertex 1, label this dodecahedron so that the journey is possible by going from vertex 1 to 2 to 3 and so on to 20, and back to 1.

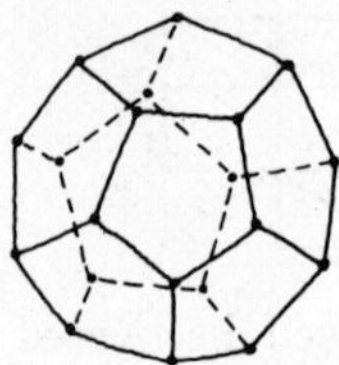

(Although Hamilton circuits have many applications in modern graph theory, unfortunately for Hamilton, the game was not successful.)

13 Consider the method of Analogy of the Sphere in *Flatland.* A point has one vertex. A line has two vertices. A square has four vertices. Therefore, a cube has eight vertices. (*a*) How many vertices should a hypercube have? Has it that many? (*b*) How many vertices should a five-dimensional hyper-hypercube have? A point has zero "sides." A line has two sides, that is, two end points. A square has four sides, that is, four equal line segments. Therefore, a cube has six sides, that is, six square faces. (*c*) How many sides, that is, cells, should a hypercube have? What figure are they? Has it that many? (*d*) How many sides should a five-dimensional hyper-hypercube have? What figure should they be?

***14** Suppose that N_0 is the number of zero-dimensional parts (vertices), N_1 the number of one-dimensional parts (edges), N_2 the number of two-dimensional parts (faces), N_3 the number of three-dimensional parts (cells), N_4 the number of four-dimensional parts, and so on, of a multidimensional figure. Then,

$$N_0 - N_1 + N_2 - \cdots + N_n = 2 \qquad \text{if } n \text{ is even}$$
$$N_0 - N_1 + N_2 - \cdots - N_n = 0 \qquad \text{if } n \text{ is odd}$$

These are the *Euler-Schläfli formulas.* (*a*) Show that the first Euler-Schläfli formula agrees with Euler's formula for polyhedra if $n = 2$. (*b*) Show that the second Euler-Schläfli formula agrees with the extension of Euler's formula if $n = 3$. Consider a hypercube moved in a fifth direction, that is, a five-dimensional hyper-hypercube. (*c*) Its vertices are the vertices of the original hypercube plus those of the new hypercube. How many vertices does it have? (*d*) Its edges are those of the two hypercubes, plus those formed by the movement of the vertices of the original hypercube. How many edges does it have? (*e*) Its faces are those of the two hypercubes, plus those formed by the movement of the edges of the original hypercube. How many faces does it have? (*f*) Its cells are those of the two hypercubes, plus those formed by the movement of the faces of the original hypercube. How many cells does it have? (*g*) Its four-dimensional sides, "hypercells," are the original hypercube, the new hypercube, and those formed by the movement of the cells of the original hypercube. How many four-dimensional hypercells does it have? (*h*) Use these results for the hyper-hypercube to verify the Euler-Schläfli formula for $n = 4$.

Branching Out

1 Plato is best known as one of the ancient Greek philosophers for his *Dialogues*. Although he was not a mathematician in the sense of one who created mathematics, he was an expositor of some of the mathematics of his time and also some of the mysticism. His deep respect for geometry is revealed by the inscription over the entrance to his Academy, "Let no one unversed in geometry enter here." His dialogue called the *Timaeus* contains the descriptions of the Platonic solids. It also contains the mystic association of the five solids with the four elements of the Greeks and the quintessence. Plato also discusses the importance of geometry in his famous dialogue, the *Republic*. You might read these two dialogues of Plato and also what other writers say about his mathematics and his mysticism.

2 As we have mentioned, Leonhard Euler was the most prolific mathematician of all time. Although Swiss by birth, Euler spent much of his life in Russia, with an interlude of several years in Germany (then called Prussia). His work touched on all fields of mathematics. He wrote rapidly, and his writings fill dozens of volumes, including much done in his head and from memory after he had become totally blind. E. T. Bell calls him a "mathematical universalist." Bell's *Men of Mathematics* is a good place to start for stories of Euler's life and highlights of his work. Although Bell does not include a discussion of the famous Königsberg bridges, it can be found in many other books.

3 Graph theory is a part of topology closely allied to Rubber Sheet Geometry. It is a bit more difficult, and interest in it as a separate topic is somewhat recent, so that discussion about it is hard to find in nontechnical books. If you can find such a book, however, you can find out about such things as isomorphic graphs, digraphs, and applications of Euler and Hamilton circuits. Beware of books on graph theory meant for mathematics specialists, however.

4 Rubber Sheet Geometry is one of the most fascinating areas of mathematics, involving problems which can be interpreted in the most sophisticated terms of research mathematicians, but which are also capable of description in the language of the layman. Such problems range from the Königsberg bridges to things called knots and pretzels to a famous problem called the four-color problem. They include problems of inside and outside and of things which have no inside and outside. There is an excellent chapter on Rubber Sheet Geometry in Kasner and Newman's *Mathematics and the Imagination.* Gamow takes some similar problems and extends them into the fourth dimension in *One Two Three . . . Infinity.*

5 The Egyptian pyramids are magnificent pieces of ancient engineering. The fact that they were built at all and the methods by which they were constructed are interesting topics. Even more interesting, in terms of the level of development of ancient mathematics, is the precision of their construction. Their bases are almost perfect squares and their angles, almost exact right angles. Moreover, their orientation north-south and east-west is also nearly perfect. The pyramids are described in many books of many types. Look in the archaeology section as well as in histories of mathematics.

6 The fourth dimension has been treated in many ways and many styles. There are serious discussions, written for the layman, of Einstein's interpretation of time as the fourth dimension. There are humorous but instructive approaches such as Gamow's in *One Two Three . . . Infinity.* Among the most entertaining, although still often of mathematical value, are those which are strictly science fiction. If you are a science fiction buff, see if you can find some with genuine mathematical content. For starters, a good one is a short story by Heinlein, "And He Built a Crooked

House," which you will find in anthologies of Heinlein's stories, or in Clifton Fadiman's *Fantasia Mathematica.* (Read some of the other stories in this collection, too.) Another story, with an interpretation of the fourth dimension closer to that of *Flatland,* is "The Appendix and the Spectacles," by Miles J. Breuer, M.D., in Fadiman's *The Mathematical Magpie.*

I'll meet
you at the...

8

Geometries with a Twist

1

Euclid's Elements

The distinctive feature of mathematics is the *deductive system*. Regardless of the type of objects studied in any branch of mathematics, the deductive system starts with some assumptions about the objects and draws a series of conclusions from the assumptions. In this respect, the *Elements* of the Greek mathematician Euclid (about 300 B.C.) is the oldest distinctly mathematical work to come down to us virtually intact.

This is not to say that we have an actual manuscript written by Euclid. The earliest manuscripts still in existence date from about the tenth century, 1,200 or 1,300 years after Euclid, and are based on a copy made by Theon of Alexandria (fourth century A.D.) some 700 years after Euclid. There is also a manuscript in the Vatican library which dates from the tenth century, but which was probably made from a copy earlier than Theon's.

In any event, the manuscripts we have were copied and recopied over many centuries. Corrections, errors, and other changes were introduced. Through careful comparison of the various copies, however, and use of commentaries by earlier scholars, contemporary scholars have been able to construct texts which are probably quite like Euclid's. Translations of these texts have been made into many languages. Indeed, parts of the *Elements* in English translation were used as textbooks in this country until very recently, and modified versions are still in use. The *Elements*, in all its copies, translations, and text forms, is generally considered to be the second best-seller of all time, topped only by the Bible.

By comparison, almost nothing is known of Euclid himself. What little we do know comes from the commentaries of the early scholars, primarily from the *Commentary* of Proclus (A.D. 410–485), but even Proclus could not reconstruct from the little evidence he had the dates of Euclid's birth and death or his place of birth. He was able to conclude that Euclid lived in Alexandria about 300 B.C., because there is a famous story about Euclid and Ptolemy I, ruler of Egypt from 306 to 283 B.C. The story, as told by Proclus, is translated here by Thomas L. Heath,[1] among the foremost of contemporary writers on Greek mathematics.

> This man lived in the time of the first Ptolemy. For Archimedes, who came immediately after the first (Ptolemy), makes mention of Euclid: and, further, they say that Ptolemy once asked him if there was in geometry any shorter way than that of the elements, and he answered that there was no royal road to geometry. He is then younger than the pupils of Plato but older than Eratosthenes and Archimedes; for the latter were contemporary with one another, as Eratosthenes somewhere says.

There is little reason to doubt that Euclid lived after Plato but before Archimedes, but it is hardly proved by the often-told story of the first Ptolemy, for there is a similar version involving Alexander and Menaechmus. In this version, Alexander is presumably Alexander the Great (356–323 B.C.) and Menaechmus is a Greek mathematician of about that time. Sir Thomas Heath tells the second version in a footnote, attributing it to Stobaeus, a commentator of about the same time as Proclus.

[1] *Euclid's Elements*, 2d ed. (Cambridge: The University Press, 1926), vol. I, p. 1.

Alexander is represented as having asked Menaechmus to teach him geometry concisely, but he replied: "O king, through the country there are royal roads and roads for common citizens, but in geometry there is one road for all."[2]

Prior to his story of Euclid and the first Ptolemy, Proclus says the distinction of Euclid's *Elements* is neither that it is about new objects nor that it is about new conclusions concerning existing objects. It involves the same geometry which must have been used by the Egyptians for centuries. It includes the work of earlier Greeks such as Thales and the Pythagoreans (Chapter 1) and ends with the Platonic solids (Chapter 7). What is new about the *Elements* is that Euclid organized the mathematics of his time into a deductive system.

The *Elements* contains thirteen books encompassing hundreds of conclusions. Contemporary mathematicians call such conclusions *theorems.* (Euclid, in Heath's translation, calls them *propositions.*) Euclid based these hundreds of theorems on only ten assumptions. Although many errors and flaws have been found in the reasoning, we should rather emphasize the soundness of the work, considering its age.

One strong point of Euclid's system is in his set of assumptions, the statements which must be given without proof and from which presumably all the other statements are proved. He divides his assumptions into two parts. The first part he calls the *postulates.* This is Heath's translation of Euclid's postulates.

Let the following be postulated:

1. To draw a straight line from any point to any point.
2. To produce a finite straight line continuously in a straight line.
3. To describe a circle with any centre and distance.
4. That all right angles are equal to one another.
5. That, if a straight line falling on two straight lines makes the interior angles on the same side less than two right angles, the two straight lines, if produced indefinitely, meet on that side on which are the angles less than the two right angles.[3]

[2] Ibid.

[3] Ibid., vol. I, pp. 154–155.

Euclid's first assumption is that there is a line connecting any two distinct points. He also implicitly assumes, without stating it, that there is only one line connecting any two points. His second postulate means that a line can be extended indefinitely in each direction, which Euclid also assumes to mean that a line is infinite in extent. Since the fifth postulate is noticeably more complicated than the others, a diagram is helpful to clarify it.

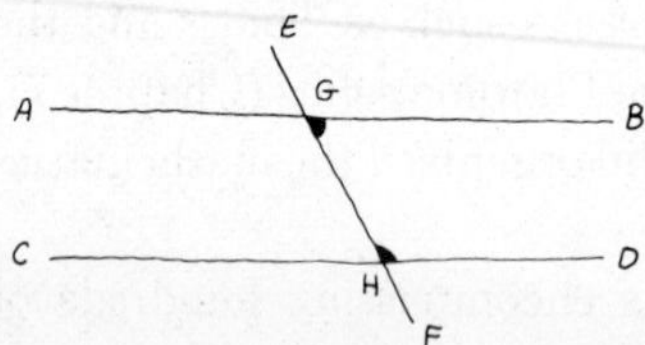

The two marked angles may be named $\measuredangle BGH$ and $\measuredangle GHD$, the middle letter indicating the vertex of the angle and the other two its sides. The fifth postulate then says that if $\measuredangle BGH + \measuredangle GHD$ is less than two right angles, the lines AB and CD will meet if extended beyond B and D.

The second part of Euclid's set of assumptions is called the *common notions*, again given in Heath's translation.

1. Things which are equal to the same thing are also equal to one another.
2. If equals be added to equals, the wholes are equal.
3. If equals be subtracted from equals, the remainders are equal.
4. Things which coincide with one another are equal to one another.
5. The whole is greater than the part.[4]

It is thought that the common notions were separated from the postulates because Euclid considered the common notions to be common to all mathematics and the postulates particular to geometry.

The list of basic assumptions, the five postulates and the five common notions, is not complete. As was pointed out before, Euclid assumed that there is just one line on every two points and that lines are infinite

[4] Ibid., vol. I, p. 155.

in extent without actually including these assumptions among the postulates. There are also some subtler assumptions of concern to modern mathematics which Euclid failed to consider. What is amazing about the list is one postulate, the fifth, which is included. However different and more complicated it is compared to the other assumptions, Euclid apparently could not prove it as a theorem. Possibly he did try to avoid the assumption of the fifth postulate, for he did not use it until he reached his twenty-ninth theorem.

2
Parallels and Perspective

Two distinct straight lines are said to be *parallel* if they do not meet. Euclid's fifth postulate is often called the *parallel postulate* because it is equivalent to a common assumption about parallel lines. For a line MN and a point A not on the line, you will probably agree that there is just one line through point A parallel to line MN.

This assumption is called the *Playfair form* of the parallel postulate. The Playfair form is true if Euclid's fifth postulate is true, and vice versa.

PLAYFAIR POSTULATE Given a line and a point not on the line, there is exactly one line through the point parallel to the line.

The Playfair postulate is named for John Playfair (1748–1819), a Scottish physicist and mathematician. Playfair wrote an English translation of the first six books of Euclid which went to ten editions, but he is remembered in mathematics today for little more than the postulate named for him and even that had been used as early as the fifth century in the *Commentary* of Proclus.

Using an assumption similar to the Playfair postulate, you would

Religious News Service Photo

The Last Supper, painted on a wall of the refectory of a monastery during about 1495 to 1497, has been restored several times. Its original lines, however, show da Vinci's mastery of perspective and the vanishing point. The beams of the ceiling, the moldings at the top of the walls, the lines of the tops of the tapestries, and the lines formed by the groups of figures all meet in one vanishing point. One interesting line can be drawn from the right front corner of the table through the knuckles of the hand with its finger raised, passing along the way the two hands which gesture directly toward Christ. The vanishing point is the center of Christ's forehead.

probably agree that the two rails of a pair of railroad tracks are parallel. But if you have ever stood between the tracks and looked down them, you will have observed that the rails appear to approach each other and to meet at the horizon. Thus, if you were to draw a picture of the tracks, you would draw them as if they met at a point called the *vanishing point;* and you would say that you had drawn them not as you believe they actually are but as you see them in *perspective.* If we portray objects in perspective, two lines which seemed obviously parallel do meet at the vanishing point. The geometry of such lines and points is called *projective geometry.* Projective geometry is more basic than Euclid's geometry, although Euclid's preceded it by many hundreds of years.

Interest of mathematicians in projective geometry most likely originated in the world of art. Before the Renaissance, paintings were flat, without a feeling of depth that the use of perspective and vanishing points provides. The Renaissance artists wanted to paint things as they actually appear, so they developed the theory of perspective. Leonardo da Vinci (1452–1519) is particularly famous for the way he made lines of buildings, furniture, even groupings of people, appear to meet at a vanishing point. Besides being one of the greatest painters and sculptors of all time, da Vinci was also an intensive student of science, engineering, and mathematics.

Some of the first work in the mathematics of perspective was done soon after da Vinci's time by Gérard Desargues (1593–1662), a French architect and engineer who was probably inspired by such considerations as perspective drawing of buildings. Credit as a forerunner is also given to Blaise Pascal, equally famous for his work in probability (Chapter 6) and in geometry. Jean-Victor Poncelet (1788–1867) is often considered to be the founder of projective geometry.

Projective geometry deals in properties of objects projected in ways similar to the way a slide or movie projector works. In fact, you can do a simple experiment with a flashlight. If you shine the flashlight directly at a wall, the beam of light will form a *circle.* If you shine it at an angle, however, the light will form a type of elongated shape called an *ellipse.* In projective geometry, a circle and an ellipse can be projected into one another.

Of course, the experiment above requires three dimensions. An example which reduces projective geometry to two dimensions is given by

Jean-Victor Poncelet was an officer of the engineers in the army of Napoleon. The Russians burned Moscow so that Napoleon would have no place to quarter his troops during the winter of 1812–1813, and Napoleon returned to France leaving the army to fend for itself. Poncelet was among the troops left in Russia that winter. But because of his officer's uniform, he was taken prisoner for questioning by the Russians and was not left to starve or freeze on the Russian steppes.

Poncelet was imprisoned in Russia for more than a year in 1813 and 1814. To keep his mind occupied, he tried to reconstruct from memory all the mathematics he had learned as a student. Some say that, being without books, he constructed geometry somewhat differently and so created projective geometry. It is more likely that having relearned all he had been taught, he went on to consider new possibilities. In 1822, he published the new results he had brought back from Russia in a book which is considered to be the beginning of projective geometry as a separate field of geometry.

a very famous theorem of Desargues. Suppose we have two triangles, ABC and $A'B'C'$, where the lines joining corresponding vertices of the triangles, AA', BB', and CC', all meet in the same point O.

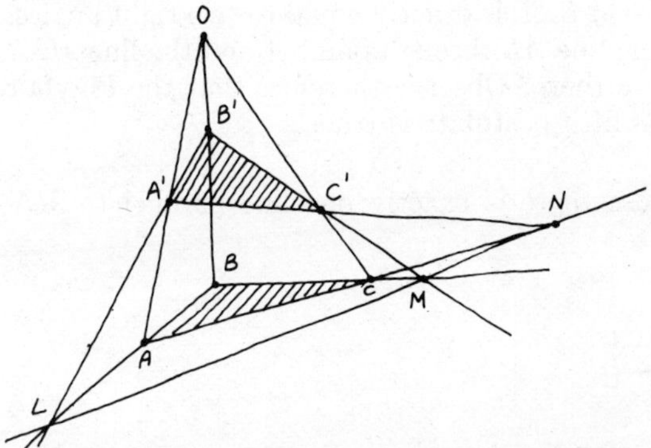

Then the triangles ABC and $A'B'C'$ are said to be *perspective* from point O. Now, we extend the sides AB and $A'B'$ of the triangles until they meet in a point L, sides BC and $B'C'$ until they meet in M, and sides AC and $A'C'$ until they meet in N. *Desargues' theorem* says that the three points L, M, and N will all be on the same straight line.

Although Desargues' theorem is in two dimensions, its proof is very easy if again we are allowed three dimensions. Triangle ABC is simply the shadow made by triangle $A'B'C'$ using a flashlight placed at point O. Points L, M, and N are in the plane of triangle ABC and also in the plane of triangle $A'B'C'$. But two planes which meet each other meet in a line, such as the line formed by two facing pages of this book. Since points L, M, and N are in two planes, they must be on the line where the planes meet.

EXERCISE 1 Recall Euclid's fifth postulate: If a straight line falling on two straight lines makes the interior angles on the same side less than two right angles, the two straight lines, if produced indefinitely, meet on that side on which are the angles less than the two right angles.

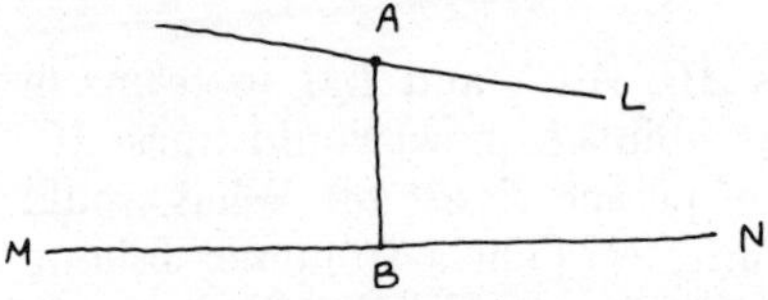

(*a*) Suppose that $\sphericalangle ABN + \sphericalangle BAL$ is less than two right angles. What can you say about the line AL through point A and the line MN? (*b*) Suppose that $\sphericalangle ABN + \sphericalangle BAL$ is greater than two right angles. What can you say about the line AL through point A and the line MN? (*c*) Suppose that $\sphericalangle ABN + \sphericalangle BAL$ is exactly equal to two right angles. What can you say about the line AL through point A and the line MN? (*d*) How many lines AL are there? Observe therefore that the Playfair postulate is true if Euclid's fifth postulate is true.

EXERCISE 2 (*a*) Suppose there is exactly one line parallel to MN through point A.

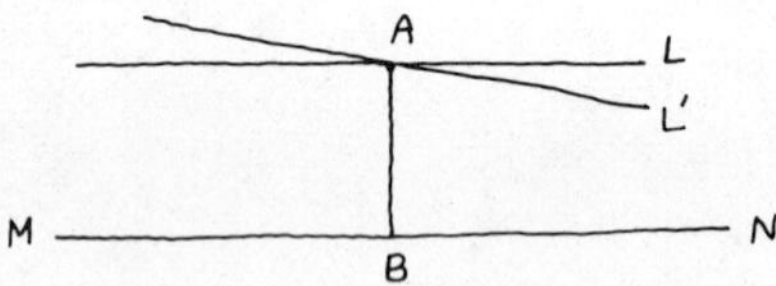

What is the sum of the angles $\sphericalangle ABN + \sphericalangle BAL$? (*b*) Suppose another line AL' is drawn through A, as shown. What can you say about $\sphericalangle ABN + \sphericalangle BAL'$? (*c*) Since AL is the only parallel to MN through point A, what must happen to AL' and MN? Observe therefore that Euclid's fifth postulate is true if the Playfair postulate is true.

EXERCISE 3 In Desargues' theorem, suppose sides AC and $A'C'$ of the triangles are parallel. Draw point L, the intersection of AB and $A'B'$, and point M, the intersection of BC and $B'C'$.

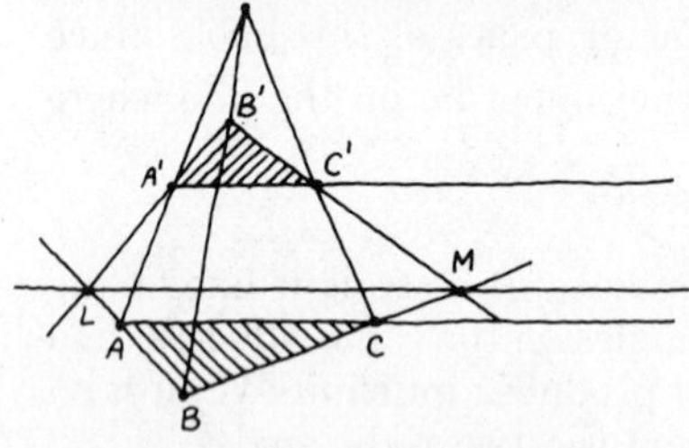

(*a*) What can you say about the lines AC, $A'C'$, and LM in terms of parallels? (*b*) If you were standing at point L, how would lines AC, $A'C'$, and LM appear to you in terms of perspective? (*c*) What would be the name that artists give to the "point" N? [The astronomer Johann Kepler (1571–1630) called such a point a "point at infinity."]

3
Absolute Geometry

The twenty-ninth theorem of Euclid is a turning point of the *Elements*, for it is there that he first uses the complicated fifth postulate. For this reason, theorems which use only the first four postulates and the five common notions are said to belong to *absolute geometry*, and theorems which use the fifth postulate belong to *Euclidean geometry*. First we will look at four theorems of absolute geometry.

Two triangles are called *congruent* if they are equal, that is, if each angle of the first is equal to an angle of the second, and if each side of the first is equal to the corresponding side of the second. A theorem of absolute geometry says that two triangles are congruent if *two sides and the included angle* of the triangles, that is, the angle formed by the two sides, are equal.

Suppose the triangles are ABC and DEF, that the sides $AB = DE$ and $AC = DF$, and that the included angles $\measuredangle BAC = \measuredangle EDF$.

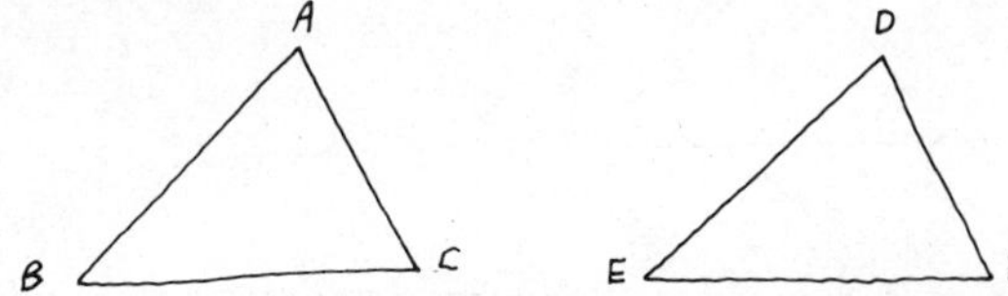

If point A is placed on point D so that AB coincides with DE, then point B will coincide with point E since $AB = DE$. Also, point C will coincide with point F since $\measuredangle BAC = \measuredangle EDF$ and $AC = DF$. Since B coincides with E and C with F, BC must coincide with EF. There are several hidden assumptions in this proof, but in particular we observe that it is here Euclid makes the assumption that there is just one line on any two points. If this assumption is allowed, then $BC = EF, \measuredangle ABC = \measuredangle DEF$, and $\measuredangle ACB = \measuredangle DFE$, so the two triangles are congruent.

An *isosceles* triangle is a triangle which has two of its sides equal. If a triangle is isosceles, that is, if it has two of its sides equal, then the angles opposite these sides are also equal. This very famous statement is known as *pons asinorum*, quite literally "the bridge of asses" in Latin. H. S. M. Coxeter[5] explains.

[5] *Introduction to Geometry* (New York: John Wiley & Sons, Inc., 1961), p. 6.

The name pons asinorum for this famous theorem probably arose from the bridgelike appearance of Euclid's figure (with the construction lines required in his rather complicated proof) and from the notion that anyone unable to cross the bridge must be an ass. Fortunately, a far simpler proof was supplied by Pappus of Alexandria about 340 A.D.

One reason that Euclid's figure and proof are complicated is that his statement of the theorem includes the equality of other pairs of angles besides those opposite the equal sides of the isosceles triangle. For the figure to which Coxeter refers, and the proof as Euclid wrote it, you may refer to Heath's translation of the *Elements.*

In his proof mentioned by Coxeter, Pappus appears to use only the congruent triangle theorem. Suppose that ABC is an isosceles triangle with equal sides $AB = AC$. We may think of the one triangle as two triangles, where the second is the original reversed.

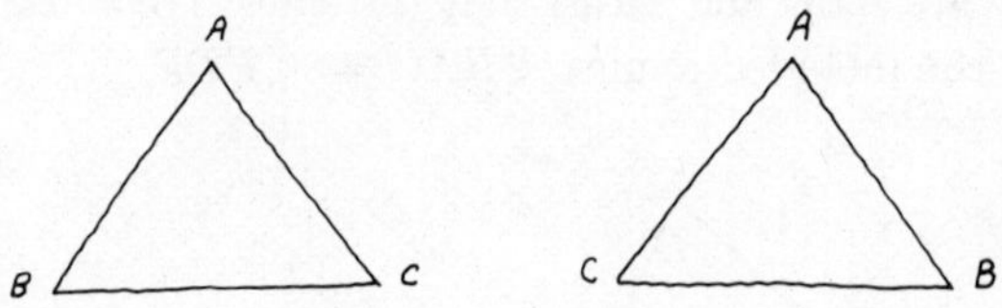

Since $AB = AC$ and $AC = AB$, two sides of the two triangles are equal, respectively, as are the included angles, $\measuredangle BAC$ and $\measuredangle CAB$, since they are the same angle. Therefore, the two triangles are congruent and all their corresponding parts are equal. In particular, $\measuredangle ABC = \measuredangle ACB$. We observe that this proof uses a hidden assumption that the triangle can be reversed without changing its size or shape.

The isosceles triangle theorem may be stated in the following form: If a triangle has two sides equal, then the angles opposite the two sides are equal. The *converse* of a theorem is another theorem which reverses the if and then parts (Chapter 3, Section 4). The converse of the isosceles triangle theorem is: If the angles opposite two sides of a triangle are equal, then the two sides are equal.

Suppose that ABC is a triangle with two equal angles, $\measuredangle ABC = \measuredangle ACB$. If AB is not equal to AC, then one of them is larger than the other. Suppose AB is the larger and place D on AB so that $BD = AC$.

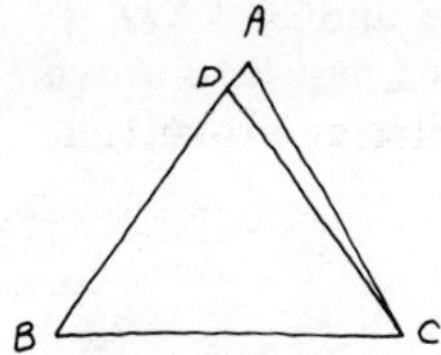

Since $BD = AC$, BC equals itself, and the included angles $\sphericalangle DBC = \sphericalangle ACB$, triangle DBC is congruent to triangle ACB. But triangle DBC is part of triangle ACB, so it cannot be equal to triangle ACB by common notion 5. Therefore AB cannot be larger than AC. Similarly, AC cannot be larger than AB. Thus AB must be equal to AC. This proof is a type of indirect proof called *reductio ad absurdum*, Latin for "reduction to the absurd."

EXERCISE 4 A third proof of the isosceles triangle theorem may be given if it is allowed that we can *bisect* an angle, that is, divide it into two equal angles. Suppose that ABC is an isosceles triangle with $AB = AC$, and let D be placed on BC so that AD bisects $\sphericalangle BAC$.

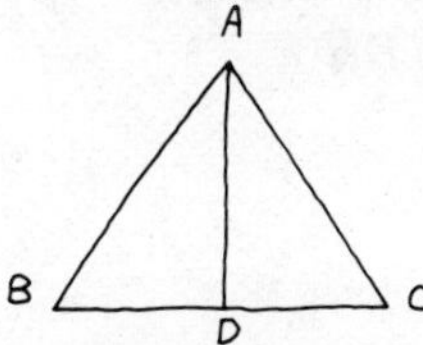

(*a*) List the parts of the two triangles ABD and ACD which are known to be equal from the information given. (*b*) Use the congruent triangle theorem to prove that $\sphericalangle ABD = \sphericalangle ACD$.

EXERCISE 5 Another theorem of absolute geometry says that two intersecting lines have vertical angles equal. Suppose the lines AB and CD meet in the point E.

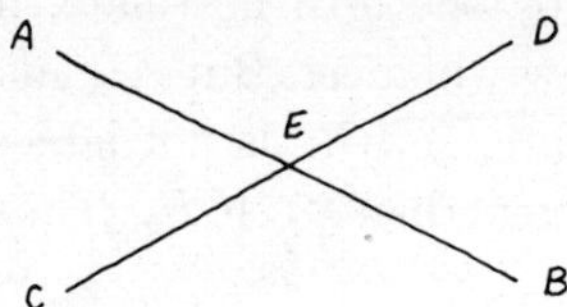

Observe that $\measuredangle AEC + \measuredangle AED =$ two right angles and $\measuredangle AED + \measuredangle DEB =$ two right angles. (*a*) Use common notions 1 and 3 to prove that the vertical angles $\measuredangle AEC = \measuredangle DEB$. (*b*) Similarly, prove that the vertical angles $\measuredangle AED = \measuredangle BEC$.

4
Euclidean Geometry

We turn now to Euclid's twenty-ninth theorem, the first to use the fifth postulate. All theorems of absolute geometry are also theorems of Euclidean geometry. Because it uses the fifth postulate, however, the twenty-ninth theorem is a theorem of Euclidean geometry but not of absolute geometry. The most famous part of Euclid's twenty-ninth theorem is: If two parallel lines are cut by a transversal, then the alternate interior angles are equal. Suppose lines AB and CD are parallel. A line such as EF which crosses both the parallel lines is a *transversal*. *Alternate interior angles* are angles between the parallel lines and on opposite sides of the transversal, such as $\measuredangle AGH$ and $\measuredangle DHG$.

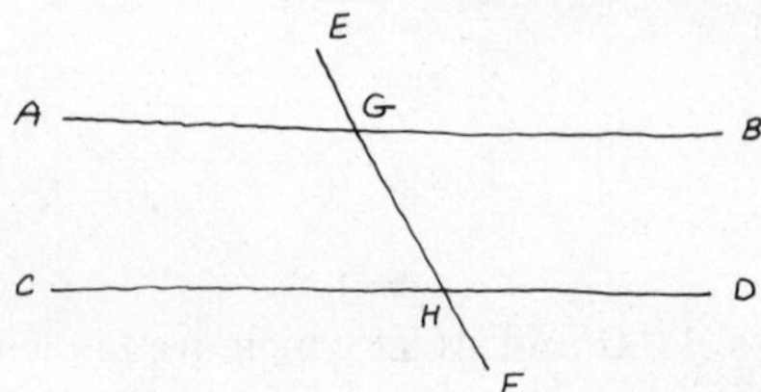

The proof which uses the fifth postulate is another proof by *reductio ad absurdum*. Suppose that $\measuredangle AGH$ is not equal to $\measuredangle DHG$. For instance, suppose $\measuredangle AGH$ is larger than $\measuredangle DHG$. We add $\measuredangle BGH$ to each. Then $\measuredangle AGH + \measuredangle BGH$ is equal to two right angles. Therefore, $\measuredangle DHG + \measuredangle BGH$ is less than two right angles. Thus, by the fifth postulate, if AB and CD are extended beyond B and D, they will meet. But AB and CD are parallel, so they cannot meet. Therefore, $\measuredangle AGH$ is not larger than $\measuredangle DHG$. Similarly, $\measuredangle DHG$ is not larger than $\measuredangle AGH$. Thus $\measuredangle AGH = \measuredangle DHG$.

Finally, we use the parallel theorem to prove the well-known theorem of Euclidean geometry that the sum of the angles of a triangle is equal to two right angles. Suppose that ABC is any triangle, and extend side BC to D. Draw CE through the point C and parallel to AB. We observe that the construction of line CE uses the Playfair postulate.

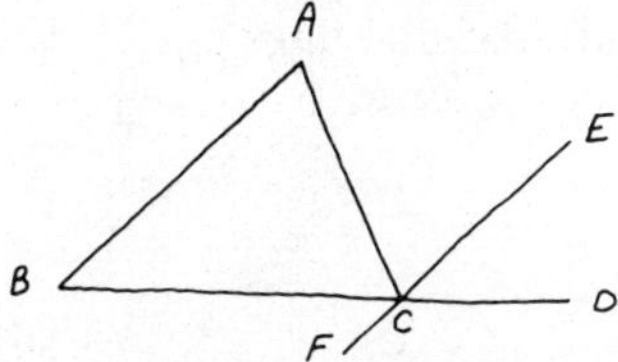

AB is parallel to CE and AC is a transversal; therefore, by the parallel theorem, $\measuredangle BAC = \measuredangle ACE$. Also, BC is a transversal, so $\measuredangle ABC = \measuredangle FCB$. But $\measuredangle FCB = \measuredangle ECD$; therefore, $\measuredangle ABC = \measuredangle ECD$. $\measuredangle ACD = \measuredangle ACE + \measuredangle ECD$, thus $\measuredangle ACD = \measuredangle BAC + \measuredangle ABC$. But $\measuredangle ACD + \measuredangle ACB =$ two right angles, so $\measuredangle BAC + \measuredangle ABC + \measuredangle ACB =$ two right angles.

EXERCISE 6 Each of the following statements is true either in both absolute and Euclidean geometries or in Euclidean geometry only. Say which ones are statements of absolute and Euclidean geometries and which are Euclidean only. (*a*) A line can be drawn on any two points. (*b*) All right angles are equal to one another. (*c*) If equals be subtracted from equals, the remainders are equal. (*d*) Given a line and a point not on the line, there is exactly one line through the point parallel to the line. (*e*) If two sides of a triangle are equal, then the angles opposite those sides are equal. (*f*) The sum of the angles of a triangle is equal to two right angles. (*g*) If two lines intersect, then the vertical angles are equal. (*h*) If two parallel lines are cut by a transversal, then the alternate interior angles are equal.

EXERCISE 7 Two lines are *perpendicular* if the angle at which they meet is a right angle. Suppose that AB and CD are parallel, and GH is perpendicular to CD.

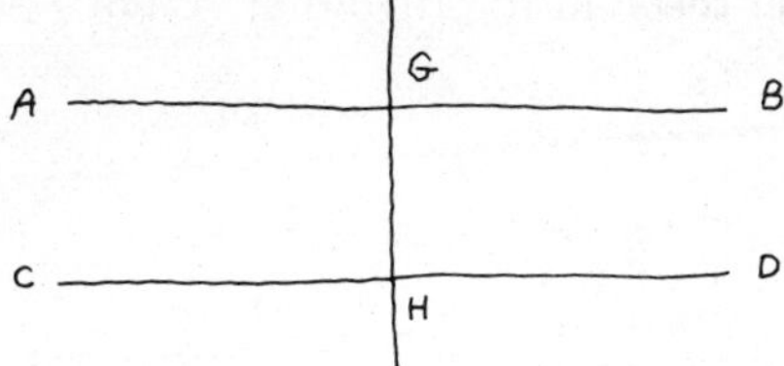

(*a*) Why is $\measuredangle AGH$ a right angle? (*b*) Why is $\measuredangle BGH$ a right angle? Observe that if a line is perpendicular to one of two parallel lines, then it is also perpendicular to the other.

EXERCISE 8 Suppose that AB and CD are parallel lines cut by the transversal EF.

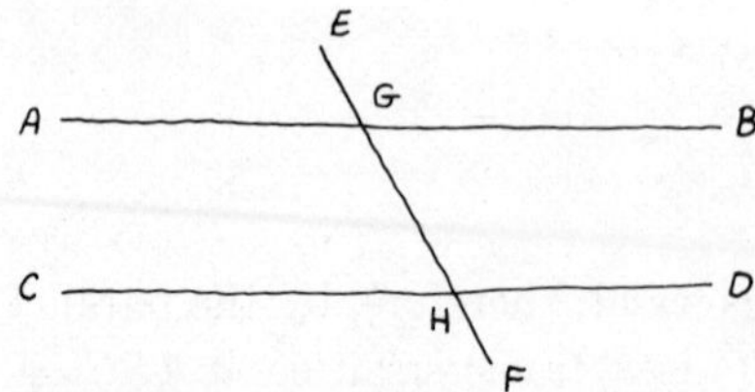

(*a*) The angles $\measuredangle BGH$ and $\measuredangle CHG$ are also alternate interior angles. Prove $\measuredangle BGH = \measuredangle CHG$ using common notion 3 and the fact that $\measuredangle AGH = \measuredangle DHG$. (*b*) A second part of the twenty-ninth theorem says that the exterior angle is equal to the opposite interior angle. Prove, for example, that $\measuredangle EGB = \measuredangle DHG$. (*c*) The last part of the twenty-ninth theorem says that interior angles on the same side of the transversal equal two right angles. Prove, for example, that $\measuredangle BGH + \measuredangle DHG =$ two right angles.

EXERCISE 9 A proof similar to that in the text, that the sum of the angles of a triangle is equal to two right angles, is credited to Eudemus (about 335 B.C.). Suppose that ABC is any triangle and draw DE through A and parallel to BC. (*a*) What postulate is assumed in the construction of line DE?

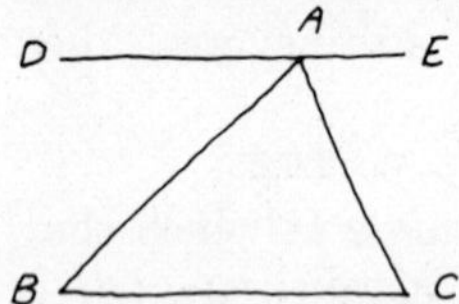

(*b*) Use this diagram and the parallel theorem to prove that $\measuredangle ABC + \measuredangle BAC + \measuredangle ACB =$ two right angles.

5
The Fifth Postulate

At about the same time that Desargues and Pascal were studying the problem of perspective, an amateur in mathematics was studying the fifth postulate. Because this postulate was so obviously different from and more complex than the other assumptions of Euclid, for centuries mathematicians tried to prove it as a theorem. If it were a theorem, it could be proved from the other postulates and the common notions.

DEFINITION A statement is *dependent* on a set of statements if it can be proved or disproved using only those statements and statements proved from those statements.

The theorem that the sum of the angles of a triangle is equal to two right angles is dependent on the Playfair postulate and also on Euclid's parallel theorem, which are in turn dependent on Euclid's fifth postulate and some of the other postulates and common notions. Even the earliest commentators on the *Elements* thought that the fifth postulate itself might be dependent on the other postulates and the common notions.

One of the first attempts to prove the fifth postulate to be dependent on the other assumptions of Euclid was made by Claudius Ptolemy (about 150 A.D.). As reported by Proclus, Ptolemy started with two parallel lines AB and CD, and a transversal FG. The angles $\sphericalangle AFG + \sphericalangle FGC$ will be either larger than, smaller than, or equal to two right angles. Ptolemy assumes that whichever case holds for one pair of parallels will also hold for every other pair.

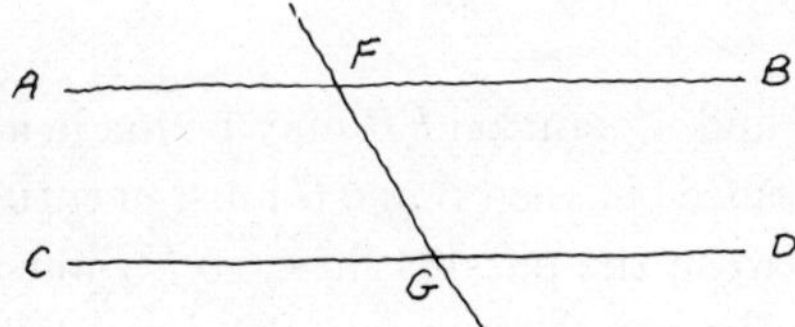

FA and GC are one pair of parallels. Suppose that $\sphericalangle AFG + \sphericalangle FGC$ is larger than two right angles. Then FB and GD are another pair of parallels, and the same case holds, so $\sphericalangle BFG + \sphericalangle FGD$ is also larger than two right angles. But then $\sphericalangle AFG + \sphericalangle FGC + \sphericalangle BFG + \sphericalangle FGD$ is larger than four right angles, which is absurd. Similarly, $\sphericalangle AFG + \sphericalangle FGC$ cannot be smaller than two right angles. Therefore, Ptolemy concludes, $\sphericalangle AFG + \sphericalangle FGC$ is equal to two right angles, from which the fifth postulate can be proved. However, Ptolemy's assumption that one case always holds, even on the two sides of a transversal, is itself absurd.

Proclus attempted a proof of the fifth postulate, assuming that the distance between two points on two intersecting lines may be as great as one pleases by extending the lines sufficiently.

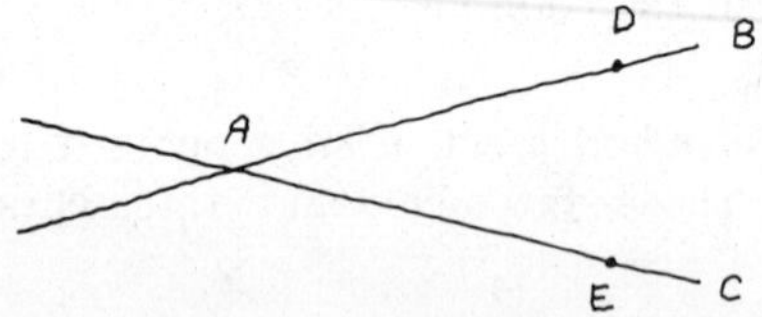

On lines AB and AC, the distance between points D and E can be made as large as one pleases by extending AB and AC beyond B and C and moving D and E out along the lines. This assumption is not dependent on the fifth postulate and can be proved as a theorem of absolute geometry.

Using this assumption, Proclus then attempts to prove that a line which meets one of two parallels must also meet the other. Suppose that AB and CD are parallel lines and that EG is a line which meets AB at F.

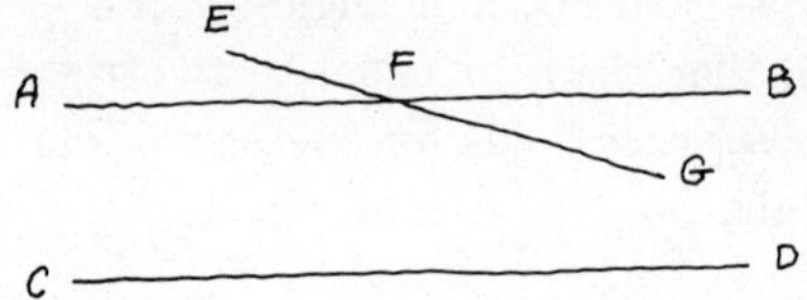

The distance between a point on FG and a point on FB may be made as large as we please. Therefore, the distance between B and G must eventually be greater than the distance between the parallel lines, so EG must meet CD. The argument appears to be convincing, but Proclus introduces a second assumption: that the distance between parallel lines is always a finite amount. This assumption, it turns out, can only be proved using the fifth postulate.

An ambitious attempt to prove the fifth postulate dependent on the other assumptions was made by Girolamo Saccheri (1667–1733) in his book *Euclides ab omni naevo vindicates* (Euclid Freed of Every Flaw). Saccheri, an Italian Jesuit priest, was fascinated by the method of *reductio ad absurdum*. Others had tried to prove the fifth postulate directly from the other assumptions, perhaps using *reductio ad absurdum* in the process. Saccheri applied *reductio ad absurdum* to the entire problem. He assumed the fifth postulate itself to be false, and from that assumption tried to reach an absurdity.

Saccheri used a two-right-angled isosceles quadrilateral, now called the *Saccheri quadrilateral*. The Saccheri quadrilateral has two adjacent right angles, and two equal sides containing those angles.

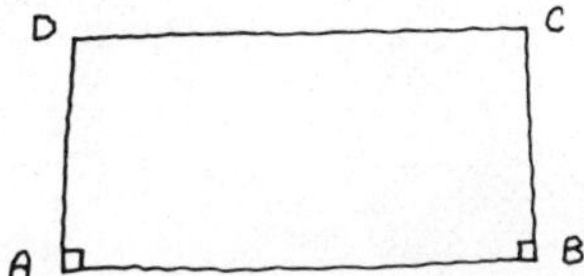

In the Saccheri quadrilateral $ABCD$, the small squares in $\measuredangle DAB$ and $\measuredangle CBA$ indicate that these angles are right angles. The equal sides are then sides AD and BC. From this figure it can be proved as a theorem of absolute geometry, without using the fifth postulate, that the angles $\measuredangle ADC = \measuredangle BCD$. It cannot be proved as a theorem of absolute geometry, however, that $\measuredangle ADC$ and $\measuredangle BCD$ are also right angles. This proof depends on the fifth postulate, and so belongs to Euclidean geometry. There are three possibilities for the angles $\measuredangle ADC$ and $\measuredangle BCD$ which Saccheri named:

1 *The hypothesis of the right angle*: The angles are both right angles.
2 *The hypothesis of the obtuse angle*: The angles are equal and each more than a right angle.
3 *The hypothesis of the acute angle*: The angles are equal and each less than a right angle.

Under either of the second two hypotheses, the fifth postulate would be false.

Saccheri succeeded relatively easily in eliminating the hypothesis of the obtuse angle. It is important, however, that in so doing he used the assumption that a line is infinite in extent. From the hypothesis of the acute angle, Saccheri derived a long series of theorems without reaching an absurdity. His final conclusions concern two lines which may be said to meet *at infinity*, a concept related to the one for which we earlier used the artist's term of a vanishing point. Saccheri proved that the lines would then have a common perpendicular at infinity. On this basis, he dismissed the hypothesis of the acute angle. In his anxiety to eliminate the cases which did not agree with the fifth postulate, Saccheri failed to realize that his theorems were the basis of a whole new geometry.

EXERCISE 10 Consider the Saccheri quadrilateral. Suppose the point E bisects the side AB of the Saccheri quadrilateral; that is, point E is placed on AB so that $AE = EB$.

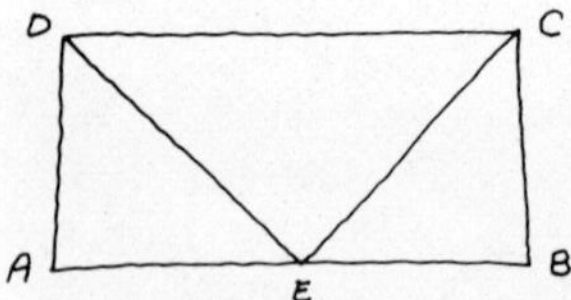

(*a*) Prove that triangle ADE is congruent to triangle BCE. Thus $\measuredangle ADE = \measuredangle BCE$. Also, $DE = CE$. (*b*) What can you conclude about $\measuredangle CDE$ and $\measuredangle DCE$? (*c*) How can you conclude that $\measuredangle ADC = \measuredangle BCD$ without using the fifth postulate?

EXERCISE 11 Consider the "proof" of Pappus. Suppose the fifth postulate is not assumed; that is, AB is not necessarily the only line through F parallel to CD.

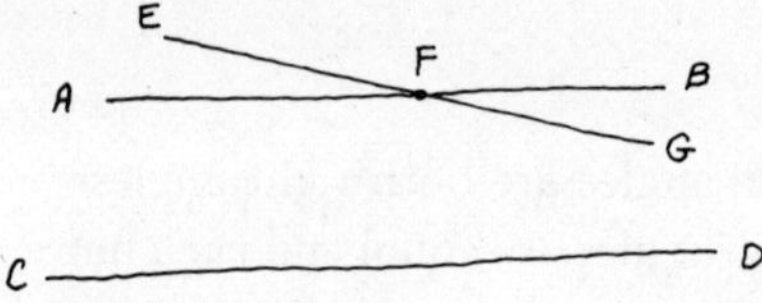

What could be true of line EG in terms of the geometry of perspective?

EXERCISE 12 Consider the "proof" of Saccheri. Suppose lines EG and CD meet at infinity; that is, at a vanishing point.

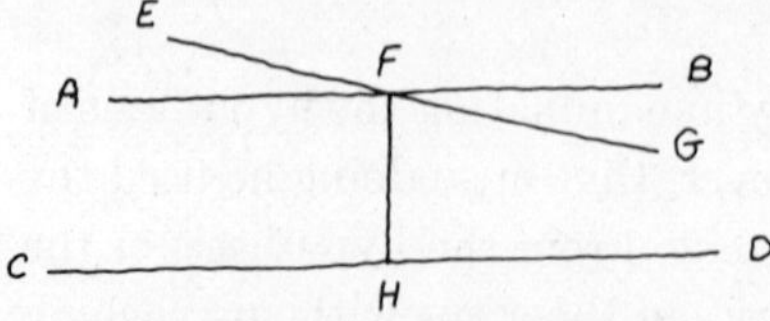

(a) If FH is perpendicular to CD and to AB, can FH also be perpendicular to EG? Suppose you are standing at point H looking out toward D. (b) If the lines EG and CD do not cross, how would they appear to you? (c) If a line like FH were drawn at infinity, how would it appear relative to EG and CD?

6

Lobachevskian Geometry

Realization that a consistent geometry can be created from Saccheri's hypothesis of the acute angle took until the nineteenth century. In that century it was discovered by three different mathematicians. The first to begin construction of a geometry other than Euclid's was the great German mathematician Gauss, who gave the construction the name *non-Euclidean geometry*. Gauss did not publish his work, however, and told only a few trusted friends about it.

Johann Bolyai (1802–1860), a Hungarian mathematician whose father was also a mathematician and a close friend of Gauss, was another who discovered the new geometry. Bolyai used the Playfair form of the fifth postulate in an approach similar to Saccheri's. To deny the Playfair postulate would be to assume that through a given point there is either no parallel to a given line or more than one parallel to a given line. Bolyai disposed of the first case, that of no parallel, as easily as Saccheri had that of the hypothesis of the obtuse angle. The second, that of more than one parallel or, in particular, two parallels, led to a series of theorems as did Saccheri's hypothesis of the acute angle. Bolyai realized, however, that he had found a new geometry. He wrote to his father, "I have created a new universe from nothing."

Bolyai published his new geometry as an appendix to a work on geometry published by his father in 1832–1833. Unfortunately for Bolyai, a Russian mathematician, Nicolai Ivanovitch Lobachevsky (1793–1856), had published a similar new geometry in Russian in 1829–1830. Lobachevsky's work did not spread to western Europe until the 1840s, when he published it again in German, and it finally came to Bolyai's attention. Although the geometry is sometimes called the geometry of Gauss-Lobachevsky-Bolyai, in honor of all three discoverers, it is more

David Eugene Smith Collection, Columbia University Library

The discovery of non-Euclidean geometry is considered to be one of the great turning points in the history of mathematics. Mathematics was no longer merely a description of the physical world. The discovery opened a world of theoretical mathematics in which a system need only be free of contradictions within itself. That Gauss' genius thought of such a theoretical world is not too surprising. That the same world was created simultaneously far across Russia is remarkable.

Nicolai Ivanovitch Lobachevsky spent his entire life at the University of Kazan, hundreds of miles east of Moscow, as student, professor, dean, and rector. Founded just two years before Lobachevsky entered as a student in 1807, Kazan was far from the great European centers of learning in more than distance. Lobachevsky had the support of the professors for his good administration of the University, but his new geometry, which he published twice between 1829 and 1837, was ignored throughout Russia. He published it in German in 1840, was elected a foreign correspondent of the Royal Society of Göttingen through the influence of Gauss, and his fame spread throughout Europe.

often shortened to *Lobachevskian geometry* in honor of the first person to publish it. An even more common name is *hyperbolic geometry*, a name coined many years later.

In Lobachevskian geometry there are two lines through a given point parallel to a given line.

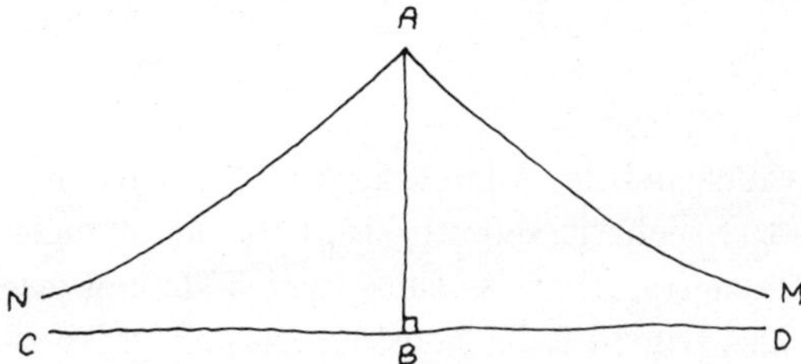

Suppose AM and AN are both parallel to CD. If we bisect $\measuredangle NAM$, the line AB is perpendicular to CD. We see that $\measuredangle BAM$ is less than a right angle. The size of $\measuredangle BAM$ depends on the length of AB. If AB is small, $\measuredangle BAM$ is close to a right angle.

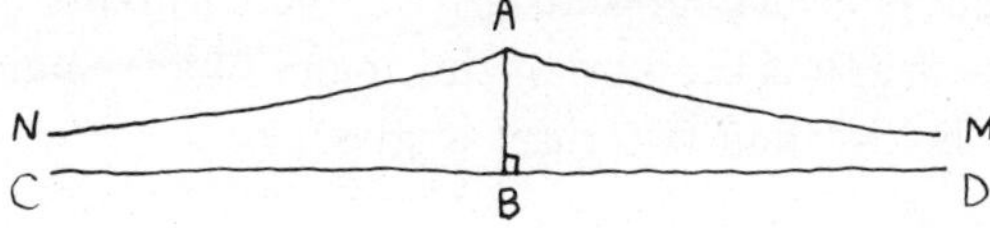

If AB is large, $\measuredangle BAM$ is close to zero.

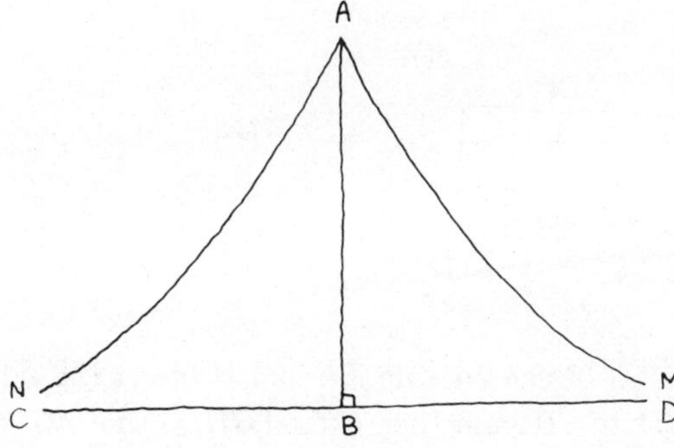

Lobachevsky called $\measuredangle BAM$ the *angle of parallelism* corresponding to the length AB. Any line through A which makes an angle with AB greater than the angle of parallelism is called an *ultraparallel;* thus, if $\measuredangle BAP$ is greater than $\measuredangle BAM$, then AP is an ultraparallel to CD.

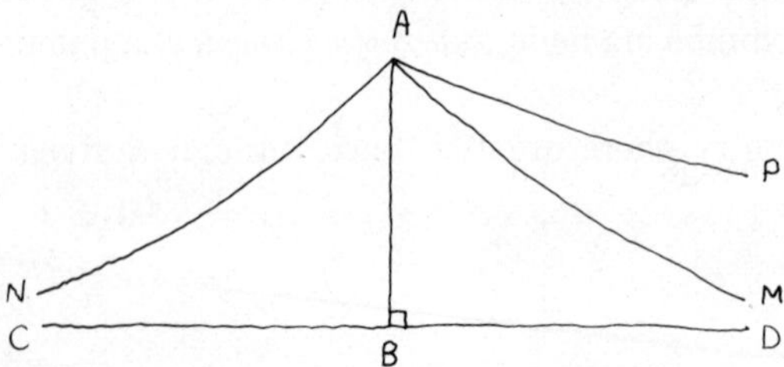

Because the angle of parallelism is less than a right angle, Lobachevskian geometry corresponds to Saccheri's hypothesis of the acute angle. All theorems of absolute geometry, that is, those which do not use Euclid's fifth postulate, are also true in Lobachevskian geometry. Theorems depending on the fifth postulate, however, are either untrue or different in Lobachevskian geometry. For example, in Lobachevskian geometry, the sum of the angles of a triangle is not equal to but less than two right angles.

If the correspondence between Lobachevskian geometry and the hypothesis of the acute angle is accepted, and also all the theorems of absolute geometry, we can prove that the sum of the angles of a triangle in Lobachevskian geometry is less than two right angles.

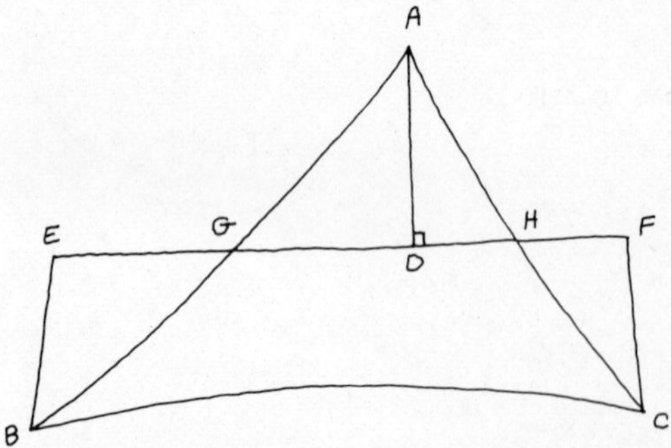

In the triangle ABC, let G bisect the side AB and H bisect the side AC. Draw AD perpendicular to GH, and then extend GH so that $EG = GD$ and $FH = HD$. Draw EB and FC. Then in triangle AGD and triangle BGE, $AG = GB$, $EG = GD$, and the included angles $\measuredangle AGD = \measuredangle BGE$, since they are vertical angles (Exercise 5); therefore, triangle AGD is congruent to triangle BGE. Similarly, triangle ADH is congruent to triangle CFH. Consequently, $\sphericalangle BEG = \sphericalangle ADG$ and $\sphericalangle CFH = \sphericalangle ADH$, so the angles $\sphericalangle BEG$ and $\sphericalangle CFH$ are also right angles. Furthermore, $EB = AD$ and $CF = AD$, so $EB = CF$. We thus recognize the quadrilateral $EBCF$ as a Saccheri quadrilateral. Since Lobachevskian geometry corresponds to the hypothesis of the acute angle, $\sphericalangle EBC + \sphericalangle FCB$ is less than two right angles, for each angle is less than a right angle. But $\sphericalangle EBG = \sphericalangle DAG$ and $\sphericalangle FCH = \sphericalangle DAH$, so $\sphericalangle BAC + \sphericalangle ABC + \sphericalangle ACB = \sphericalangle EBG + \sphericalangle FCH + \sphericalangle ABC + \sphericalangle ACB$, which is less than two right angles.

EXERCISE 13 Each of the following statements is true in either Euclidean or Lobachevskian geometry, but not in both. Say which ones are Euclidean and which ones are Lobachevskian. (*a*) There is exactly one line through a given point parallel to a given line. (*b*) There are two lines through a given point parallel to a given line. (*c*) The angle of parallelism is less than a right angle. (*d*) The Saccheri hypothesis of the right angle is true. (*e*) The Saccheri hypothesis of the acute angle is true. (*f*) The sum of the angles of a triangle is equal to two right angles. (*g*) The sum of the angles of a triangle is less than two right angles. (*h*) If two parallel lines are cut by a transversal, the alternate interior angles are equal.

EXERCISE 14 Suppose that AB is perpendicular to CD, and also that AP is perpendicular to AB.

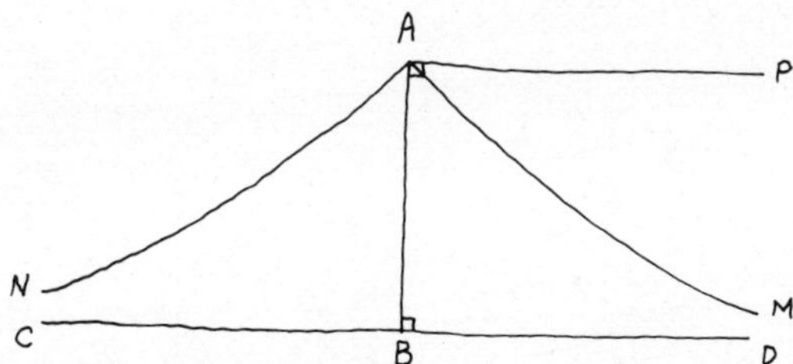

(*a*) Why must $\sphericalangle BAM$ be less than $\sphericalangle BAP$? (*b*) Why must AP be an ultraparallel to CD? (*c*) Why is it not true in Lobachevskian geometry that if two parallel lines are cut by a transversal, then the alternate interior angles are equal? (*d*) Why is it not true in Lobachevskian geometry that if a line is perpendicular to one of two parallel lines, it is perpendicular to the other? (*e*) If a line is perpendicular to two lines, what is the relationship of the two lines in Lobachevskian geometry?

EXERCISE 15 (*a*) A kind of triangle called an *asymptotic triangle* exists in Lobachevskian geometry although not in Euclidean geometry. An asymptotic triangle in Lobachevskian geometry has two of its sides parallel. The "angle" at M is considered to be zero.

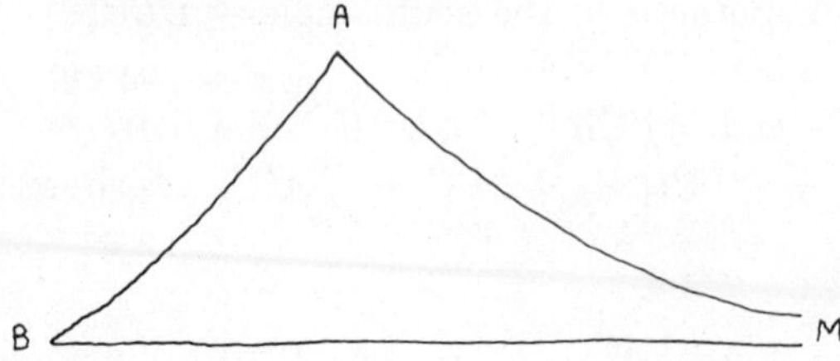

Explain why $\measuredangle ABM + \measuredangle MBA$ is less than two right angles. (*b*) A triangle with two sides both parallel to the third side is said to be *doubly asymptotic.*

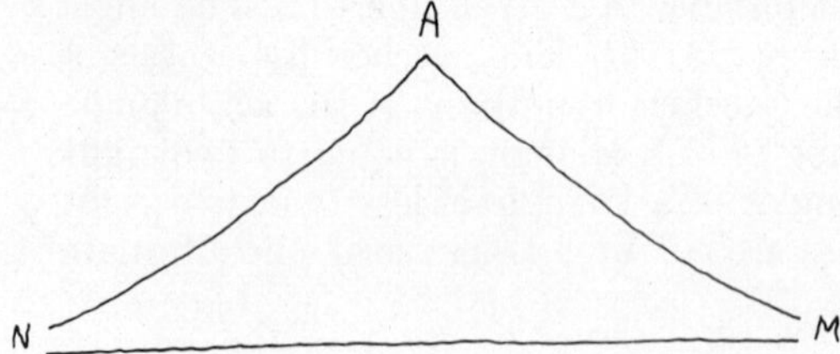

Explain why $\measuredangle NAM$ is less than two right angles. (*c*) A triangle with all its sides parallel is *trebly asymptotic.*

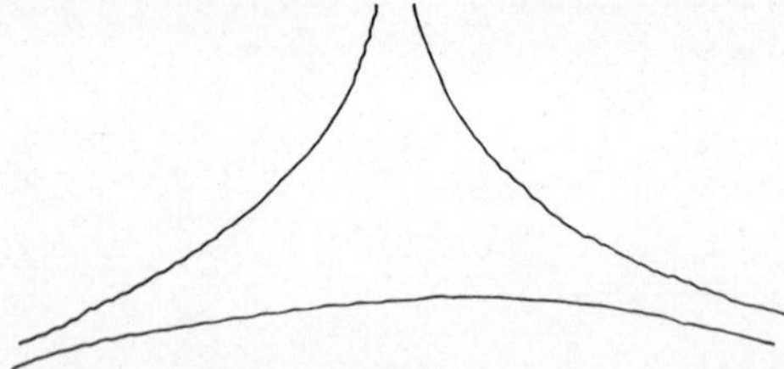

What is the sum of the angles of a trebly asymptotic triangle?

You may have found it disturbing that we must use "curved" lines to draw Lobachevskian "straight" lines. The reason is that we are making our drawings in a Euclidean plane and not a Lobachevskian plane. We think of the Euclidean plane as a flat surface, such as the surface of this page, and say that it has *zero curvature.*

To construct a surface which may represent the Lobachevskian plane, we imagine that W is a weight attached to P by a string.

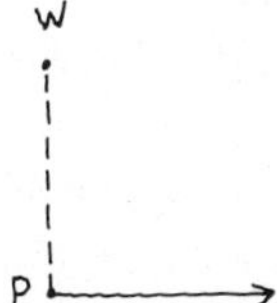

If we pull P in a direction perpendicular to WP, the resulting curve traced by W is called a *tractrix*.

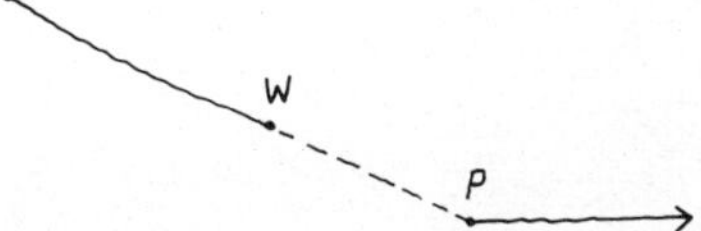

We construct a similar tractrix in the opposite direction, and revolve them both around the line along which P was pulled. The resulting surface is called a *pseudosphere*. The pseudosphere has *negative curvature*. It represents a finite part of the *Lobachevskian plane* in the same way that this page represents a finite part of the Euclidean plane. The pseudosphere shown here has a doubly asymptotic triangle AMN drawn on it.

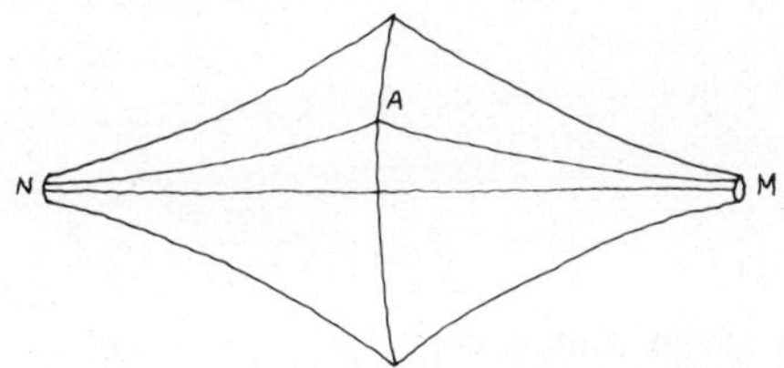

7
Riemannian Geometry

Saccheri eliminated the hypothesis of the obtuse angle and Bolyai eliminated the case in which there is no parallel to a given line through a given point by assuming that a line is infinite in extent. Bernhard

Riemann (1826–1866) realized that there is a difference between *infinite* and *unbounded.* Thus Riemann was able to construct a third geometry, different from both Euclidean and Lobachevskian, which corresponds to the case in which there is no parallel to a given line through a given point, and also to the hypothesis of the obtuse angle.

Euclid's second postulate says that a straight line can be produced indefinitely. We have already seen that in a non-Euclidean geometry a "straight" line may appear curved when drawn on the Euclidean plane. A circle in the Euclidean plane may be produced indefinitely simply by going round and round it.

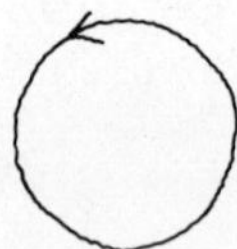

In this sense, the circle is *unbounded.* The circle is clearly *finite in extent,* however.

An ordinary sphere has *positive curvature,* and represents the *Riemannian plane.* The lines in *Riemannian geometry* are the great circles of the sphere, the circles which bisect the sphere.

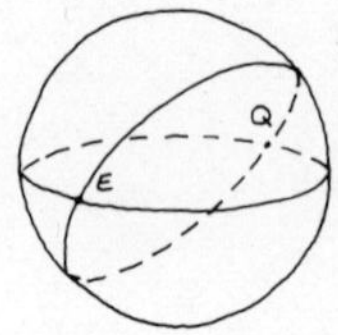

No line can be drawn parallel to a given line. Every two lines meet in at least one point; in fact, in the two points P and Q at opposite poles of the sphere.

We recall that Euclid and others who investigated the problem of the fifth postulate assumed as part of the second postulate that a line is infinite in extent, although that is not stated in the postulate. Euclid and the others also assumed that there is only one line on two given points, although the first postulate says merely that there exists a line on two given points. Thus, it is possible to have a second line, or as many lines as we please, on two points P and Q. Furthermore, the lines need not be infinite in extent as long as they are unbounded. Rie-

Historical Pictures Service, Chicago

Bernhard Riemann spent his life in the heart of the intellectual world of nineteenth century Europe. He studied at the University of Berlin, and presented his doctoral dissertation at the University of Göttingen under the direction of Gauss. Gauss also chose the subject for Riemann's probationary lecture, a memorable paper on the foundations of geometry, delivered in 1854.

This paper described far more than what is presented as Riemannian geometry in this section. It took up the entire concept of space, including dimension and curvature. Although Riemann specifically described spaces of positive curvature and also zero curvature (Euclidean), his theory is general enough to encompass the negative curvature of Lobachevskian space as well. It is on the basis of Riemann's concepts of space that the pseudosphere was discovered to be a model of Lobachevskian geometry by Eugenio Beltrami (1835–1900). It is on the mathematics of Riemann's general mathematical theory that Einstein based his general theory of relativity (Section 9).

mann realized that the great circles of a sphere could be interpreted as lines on a surface of positive curvature without violating the basic postulates.

In Riemannian geometry the sum of the angles of a triangle is greater than two right angles. Suppose, for example, we take the north pole to be A, and draw BC on the equator of a sphere patterned after the earth.

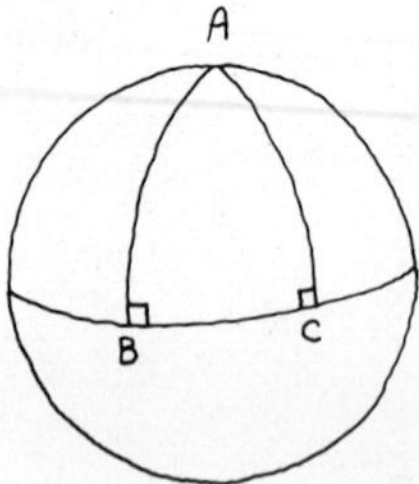

Then $\measuredangle ABC$ and $\measuredangle ACB$ are both right angles. Thus $\measuredangle ABC + \measuredangle ACB + \measuredangle BAC$ is greater than two right angles. Riemannian geometry then may be assumed to correspond to the Saccheri hypothesis of the obtuse angle.

Riemann completed the work begun by Gauss, Lobachevsky, and Bolyai, in finally proving that the fifth postulate is completely independent of the other assumptions, and that a new and consistent geometry can be constructed from each case of its denial. We get the three distinct geometries, Euclidean, Lobachevskian, and Riemannian.

The variation of Riemannian geometry we constructed on the sphere is called *spherical geometry*, which was studied well before the time of Riemann. It was not connected with a case of the fifth postulate in earlier studies, because it has two properties which do not correspond to other assumptions of Euclid. First, as we have seen, every two lines, that is, great circles on the sphere, meet not once but twice in opposite poles. Second, two points determine not only one line but any number of lines if the two points happen to be opposite poles.

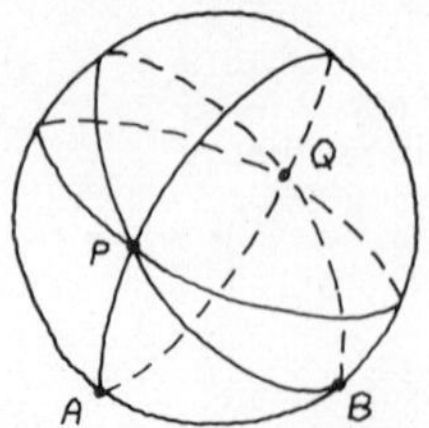

Although these properties do not violate the stated postulates of Euclid, they do violate some of the subtler assumptions made in absolute geometry, in particular, in the proof of the congruent triangle theorem.

Felix Klein (1849–1929) discovered a way to construct a different variation of Riemannian geometry which does correspond to Euclid's assumptions in the two ways given above. In Klein's geometry, every two lines meet in exactly one point, and every two points determine exactly one line. To construct Klein's geometry, we consider every pair of opposite poles to be the same point. Thus P and Q above are the same point, say P. The lines AP and BP meet in only the one point P, and the two points A and P determine exactly one line. We ignore lines meeting in Q and lines determined by P and Q, since P and Q are considered to be the same point.

Klein called his variation of Riemannian geometry *elliptic geometry*. At the same time he gave the names *hyperbolic geometry* to Lobachevskian geometry and *parabolic geometry* to Euclidean geometry. Today, the commonly used names are elliptic and hyperbolic geometry for the non-Euclidean geometries, and Euclidean (not parabolic) geometry for Euclid's geometry. All three of these geometries share the property that two points determine exactly one line, whereas spherical geometry does not. Euclidean and hyperbolic geometry share the property that a line is infinite in extent, whereas the line is finite in extent in both elliptic and spherical geometry.

EXERCISE 16 Each of the following statements is true in either Euclidean or Riemannian geometry, but not in both. Say which ones are Euclidean and which are Riemannian. (*a*) There is exactly one line through a given point parallel to a given line. (*b*) Every two lines meet in at least one point. (*c*) Lines are infinite in extent. (*d*) Lines are finite in extent but unbounded. (*e*) The Saccheri hypothesis of the right angle is true. (*f*) The Saccheri hypothesis of the obtuse angle is true. (*g*) The sum of the angles of a triangle is equal to two right angles. (*h*) The sum of the angles of a triangle is greater than two right angles. (*i*) There exist triangles which have two right angles. (*j*) The plane has positive curvature.

EXERCISE 17 (*a*) Suppose triangle ABC is a *right triangle* in Riemannian geometry; that is, triangle ABC has one right angle.

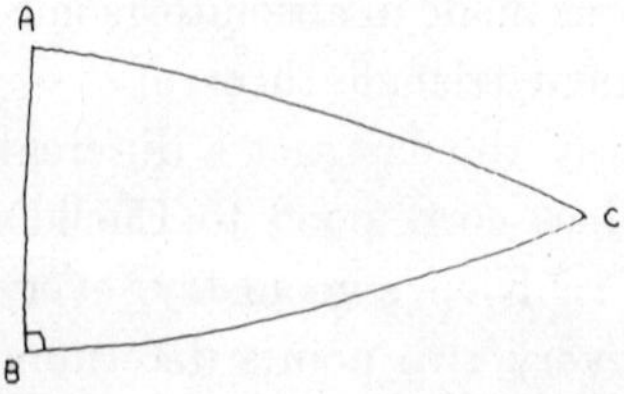

Explain why $\sphericalangle CAB + \sphericalangle BCA$ is greater than a right angle. (*b*) A *doubly right-angled triangle* exists in Riemannian geometry, although not in Euclidean geometry. It has two right angles. Suppose triangle ABC is a doubly right-angled triangle.

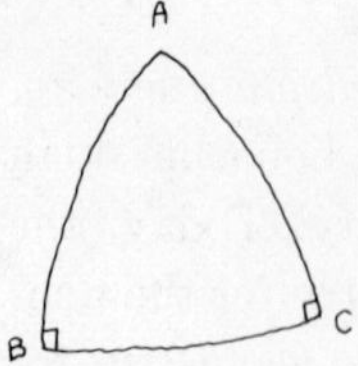

Explain why $\sphericalangle CAB$ must be greater than zero. (*c*) Suppose triangle APC is a doubly right-angled triangle in elliptic geometry.

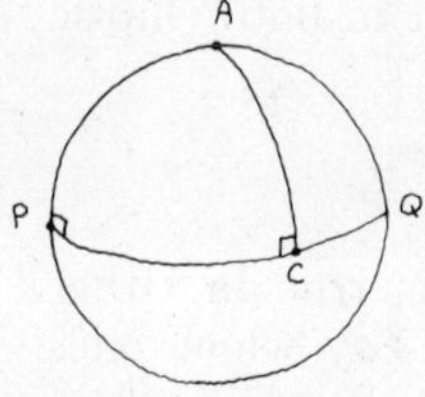

$\sphericalangle APC + \sphericalangle ACP + \sphericalangle PAC$ is greater than two right angles; but what amount must $\sphericalangle APC + \sphericalangle ACP + \sphericalangle PAC$ be less than?

EXERCISE 18 Complete this proof that the sum of the angles of a triangle in elliptic geometry is greater than two right angles:

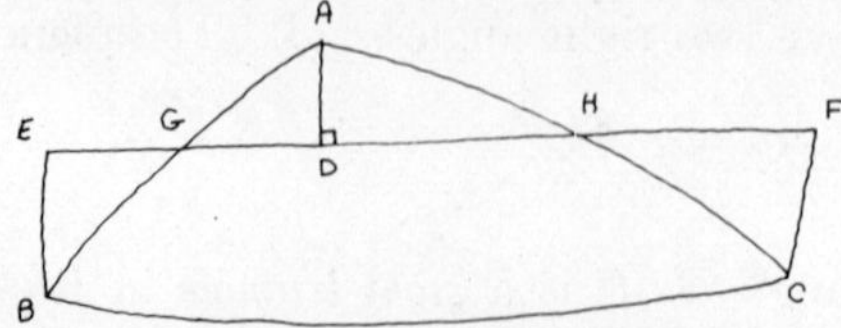

In the triangle ABC, let G bisect the side AB and H bisect the side AC. Draw AD perpendicular to GH. Extend GH so that $EG = GD$ and $FH = HD$. Draw EB and FC. (*a*) Prove that triangle AGD is congruent to triangle BGE. (*b*) Prove that triangle ADH is congruent to triangle CFH. (*c*) Why are angles $\measuredangle BEG$ and $\measuredangle CFH$ right angles? (*d*) Why is $EB = CF$? (*e*) Why is $EBCF$ a Saccheri quadrilateral? (*f*) Explain why, if elliptic geometry corresponds to the hypothesis of the obtuse angle, $\measuredangle BAC + \measuredangle ABC + \measuredangle ACB$ is greater than two right angles.

8

The Möbius Strip

We can discover some further properties of elliptic geometry by studying a surface called the *Möbius strip*. The Möbius strip is named for Augustus Ferdinand Möbius (1790–1868), a German mathematician. To construct a Möbius strip, you take a long, narrow piece of paper.

Twist the paper once and tape the ends, so that corner A is joined to corner D rather than B, and C to B.

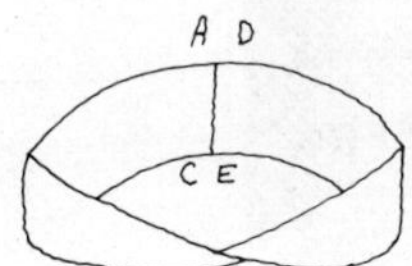

Start to draw a line around the outside of the strip. You will find that when you reach the starting point again, you are on the inside of the strip, but without having crossed the edge.

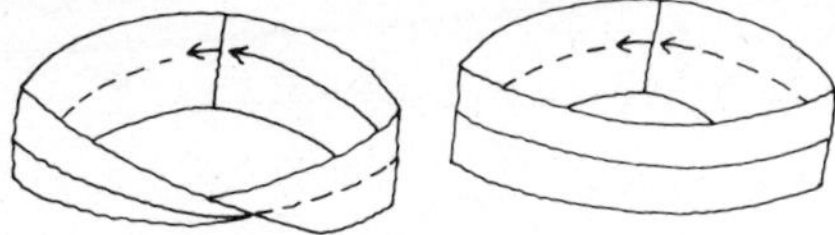

Escher Foundation, Haags Gemeentemuseum, The Hague

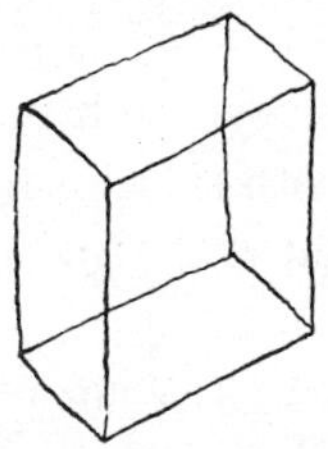

Does this drawing represent a two-dimensional or a three-dimensional figure? If three-dimensional, which edge is in front and which in back? The second question is represented by the figure on the piece of paper lying on the ground in Escher's Belvedere. The boy on the bench is contemplating a three-dimensional representation of the same question. Which is the front edge and which the back edge of those that appear to cross in the cubelike object he holds?

Now we examine the building itself. (We have gone in reverse order—it is to be hoped you looked first at the building, then the cubelike object, then the drawing.) Two people are climbing a ladder which has its foot inside the building but its top outside. The woman and the man at the far right are directly above one another and looking through identically oriented archways, but one looks toward us and the other away. As Escher says in The Graphic Work of M. C. Escher: "Is it any wonder that nobody in this company can be bothered about the fate of the prisoner in the dungeon who sticks his head through the bars and bemoans his fate?"

The building is of the same impossible construction as the cubelike object; possible, that is, in two dimensions but not in three. The columns in the lower tier have the same arrangement as the crossed edges of the object. The columns starting from the arches at the back join the railing of the front and those from the arches at the front join the railing at the back.

The ordinary ring is *two-sided* like the ordinary plane, that is, you must cross the edge to get from one side to the other. The Möbius strip, however, is *one-sided.* A conveyer belt made in the shape of a Möbius strip will wear equally on both sides instead of only one. Coxeter gives the interesting sidelight that a patent for such a belt is held by the Goodrich Company.

Moreover, the ordinary ring is cut into two separate rings by a line around it. If you cut the Möbius strip along the line around it, however, it will remain one strip, although now with two twists in it. The Möbius strip, unlike the ordinary ring and also the ordinary plane, is not divided into two parts by a line around it.

EXERCISE 19 (*a*) Construct a Möbius strip as described. (*b*) Draw a line around the strip and observe that when you reach your starting point you are on the other side. (*c*) Cut the strip along the line you have drawn, being careful not to cut the edge, and observe that the strip remains in one piece but with two twists. (*d*) Cut the doubly twisted strip around its center line again. Is the Möbius strip divided into two parts by two lines around it?

The elliptic plane shares the properties of the Möbius strip because the Möbius strip is topologically equivalent to the elliptic plane with one point removed. We recall that two geometric figures are topologically equivalent if one can be made into the other by stretching or shrinking. Let the sphere represent the elliptic plane, with every pair of opposite poles of the sphere considered to be the same point. Remove one point, say the north pole. Since the south pole is opposite, it too is removed. Stretch the hole left by removing the point so that the north cap of the sphere is open. Again, since opposite points are identical, an identical cap at the south pole is removed. Also, on the sphere with two caps removed, since opposite points are identical, we can consider only the front half, bounded by AB, AC, CD, and BD. Points A and D are opposite, so they are identical, as are points C and B. But this identification is exactly the construction of the Möbius strip; thus, the sphere with one point removed and opposite points identical is topologically equivalent to the Möbius strip.

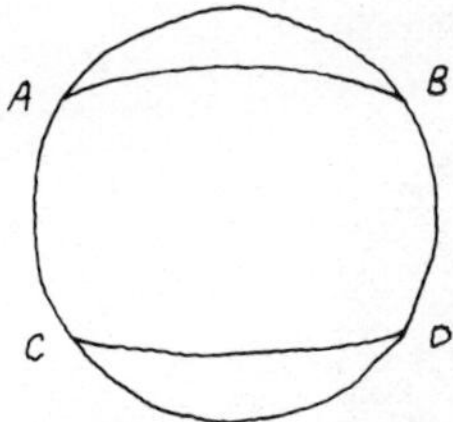

Because of the topological equivalence of the Möbius strip and the elliptic plane with one point removed, we conclude that the elliptic plane has the properties we have observed for the Möbius strip. In particular, the elliptic plane is one-sided, and it is not cut into two parts by a line in it.

EXERCISE 20 Each of the following statements is true in the spherical form of Riemannian geometry, in the elliptic form of Riemannian geometry, or in both. Say which ones are spherical, which ones are elliptic, and which ones are both. (*a*) Two points may determine more than one line. (*b*) Two points determine exactly one line. (*c*) The line is finite in extent. (*d*) The plane has positive curvature. (*e*) The plane is two-sided. (*f*) The plane is one-sided. (*g*) The plane is cut into two parts by a line in it. (*h*) The plane is not cut into two parts by a line in it.

EXERCISE 21 Each of the following statements is true in one or more of Euclidean, hyperbolic, and elliptic geometry. List one, two, or three geometries in which each statement is true. (*a*) Lines are finite in extent. (*b*) Lines are infinite in extent. (*c*) There are no parallel lines. (*d*) A line has one or more parallels through a given point. (*e*) Every two points determine exactly one line. (*f*) The plane has zero curvature. (*g*) The plane has negative curvature. (*h*) The plane has positive curvature. (*i*) The plane is two-sided. (*j*) The plane is one-sided.

9

The Shape of Space

For almost a century the non-Euclidean geometries were considered to be merely interesting curiosities. Then, in 1905, the great German-born American physicist Albert Einstein (1879–1955) announced the special

Copyright by Philippe Halsman

Albert Einstein was one of the towering figures of the twentieth century. His fame went beyond the world of science; and when he died in 1955, his name was as familiar to laymen as to physicists and mathematicians. It is perhaps surprising to learn that Einstein did not show the makings of a genius as a child, as Gauss did, or proceed smoothly through his doctorate to a lectureship and professorship as did Riemann. The special theory of relativity was published in 1905 by a twenty-six-year-old clerk in the patent office of Berne, Switzerland.

In 1909, Einstein was invited to become a professor at the University of Zurich. From there his fame spread; and by the time the general theory of relativity was published, he had been elected to the Prussian Academy of Sciences in Berlin. In 1919, the Royal Society of London sent out two expeditions to test the general theory of relativity during a total eclipse of the sun. They verified that the sun's gravitational field bent light from stars by exactly the amount Einstein had predicted, and Einstein became famous throughout the world.

theory of relativity. Eleven years later, in 1916, Einstein published the general theory of relativity. The general theory of relativity is based on mathematics developed from Riemannian geometry. It suggests that space itself is curved. Today it is generally agreed that space is indeed curved; and if it is, then the geometry of space would be non-Euclidean. However, whether the curvature is positive or negative, that is, which non-Euclidean geometry describes space, is still in doubt.

Before we can consider the curvature of space, we must examine how we actually perceive space. Dionys Burger, a contemporary Dutch physicist and mathematician, has written a book which is a sequel to Abbott's *Flatland* (Chapter 7, Section 8). Burger's sequel, in which Flatland turns out to be a positively curved plane, is appropriately called *Sphereland.* In the story, two Flatlanders are perplexed by discoveries about their land. Their introduction to the problem can serve as our introduction to the corresponding problem of space. The characters are a Hexagon, the grandson of the Square who narrates *Flatland,* and a stately Octagon, recently dismissed director of the Trigonometric Service of Flatland. As we begin, the Octagon has just revealed to the Hexagon, who is the narrator of *Sphereland,* that it is a problem in his measurements which caused his dismissal.

"Can I learn a little more?" I asked when we took leave of each other. "Couldn't you lift a tip of the veil for me, so that I can know in what direction to look for the problem?"

"All right," he said. "Tell me what the sum of the angles of a triangle is."

That unexpected question did surprise me, I must admit, but I answered: "180°, of course."

"Always?" he asked, and he left.

Amazing Results

This last comment of the former director gave me food for thought. I had been talking with him for quite a while and thought I was dealing with an intelligent man, but after this question, this doubt of his . . . was the man really in his right mind? Was he all there? Whoever doubted that the sum of the angles of a triangle is 180°? Who has ever seen any other kind of triangle? Or can conceive of such a triangle? Besides, it is easy to prove that the angles total 180°. Hadn't I better withdraw from the case? It would be difficult to get rid of the man later, and just try to talk a maniac out of some senseless truth or other once he has it in his head!

These reflections caused me a sleepless night. I considered inventing some sort of excuse, an illness, an unexpected summons—if need be, a death in the family—in order to get out of it, but when he stood in front of my door early the next morning, he inspired me again with such confidence that I decided to go along with him, and to the present day I have never regretted it.

He brought me quite a way out of the city. Here a three-cornered net had been set out with observation posts at considerable distances from each other. We made our way to post A. By means of an accurate trigonometric instrument it was possible to determine the angles between the directions AB, AC, AD, AE, and AF. My guide asked me to try using the instrument and measure the angle between the direction to C and the one to D. I found myself enjoying it, and even measured the other four angles.

"Together these five values will naturally have a sum of 360°," I declared.

"Let us try," responded my new friend—for why shouldn't I call him that—"to put it to the test."

A net of triangles is laid out.

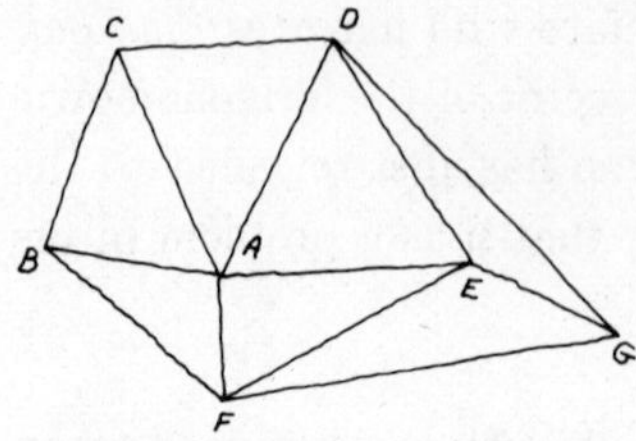

This we did and the result did turn out to be a pretty accurate 360°, but not precisely 360°!

"Of course one will never get exactly 360°," my guide said. "Every observation that is carried out, with the utmost accuracy even, entails a small error, an observational error. Because of this the final sum will never be quite correct, but a little too much or too little. The trained observer will get a smaller deviation than the untrained man, but even someone working with the closest accuracy possible will have a small error in his result."

"You don't need to tell me that," I said, "I have long been familiar with observational errors."

"Then you will also understand," the other resumed, "that with the successive repetition of the same measurements the result will sometimes be too little and sometimes too much."

"Of course," I said, "but one can also have a result that is too much, several times in succession."

"Or, on the other hand, one that is too small on successive occasions," my friend finished for me.

"So far we understand each other," he said, "but now the strange thing appears. If we go to point C and measure the angle between lines CD and CA, after which we go to point D and measure the angle between DC and DA, finally adding the values discovered for the size of the three angles of triangle ACD. . . ."

"Then we have to get 180°," I stated. "That is to say, we will never get precisely 180° but always just a little more or a little less."

"Exactly, that should be so," he said, "but it is not the case. I am always getting too large a result and the deviation I find is just a little too great to be attributable to observational error."

"This means," I concluded, "that the deviation cannot be the result of observational error, but that it is a true deviation, that therefore the sum of the angles of this triangle is really greater than 180°."

My conclusion was right, it could not be refuted logically, but when I had said it, I was shocked by my own words. For how could that be? What had I just said? The sum of the angles of a triangle greater than 180°! That is impossible! Just how could that be?[6]

Although it gets a bit ahead of our look at our own space, perhaps we should take a peek at the Flatland solution to the problem. It is suggested by the Hexagon, again the narrator, several meetings later. The Octagon is Mr. Puncto.

I was looking forward in high spirits to my next meeting with Mr. Puncto. From the moment we greeted each other he knew there had to be a reason for my good mood. "Did you find a solution?" he asked.

"No," I replied, "I can't shout Eureka yet, but do believe that I am very close to discovering the basis of the problem. I think that our curious phenomenon—the sum of the angles of a triangle exceeding 180°—has to be explained by assuming that the sides of the triangle are curved but that this curvature is not visible, I mean: not visible to us. It occurs in a direction perpendicular to our world. A three-dimensional creature must be able to see the curvature; we cannot."[7]

[6] From *Sphereland,* © 1965 by Dionys Burger with permission of Thomas Y. Crowell Company, Inc., pp. 128–131.
[7] Ibid., p. 154.

The Hexagon has told us our difficulty in understanding space. The problem is, we cannot see space as a whole. We are within it. For example, you can hold an apple in your hand and see its shape. An apple is outside and separate from us, and we can see the whole of it.

We are outside the earth also, but we cannot see the whole of it (except recently from space ships). Suppose you are an ancient Greek and that you draw a triangle on the surface of the earth and measure its angles. You find that the sum of the angles of the triangle is approximately equal to two right angles. However, this does not mean that the surface of the earth is flat, that is, that it has zero curvature and extends infinitely in all directions. Rather, it means that your triangle is very small compared with the size of the earth, so the deviation of the sum of the angles of your triangle from two right angles is too small to be detected.

Suppose instead we can draw a triangle on the earth which has one angle at the north pole, the second in the Philippines, and the third in northern Brazil. This triangle covers about a quarter of the northern hemisphere, each of its angles is approximately a right angle, and the sum of its angles is about three right angles. Thus our perception of a surface depends on how much of the surface we can see.

Similarly, our perception of space depends on how much of space we can see. Moreover, we must bear in mind that we are within space. The surface of the earth, although curved, is two-dimensional. The space we are in is three-dimensional and our travel in space is not essentially different from the Spherelanders' traveling about on their two-dimensional surface. If space were, for example, positively curved so that its geometry were elliptic geometry and finite in extent, the earth would be like a plane in the elliptic space. Space, then, would not be simply the interior of another sphere further out, but would be related to the interior of a hypersphere, the four-dimensional analog of the three-dimensional interior of an ordinary sphere. If space were negatively curved so that its geometry were hyperbolic geometry, space would be the four-dimensional analog of the three-dimensional interior of a pseudosphere.

Beside our problem of being within space and unable actually to visualize its shape, our ability to measure space is very limited indeed. The largest triangle we can draw in space would have one angle at a distant star and its other two angles at opposite sides of the earth's orbit, about 186 million miles apart.

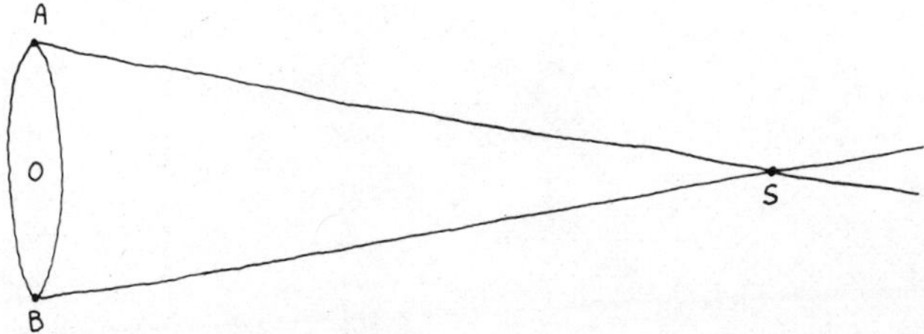

Astronomers determine the angle at the star by the positions of the star relative to background stars. These positions are called the *parallax* of the star. The word *parallax* comes not from "parallel" but from the Greek words for "to change position."

The angle at the star $\measuredangle ASB$ is determined by the parallax of the star. A and B are two positions of the earth at opposite sides of its orbit around the sun. If $\measuredangle BAS$ and $\measuredangle ABS$ were measured and the sum of the angles of triangle ABS equal to two right angles, then space would have zero curvature and extend infinitely in all directions. If the sum of the angles were less than two right angles, then space would have negative curvature. If the sum were more than two right angles, then space would have positive curvature and would be finite in extent although unbounded. Such measurements of space, however, are like drawing a tiny triangle on the surface of the earth. The parallaxes of distant stars are so slight that these variations cannot be detected.

In the early part of the current century a new phenomenon, called the *red shift*, was observed. There is a similar phenomenon called the Doppler effect which was already well known. You have surely noticed that sounds such as that of a siren, or a car horn sounding continuously, change in pitch as the vehicle goes past you. This is an example of the phenomenon of sound waves called the Doppler effect. As the vehicle approaches, the frequency of the sound wave is high relative to your position and the sound is high-pitched. As the vehicle goes away from you, beginning the moment it passes, the frequency drops relative to your position and the sound immediately has a lower pitch.

The red shift is a similar phenomenon of the light waves of stars. If a star is approaching the earth, its light waves shift to the higher

Courtesy of the National Radio Astronomy Observatory

300 FOOT RADIO TELESCOPE

Optical telescopes, from those constructed by Galileo, to the 100-inch Mount Wilson telescope used by Hubble, to the famous 200-inch telescope on Palomar Mountain, all work on the principle of collecting light waves from objects in space. This familiar type of telescope is far from obsolete (many discoveries have been made by Allan Sandage, who was once an assistant of Hubble, using the Palomar Mountain telescope), but it is being joined by new types of telescopes. One of the most important of the new types is the radio telescope, which collects radio energy rather than light energy.

Most radio telescopes are shaped like a large dish in which radio waves are collected. The dish of this radio telescope at Green Bank, West Virginia, is 300 feet in diameter. When it was completed in 1962, it was the world's largest movable dish. There is now a 330-foot movable dish in Germany. The inside surface of the 300-foot dish consists of 1.8 acres of aluminum mesh

Photograph by Arecibo Observatory, Cornell University

ARECIBO RADIO TELESCOPE

panels. The rest of the structure is steel weighing 600 tons, of which 500 tons is in moving weight. According to the description distributed by the National Radio Astronomy Observatory: "It was built primarily to study the density and distribution of hydrogen gas in the spiral arms of our Galaxy. In addition the 300-ft. does extensive research in the study of Pulsars and looking for new and interesting radio sources."

Radio waves from space were first observed in the early 1930s. A 31-foot dish was built by Grote Reber in Wheaton, Illinois, in 1936–1937. It now belongs to the National Radio Astronomy Observatory. Today there are radio telescopes with dishes of about 100 feet all over the world, including several 85-foot and a 140-foot also at Green Bank. The largest radio telescope in the world has a dish 1000 feet wide. The dish, of course not movable although the platform above it is, covers a valley in Arecibo, Puerto Rico.

frequencies relative to the earth, and the higher-frequency waves give a violet color. If the star is receding from the earth, its light waves relative to the earth shift toward the reds, which are the lower frequencies. Astronomers have observed a shift toward the reds in the light spectra of distant galaxies. This is the red shift, which indicates that distant galaxies are moving away from us. In 1929, Edwin Hubble (1889–1953), measuring the red shift on the 100-inch Mt. Wilson telescope, discovered that not only are the galaxies receding, but at a startling rate given by a formula now called Hubble's law.

The Einsteinian perception of space was a finite but unbounded universe, curving in upon itself with positive curvature but with the galaxies held in equilibrium by forces created by the curvature. Einstein's theory is referred to as the *steady state theory.* With the discovery of the red shift, Georges Lemaître (1884–1966), a Belgian priest and astronomer, proposed the *theory of the expanding universe* often referred to as the *"big bang" theory.* If the curvature of space were positive, the expansion would eventually become a contraction and the universe would collapse. In the expansion-contraction theory the universe would be finite but unbounded. Sir Arthur Eddington (1882–1944), British astronomer and physicist, proposed a theory of continued expansion in which the curvature would be negative and the universe infinite, a view shared by George Gamow (1904–1968), the Russian-born American physicist and writer of popular books on science.

Gamow illustrated the theories of space curvature and the expanding universe in the fanciful stories of *Mr Tompkins in Wonderland,* updated and expanded in *Mr Tompkins in Paperback.* In this excerpt,[8] Mr Tompkins has just returned from an evening at the Cosmic Opera with the daughter of his friend, the professor.

When Mr Tompkins brought Maud back to her father's house the professor was sitting in his comfortable chair with the newly arrived issue of the Monthly Notices in his hands.

'Well, how was the show?' he asked.

'Oh, wonderful!' said Mr Tompkins, 'I was especially impressed by the aria on the ever-existing universe. It sounds so reassuring.'

'Be careful about this theory,' said the professor. 'Don't you know the

[8] George Gamow, *Mr Tompkins in Paperback* (Cambridge: The University Press, 1965), pp. 62–64.

proverb: "All is not gold that glitters"? I am just reading an article by another Cambridge man, MARTIN RYLE, who built a giant radio-telescope which can locate galaxies at distances several times greater than the range of the Mount Palomar 200-inch optical telescope. His observations show that these very distant galaxies are located much closer to each other than are those in our neighbourhood.'

'Do you mean,' asked Mr Tompkins, 'that our region of the universe has a rather rare population of galaxies, and that this population density increases when we go further and further away?'

'Not at all,' said the professor, 'you must remember that, due to the finite velocity of light, when you look far out into space you look also far back into time. For example, since light takes eight minutes to come here from the Sun, a flare on the Sun's surface is observed by terrestrial astronomers with an eight-minute delay. The photographs of our nearest space neighbour, a spiral galaxy in the constellation of Andromeda—which you must have seen in books on astronomy and which is located about a million light-years away—show how it actually looked one million years ago. Thus, what Ryle sees, or I should rather say hears, through his radio-telescope, corresponds to the situation which existed in that distant part of the universe many thousand millions of years ago. If the universe were really in a steady state, the picture should be unchanged in time, and very distant galaxies as observed from here now should be seen distributed in space neither more densely nor rarely than the galaxies at shorter distances. Thus Ryle's observations showing that distant galaxies seem to be more closely packed together in space is equivalent to the statement that the galaxies everywhere were packed more closely together in the distant past of thousands of millions of years ago. This contradicts the steady state theory, and supports the original view that the galaxies are dispersing and that their population density is going down. But of course we must be careful and wait for further confirmation of Ryle's results.'

'By the way,' continued the professor, extracting a folded piece of paper from his pocket, 'here is a verse which one of my poetically inclined colleagues wrote recently on this subject.' And he read:

'Your years of toil,'
Said Ryle to Hoyle,
   'Are wasted years, believe me.
The steady state
Is out of date.
   Unless my eyes deceive me,

My telescope
Has dashed your hope;
   Your tenets are refuted.

Let me be terse:
Our universe
   Grows daily more diluted!'

Said Hoyle, 'You quote
Lemaître, I note,
   And Gamow. Well, forget them!
That errant gang
And their Big Bang—
   Why aid them and abet them?

You see, my friend,
It has no end
   And there was no beginning,
As Bondi, Gold,
And I will hold
   Until our hair is thinning!'

'Not so!' cried Ryle
With rising bile
   And straining at the tether;
'Far galaxies
Are, as one sees,
   More tightly packed together!'

'You make me boil!'
Exploded Hoyle,
   His statement rearranging;
'New matter's born
Each night and morn.
   The picture is unchanging!'

'Come off it, Hoyle!
I aim to foil
   You yet' (The fun commences)
'And in a while,'
Continued Ryle,
   'I'll bring you to your senses!'*

'Well,' said Mr Tompkins, 'it will be exciting to see what will be the outcome of this dispute,' and giving Maud a kiss on the cheek he wished them both goodnight.

[Gamow adds a footnote]

* A fortnight before the publication date of the first printing of this book there appeared an article by F. Hoyle entitled: "Recent Developments in Cosmology"

(Nature, Oct. 9, 1965, p. 111). Hoyle writes: "Ryle and his associates have counted radio sources . . . The indication of that radio count is that the Universe was more dense in the past than it is today." The author has decided, however, not to change the lines of the arias of "Cosmic Opera" since, once written, operas become classic. In fact, even today Desdemona sings a beautiful aria before she dies, after being strangled by Othello.

Epilogue

Gamow's footnote at the end of Section 9 indicates that just as his book was going to press in 1965, new information was published, indicating that the universe is indeed expanding. This author has been similarly fortunate, for while this manuscript was under preparation, the article of which the first part is below appeared on the front page of *The New York Times*. What does it prove? Nothing yet, although perhaps the positively curved and finite universe theory, for the moment, is given a boost over the negative curvature theory. More important, it shows that the question is still very much alive!

Men Report Seeing Edge of the Universe

Astronomers believe they have seen the edge of the universe.

The announcement last week of the discovery of a quasar more distant than anything previously observed has strengthened this belief, not because it is so distant but because it is not farther away.

The situation can be likened to gazing through a forest of widely scattered trees and finding that none can be seen beyond a certain distance. The implication is that, as Einstein and others believed, the universe is finite.

Beyond its expanding volume, this theory holds, nothing exists—not even space, because, in this concept, over such great distances space curves back upon itself.

Assuming the validity of the method used for estimating distances to quasars and other very remote objects, light from the newly discovered one has probably taken 12 billion years to reach the earth, traveling at 186,000 miles a second.

A Wall to Vision

The nature of quasars is unknown, although some scientists suspect they are galaxies in an early stage of formation. If they are as distant as they seem to be, they shine far more brightly than any other celestial object, both invisible light and light at radio wave lengths.

Because of this intrinsic brightness, some of them would be expected to be visible at considerably greater distances than that of the newly found quasar.

The distance estimates are determined from the observed expansion of the universe. It has been found from a variety of observations that the galaxies, or great star systems, are flying apart much like particles of dust in an expanding cloud. When such an expansion is viewed from any one particle in the cloud, the rate at which another particle is receding is directly related to its distance. The faster the rate, the greater the distance.

By determining the rate at which a quasar is flying away from the earth, astronomers can estimate its distance in terms of the length of time its light has been on the way here. The expansion is such that the universe appears to have originated in an enormous explosion of matter—a "big bang"—some 13 billion years ago, and the earliest quasars should be visible far enough away for their light to have been on its way here for that long.

But looking across vast distances—and, hence, far back into time—man can see out only to 12 billion years (in the most distant quasars). Hence, in the words of Dr. Allan R. Sandage of the Hale Observatories in California, in these quasars we are apparently seeing "the edge of the world." To cosmologists, "world" is virtually synonomous with "universe."

For the last year or two, Dr. Sandage said in a telephone interview Friday, it has been suspected that there is some sort of "wall" preventing astronomers from seeing quasars in the region beyond 12 billion years. Using new, more powerful methods of observing with radio and optical telescopes, such objects should be visible, if they exist.

Now, he said, it is beginning to look as though the "wall" is real.[9]

[9] Walter Sullivan, *The New York Times*, April 8, 1973, p. 1, col. 4, and p. 63, col. 1. ©1973 by The New York Times Company. Reprinted by permission.

Hale Observatories

ALLAN R. SANDAGE

Looking Back

1 All the assumptions of absolute geometry are also assumptions of Euclidean geometry. However, there is one assumption of Euclidean geometry which is not an assumption of absolute geometry. (*a*) Name this assumption. All theorems of absolute geometry are also theorems of Euclidean geometry. However, there are theorems of Euclidean geometry which are not theorems of absolute geometry. Which of the following theorems of Euclidean geometry are not theorems of absolute geometry? (*b*) If the two sides and included angles of two triangles are equal, then the triangles are congruent. (*c*) If a triangle has two sides equal, then the angles opposite those sides are equal. (*d*) If a triangle has two angles equal, then the sides opposite those angles are equal. (*e*) If two lines intersect, the vertical angles are equal. (*f*) If two parallel lines are cut by a transversal, the alternate interior angles are equal. (*g*) The sum of the angles of a triangle is equal to two right angles. (*h*) There is exactly one line through a given point parallel to a given line.

2 Euclid's fifteenth theorem is: If two lines intersect, the vertical angles are equal. The sixteenth theorem is: If one side of a triangle is extended, the exterior angle is greater than either opposite interior angle.

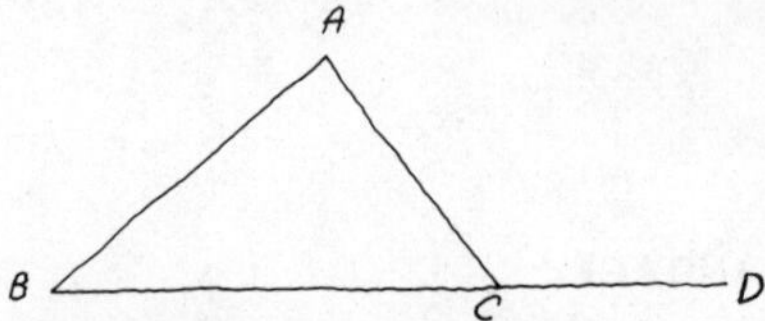

For example, in triangle ABC, if BC is extended to D, the exterior angle $\sphericalangle ACD$ is greater than either of the opposite interior angles $\sphericalangle BAC$ or $\sphericalangle ABC$. To prove that $\sphericalangle ACD$ is greater than $\sphericalangle BAC$, let E bisect AC, extend BE to F so that $BE = EF$, and draw FC.

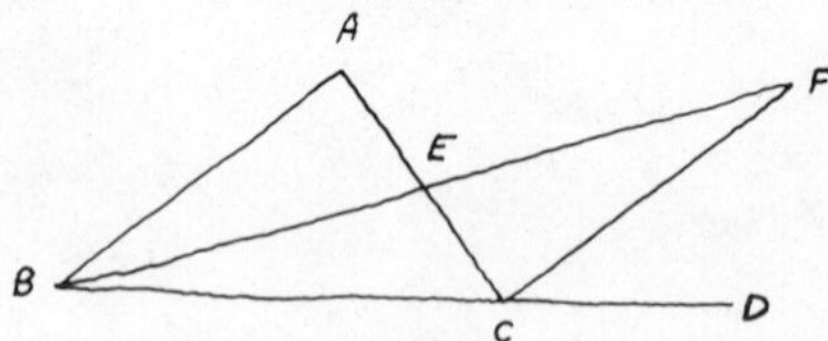

(*a*) Prove that triangle AEB is congruent to triangle CEF; thus $\sphericalangle BAE = \sphericalangle ACF$. (*b*) Why is $\sphericalangle ECF$ less than $\sphericalangle ECD$? (*c*) Why is $\sphericalangle BAC$ less than $\sphericalangle ACD$? (*d*) Is this theorem of Euclidean geometry also a theorem of absolute geometry?

3 Euclid's twenty-ninth theorem is: If two parallel lines are cut by a transversal, the alternate interior angles are equal. The converse of this theorem is: If two lines are cut by a transversal so that the alternate interior angles are equal, then the lines are parallel. The proof is by *reductio ad absurdum.*

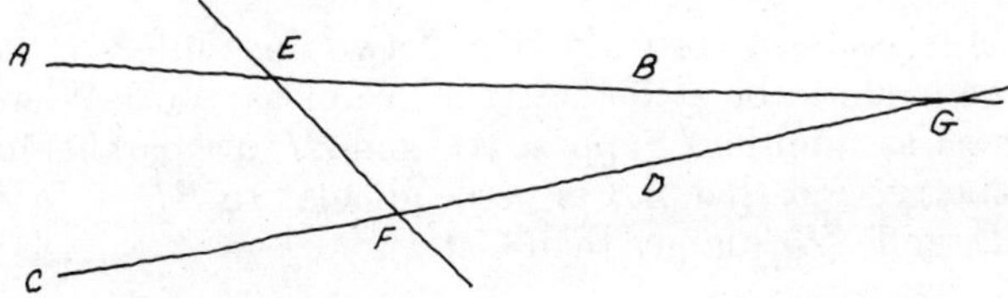

(*a*) Suppose that the lines AB and CD are not parallel but meet at G. Use the theorem in Exercise 2 to explain why $\sphericalangle CFE$ and $\sphericalangle BEF$ would not be equal. (*b*) Is this theorem of Euclidean geometry also a theorem of absolute geometry?

4 Refer to this diagram of the Saccheri quadrilateral.

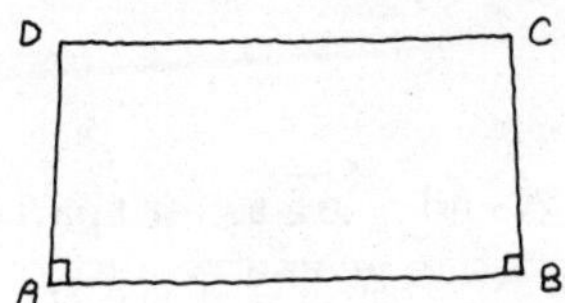

(*a*) What is true of $\sphericalangle ADC$ and $\sphericalangle BCD$ under the hypothesis of the acute angle? (*b*) What is true of $\sphericalangle ADC$ and $\sphericalangle BCD$ under the hypothesis of the obtuse angle? (*c*) What is the name of the geometry in which the hypothesis of the acute angle is true? (*d*) What is the name of the geometry in which the hypothesis of the obtuse angle is true?

5 Suppose it is given that Euclidean geometry corresponds to Saccheri's hypothesis of the right angle. In triangle ABC, let G bisect AB and H bisect AC. Draw AD perpendicular to GH. Extend GH so that $EG = GD$ and $DH = HF$. Draw EB and FC.

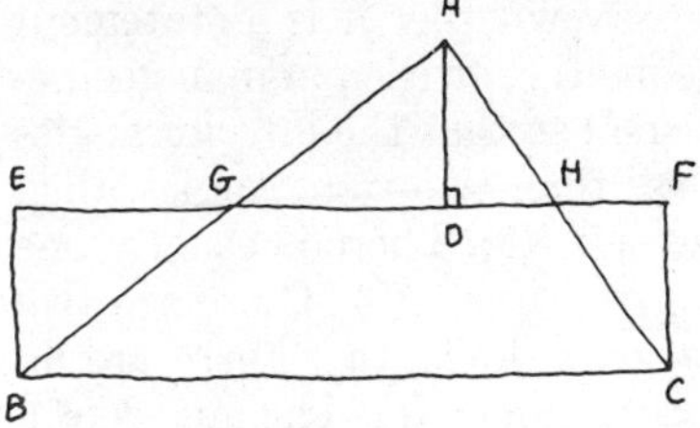

(*a*) Prove that triangle ADG is congruent to triangle BEG, and that triangle ADH is congruent to triangle CFH. (*b*) Prove that the quadrilateral $EBCF$ is a Saccheri quadrilateral with right angles $\measuredangle BEG$ and $\measuredangle CFH$ and sides $EB = FC$. (*c*) Given the hypothesis of the right angle, prove that the sum of the angles of triangle ABC is equal to two right angles.

6 Recall that Proclus assumed correctly that if two lines intersect, a point on each can be made to be as far apart as we please. (*a*) What was Proclus's incorrect assumption? Suppose AC and BD are parallel in Lobachevskian geometry, and that AB is perpendicular to BD, CD is perpendicular to BD, and AB is larger than CD.

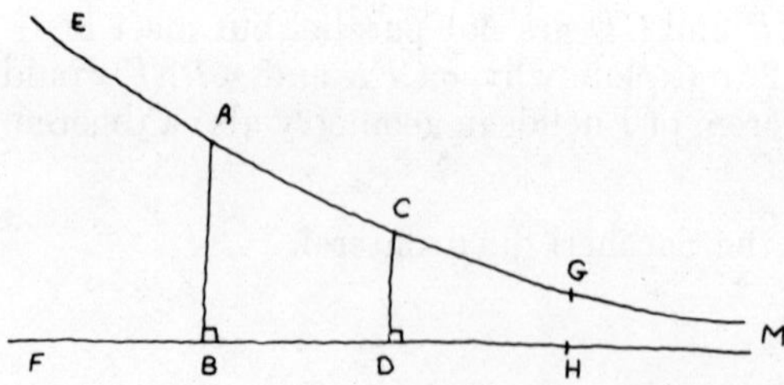

(*b*) It is possible to extend AC and BD so that E and F are as far apart as we please? (*c*) Are parallel lines in Lobachevskian geometry always no more than a given finite distance apart?

7 (*a*) What was Saccheri's incorrect assumption in the case of the hypothesis of the acute angle? Refer to the diagram in Exercise 6. (*b*) Is it possible to extend AC and BD so that G and H are as close as we please? (*c*) Is it reasonable to say that parallel lines in Lobachevskian geometry "meet at infinity"? (*d*) Of the angles of parallelism $\measuredangle BAM$ and $\measuredangle DCM$, which one is closer to a right angle? (*e*) If GH is perpendicular to BD and $\measuredangle HGM$ is drawn, of angles $\measuredangle DCM$ and $\measuredangle HGM$ which one is closer to a right angle? (*f*) Is it reasonable to say that lines AC and BD have a "common perpendicular" at infinity?

8 For each of the following statements, say whether it is a statement of Euclidean geometry, Lobachevskian geometry, or Riemannian geometry. (*a*) If two lines are cut by a transversal so that the interior angles on the same side of the transversal are less than two right angles, then the lines meet on that side of the transversal. (*b*) There is exactly one line through a given point parallel to a given line. (*c*) There are two lines through a given point parallel to a given line. (*d*) There are no lines through a given point parallel to a given line. (*e*) The angles of a

Saccheri quadrilateral all are right angles. (*f*) The angles of a Saccheri quadrilateral not given to be right angles are equal and less than right angles. (*g*) The angles of a Saccheri quadrilateral not given to be right angles are equal and greater than right angles. (*h*) The sum of the angles of a triangle is equal to two right angles. (*i*) The sum of the angles of a triangle is less than two right angles. (*j*) The sum of the angles of a triangle is more than two right angles. (*k*) There exist asymptotic triangles. (*l*) There exist doubly right-angled triangles.

9 For each of the following statements of Riemannian geometry, say whether it is a statement of spherical geometry, elliptic geometry, or both. (*a*) The line is finite in extent but unbounded. (*b*) Every two lines meet in two points which are opposite poles. (*c*) Two points may determine more than one line if the points are opposite poles. (*d*) Every pair of opposite poles is the same point. (*e*) Every two lines meet in exactly one point. (*f*) Two points determine exactly one line. (*g*) The surface is two-sided. (*h*) The surface is one-sided. (*i*) The plane is divided into two parts by a line in it. (*j*) The plane is not divided into two parts by a line in it.

Branching Out

1 Leonardo da Vinci was one of history's few true universalists, of the first rank equally in art and science. He was also an intriguing and mysterious man, illegitimately born, married to no woman and close to no man, writing his notes backward so they must be held to a mirror to be read. There is a great deal to find out about him and a great deal written about him. You might start with the chapter on da Vinci in *The Life of Science* by George Sarton, whose admiration of da Vinci's genius is obvious. From there your own interests can take you in any of several directions, ranging from da Vinci's studies of perspective to his studies of flight and predictions of modern flight and parachutes.

2 As the interests of da Vinci were not limited to the study of perspective, so too was the study of perspective not limited to da Vinci. If you examine pre-Renaissance art, you will realize that much of it displays a lack of sense of perspective and spatial relationships. For example, sizes of background figures may be out of proportion compared with foreground figures. A whimsical example is William Hogarth's post-Renaissance painting called *False Perspective*. You will find it reproduced along with other interesting examples of perspective and nonperspective in

Life Science Library's *Mathematics*. Of Renaissance artists interested in perspective, Albrecht Dürer is one of the best known. There is a nicely illustrated section including da Vinci, Dürer, Piero della Francesca, and others, in Morris Kline's *Mathematics for Liberal Arts*.

3 Although the contributions of Desargues, Pascal, and others were important, the creation of projective geometry as a branch of mathematics is usually credited to Jean-Victor Poncelet. Poncelet was another of the French mathematicians influenced by the Napoleonic era. If nothing else, his capture during the French retreat from Moscow, and subsequent imprisonment in Russia, gave him leisure to write. The story of these years is well known and credit is given for the results, but little is written about the rest of his life. Perhaps if the short chapter in Bell's *Men of Mathematics* is an indicator, it is because with the exception of the leisure of imprisonment, Poncelet's life was mostly taken up with the commonplace tasks of a military engineer.

4 Archimedes, Newton, and Gauss are considered to be the three greatest mathematicians of all time. We have met the third of these, Karl Friedrich Gauss, whom E. T. Bell calls "The Prince of Mathematicians," many times in this book. Many stories are told about Gauss as a child prodigy. As an adult, Gauss touched on just about every branch of mathematics, but he published little compared with prolific writers such as Euler. He obtained many results not published which are found in his diary, among them his discovery of non-Euclidean geometry. It is said that he published few of his discoveries because he would never publish them until he had made them perfect. His motto reflects this attitude: *Pauca sed matura* (Few but ripe). There is a great deal written about such a great mathematician, so you will have little trouble finding both anecdotes of Gauss' life and parts of his mathematical work which you can understand.

5 In this book we have not studied per se the components of a deductive system: undefined terms or primitives, definitions, axioms or postulates, and theorems. There are good discussions in many elementary books on modern mathematics. There is also a brief summary in Blumenthal's *A Modern View of Geometry*. Euclid is perhaps most often criticized in elementary books because of his definitions and his failure to recognize the necessity of undefined terms. We have seen that although he did realize that the fifth postulate was a necessary assumption, there were also omissions among the assumptions. Many mathematicians have attempted to construct complete sets of axioms for Euclidean geometry, the most prominent of them being David Hilbert. If you wish to go beyond the general concept of a deductive system and apply the deductive system to Euclidean geometry, you will also find a brief summary of Hilbert's postulates in Blumenthal.

6 Most people find the discovery of non-Euclidean geometries as startling as did the mathematicians who discovered them. However, not much in depth has been written on a completely nontechnical level about the non-Euclidean geometries. There are brief descriptions in such books as Kasner and Newman's *Mathematics and the Imagination* and Life Science Library's *Mathematics*. An excellent but technical account is provided by Roberto Bonola in *Non-Euclidean Geometry*. By skipping the proofs and concentrating on the verbal and historical parts, you can probably glean a great deal of information from Bonola. Translations of the work of Bolyai and Lobachevsky are also included as appendices.

7 The two Ptolemys we have mentioned are often confused. Ptolemy I was ruler of Egypt at the time of Euclid (about 300 B.C.). Claudius Ptolemy was an astronomer of Alexandria about A.D. 150. Aside from his attempt to prove the fifth postulate [and possibly he was the first to use the number zero (Chapter 1, Section 4)], Ptolemy of Alexandria is most famous for the Ptolemaic concept of the universe in which the earth is the center and everything else revolves about it. This concept was perpetuated, largely by the Church, for hundreds of years. It was not until the sixteenth and seventeenth centuries that the Copernican system, in which the sun is the center of at least our solar system, with the earth revolving about it, became accepted. Galileo Galilei, whom we have met in other chapters, is probably most famous for his defense of the Copernican system. Find out about the Ptolemaic system and then go on to Copernicus, Galileo, and his contemporary, Kepler. There are good articles in Newman's *World of Mathematics*, although they do not include Galileo's legendary remark, "The earth *does* move all the same." For that, see Eves' *In Mathematical Circles*.

8 We have mentioned but not described Einstein's special and general theories of relativity. The first concerns the connection between space and time; the second, gravity and space curvature. Although we have seen that Einstein's description of space, the steady state theory, is no longer considered to be accurate, his proposals of time as a fourth dimension, of space as curved, and the universe as finite revolutionized physics as completely as Lobachevskian and Riemannian geometries did geometry. It is amazing to learn that so great a scientist came to the United States as a refugee from Hitler's Germany, and there are many stories of his humanness. There are several books about Einstein and his work written for nonscientists, including Coleman's *Relativity for the Layman* and, of course, Gamow's *Mr Tompkins* books and *One Two Three . . . Infinity*. Another book, *From Euclid to Eddington* by Edmund Whittaker, is far more comprehensive and difficult, but the verbal parts can take you beyond Einstein to the expanding universe.

9 This chapter has included excerpts from some entertaining and whimsical books that will help you to understand the fourth dimension and curvature of space. *Flatland*, written before there was any idea of curved space, helps us understand the concept of a fourth dimension by showing the difficulties two-dimensional creatures have in comprehending a third dimension. *Sphereland* helps us understand curved space by showing the difficulties that two-dimensional creatures have in comprehending a curved two-dimensional space. Of course, the curved space of the two-dimensional creatures of Sphereland turns out to be a sphere, so if our space were similarly curved it would be a hypersphere, which is as hard for us to imagine as a sphere for the Spherelanders. The *Mr Tompkins* stories will help you understand the phenomena of curved space and the expanding universe. There are also stories which explain quantum theory and related aspects of atomic physics in the Mr Tompkins books.

A Brief Bibliography

For the General Reader—Use Freely

Abbott, Edwin A. *Flatland, A Romance of Many Dimensions*. 5th. ed., revised. New York: Barnes & Noble, Inc., 1963.

Bell, E. T. *Mathematics, Queen and Servant of Science*. New York: McGraw-Hill Book Company, Inc., 1951.

———. *Men of Mathematics*. New York: Simon and Schuster, 1937.

Bergamini, David. *Mathematics* (Life Science Library). New York: Time-Life Books, 1970.

Burger, Dionys. *Sphereland*. Trans. Cornelie J. Rheinboldt. New York: Thomas Y. Crowell Company, 1965.

Campbell, Stephen K. *Flaws and Fallacies in Statistical Thinking*. Englewood Cliffs, N.J.: Prentice-Hall, Inc., 1974.

Coleman, James A. *Relativity for the Layman*. New York: William Frederick Press, 1958.

Dantzig, Tobias. *Number, the Language of Science*. 4th. ed. New York: Macmillan Publishing Co., Inc., 1954.

Eves, Howard W. *In Mathematical Circles*. Boston: Prindle, Weber & Schmidt, Inc., 1969.

Fadiman, Clifton. *Fantasia Mathematica*. New York: Simon and Schuster, 1958.

———. *The Mathematical Magpie*. New York: Simon and Schuster, 1962.

Gamow, George. *Mr Tompkins in Paperback*. Cambridge: The University Press, 1965.

———. *One Two Three . . . Infinity*. New York: The Viking Press, Inc., 1947. Rev. ed., 1961.

Hardy, G. H. *A Mathematician's Apology*. Cambridge: The University Press, 1967.

Huff, Darrell. *How to Lie with Statistics*. New York: W. W. Norton & Company, Inc., 1954.

Kasner, Edward and James R. Newman. *Mathematics and the Imagination*. New York: Simon and Schuster, 1940.

King, Amy C., and Cecil B. Read, *Pathways to Probability*. New York: Holt, Rinehart and Winston, Inc., 1963.

Kline, Morris. *Mathematics for Liberal Arts*. Reading, Mass.: Addison-Wesley Publishing Company, 1967.

———, ed. *Mathematics in the Modern World: Readings from Scientific American*. San Francisco: W. H. Freeman and Company, 1968.

Newman, James R. *The World of Mathematics*. New York: Simon and Schuster, 1956.

Osen, Lynn M. *Women in Mathematics*. Cambridge, Mass.: The MIT Press, 1974.

Rapport, Samuel, and Helen Wright. *Mathematics*. New York: New York University Press, 1963.

Sarton, George. *The Life of Science*. New York: Henry Shuman, Inc., 1948.

Zaslavsky, Claudia. *Africa Counts*. Boston: Prindle, Weber & Schmidt, Inc., 1973.

Ranging in Difficulty—Try Some

Allendoerfer, Carl B., and Cletus O. Oakley. *Principles of Mathematics*. 3d. ed. New York: McGraw-Hill Book Company, Inc., 1969.

Blumenthal, Leonard M. *A Modern View of Geometry*. San Francisco: W. H. Freeman and Company, 1961.

Bonola, Roberto. *Non-Euclidean Geometry*. Trans. H. S. Carslaw. The Open Court Publishing Company, 1912. New York: Dover Publications, Inc., 1955.

Coxeter, H. S. M. *Introduction to Geometry*. 2d. ed. New York: John Wiley & Sons, 1969.

Eves, Howard. *An Introduction to the History of Mathematics*. 3d. ed. New York: Holt, Rinehart and Winston, Inc., 1969.

Eves, Howard, and Carroll V. Newsom. *An Introduction to the Foundations and Fundamental Concepts of Mathematics*. New York: Holt, Rinehart and Winston, Inc., 1958.

Heath, Thomas L. *A Manual of Greek Mathematics*. Oxford: Oxford University Press, 1931. New York: Dover Publications, Inc., 1963.

Kline, Morris. *Mathematical Thought from Ancient to Modern Times*. New York: Oxford University Press, 1972.

Meschkowski, Herbert. *Evolution of Mathematical Thought*. Trans. Jane H. Gayl. San Francisco: Holden-Day, Inc., 1965.

Nagel, Ernest, and James R. Newman. *Gödel's Proof*. New York: New York University Press, 1958.

Neugebauer, O. *The Exact Sciences in Antiquity*. 2d. ed. Providence, R.I.: Brown University Press, 1957. New York: Dover Publications, Inc., 1969.

van der Waerden, B. L. *Science Awakening*. Trans. Arnold Dresden. New York: Oxford University Press, 1961.

Weyl, Hermann. *Symmetry*. Princeton, N.J.: Princeton University Press, 1952.

Whittaker, Edmund. *From Euclid to Eddington*. Cambridge: The University Press, 1949.

Wilder, Raymond L. *Evolution of Mathematical Concepts: An Elementary Study*. New York: John Wiley & Sons, 1968.

For the Specialist—Proceed at Your Own Risk

Boole, George. *An Investigation of the Laws of Thought*. New York: Dover Publications, Inc., 1958.

———. *The Mathematical Analysis of Logic*. Oxford: Basil Blackwell, 1948. New York: Barnes & Noble, 1965.

Clifford, A. H., and G. B. Preston. *The Algebraic Theory of Semigroups*. Vol. I. Providence, R.I.: American Mathematical Society, 1961.

Coxeter, H. S. M. *Regular Polytopes*. 2d. ed. New York: Macmillan Publishing Co., Inc., 1963.

Heath, Thomas L. *Euclid's Elements*. 2d. ed. Cambridge: The University Press, 1926. New York: Dover Publications, Inc., 1956.

McKeon, Richard, ed. *The Basic Works of Aristotle*. New York: Random House, 1941.

Reichenbach, Hans. *The Philosophic Foundations of Quantum Mechanics*. Berkeley: University of California Press, 1944.

Rosser, J. Barkley, and Atwell R. Turquette. *Many-Valued Logics*. Amsterdam: North-Holland Publishing Company, 1952.

Russell, Bertrand, and Alfred North Whitehead. *Principia Mathematica*. 2d. ed. Cambridge: The University Press, 1927.

———. *Principia Mathematica to Fifty-Six*. 2d. ed. Cambridge: The University Press, 1962.

Answers to Selected Exercises

Chapter 1

1 (*a*) 𓎆𓎆𓎆𓎆 (*b*) 𓎆𓎆𓎆𓎆𓏺𓏺𓏺𓏺 (*c*) 𓍢𓍢𓍢

(*d*) 𓆼𓆼𓍢𓍢𓍢𓎆𓎆𓎆𓎆 (*e*) 𓆼𓍢𓍢𓍢𓍢𓍢𓏺𓏺𓏺𓏺

(*f*) 𓆼𓆼𓍢𓍢𓍢𓍢𓍢𓎆𓎆𓎆𓏺𓏺𓏺𓏺 (*g*) 𓍢𓍢𓍢𓍢𓍢𓍢𓍢𓍢𓍢𓎆𓎆𓎆𓎆𓎆𓎆𓎆𓎆𓎆𓏺𓏺𓏺𓏺𓏺𓏺𓏺𓏺𓏺

2 (*a*) 𓆼𓆼𓍢𓍢𓍢𓍢𓍢𓎆𓎆𓎆𓎆𓎆𓎆𓏺𓏺𓏺𓏺𓏺𓏺𓏺𓏺𓏺

$$\begin{array}{r} 1{,}325 \\ 1{,}244 \\ \hline 2{,}569 \end{array}$$

(*b*) 𓆼𓍢𓍢𓍢𓍢𓍢𓍢𓍢𓍢𓎆𓎆𓎆𓎆𓏺𓏺𓏺𓏺𓏺

$$\begin{array}{r} 1{,}603 \\ 242 \\ \hline 1{,}845 \end{array}$$

(*c*) 𓆼𓍢𓎆𓎆𓎆𓎆𓎆𓎆𓏺𓏺

$$\begin{array}{r} 685 \\ 477 \\ \hline 1{,}162 \end{array}$$

3 (*a*) CCC (*b*) D (*c*) MMDL
(*d*) MDCCCXII (*e*) MLXVI
(*f*) MCCCCLXXXXII and MCDXCII
(*g*) MDCCCCXXXXII and MCMXLII

4 (*a*) MMMDCCLXXXIII

$$\begin{array}{r} 2{,}162 \\ 1{,}621 \\ \hline 3{,}783 \end{array}$$

(*b*) MMMDCCXVI

$$\begin{array}{r} 2{,}888 \\ 828 \\ \hline 3{,}716 \end{array}$$

(*c*) MMMCCX

$$\begin{array}{r} 1{,}944 \\ 1{,}266 \\ \hline 3{,}210 \end{array}$$

5 (*a*) $\overline{\mu}$ (*b*) $\overline{\mu\delta}$ (*c*) $\overline{\tau}$ (*d*) $\overline{{}_{/}\beta\tau\mu}$
(*e*) $\overline{{}_{/}\alpha\phi\delta}$ (*f*) $\overline{{}_{/}\beta\phi\lambda\delta}$ (*g*) $\overline{\text{ϡϙ}\theta}$

6 (*a*) 36 (*b*) 59 (*c*) 102 (*d*) 854
(*e*) 44,082 (*f*) 43,242 (*g*) 216,882

7 (*a*) Egyptian, Roman, Babylonian (*b*) Greek (*c*) Simple grouping (*d*) Cipher (*e*) Egyptian, Roman, Greek (*f*) Babylonian (*g*) It is not necessary to create new symbols for larger numbers. (*h*) There are ambiguities.

9 (*a*) $\sqrt{3}$. It has not yet been shown that $\sqrt{3}$ is irrational (Exercise 12).
(*b*) (*i*) $\sqrt{5}$; (*ii*) $\sqrt{13}$; (*iii*) $\sqrt{17}$; (*iv*) $\sqrt{20}$; (*v*) 5
(*c*) The values of c in parts a and b (v) are rational.

10 $97/56 \approx 1.732\cdots$

14 (*a*) 6 inches (*b*) 3 (*c*) $2/\sqrt{3}$ inches
(*d*) $12/\sqrt{3}$ inches (*e*) $6/\sqrt{3} \approx 3.464$

15 (*a*) 72 (*b*) 3,612 (*c*) 1,244 (*d*) 72,024 (*e*) 216,944

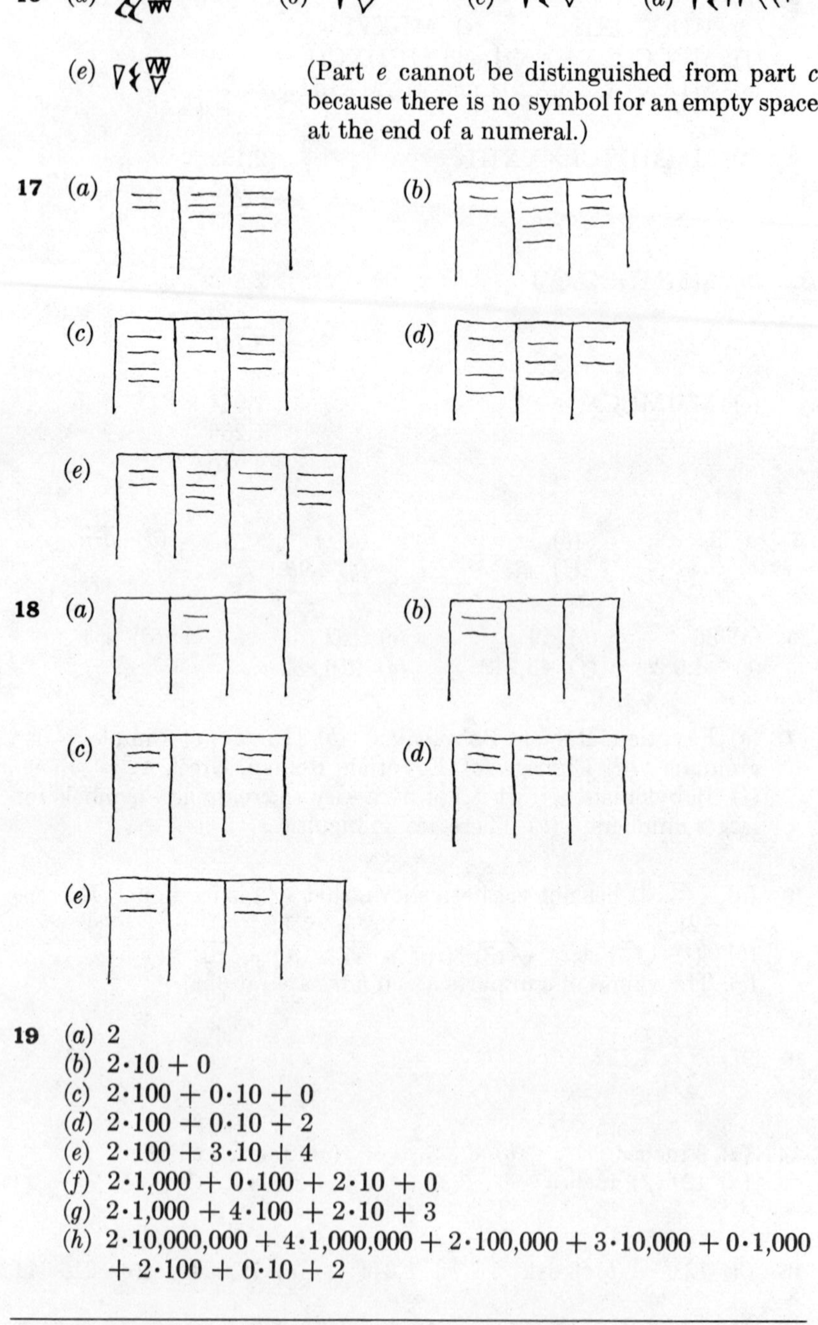

16 (*a*) (*b*) (*c*) (*d*)

(*e*) (Part *e* cannot be distinguished from part *c* because there is no symbol for an empty space at the end of a numeral.)

17 (*a*) (*b*) (*c*) (*d*) (*e*)

18 (*a*) (*b*) (*c*) (*d*) (*e*)

19 (*a*) 2
(*b*) $2 \cdot 10 + 0$
(*c*) $2 \cdot 100 + 0 \cdot 10 + 0$
(*d*) $2 \cdot 100 + 0 \cdot 10 + 2$
(*e*) $2 \cdot 100 + 3 \cdot 10 + 4$
(*f*) $2 \cdot 1{,}000 + 0 \cdot 100 + 2 \cdot 10 + 0$
(*g*) $2 \cdot 1{,}000 + 4 \cdot 100 + 2 \cdot 10 + 3$
(*h*) $2 \cdot 10{,}000{,}000 + 4 \cdot 1{,}000{,}000 + 2 \cdot 100{,}000 + 3 \cdot 10{,}000 + 0 \cdot 1{,}000 + 2 \cdot 100 + 0 \cdot 10 + 2$

20 (*a*) Babylonian, Mayan, Hindu-Arabic
(*b*) Greek
(*c*) Hindu-Arabic
(*d*) Hindu-Arabic
(*e*) It is not necessary to create new symbols for larger numbers.
(*f*) There are no ambiguities.
(*g*) The numerals are not cumbersome to write.
(*h*) The symbol for each number is not difficult to remember.

21 (*a*), (*c*), and (*e*)

22 (*a*), *i*, *ii*, *v*, and *ix* (*b*) Yes (*c*) No

23 *i*, *ii*, *iii*, *iv*, *v*, and *ix*

24 (*a*) *i*, *ii*, *iii*, *iv*, *v*, *vii*, and *ix* (*b*) Yes (*c*) No

25 (*a*) 9 (*b*) 9 (*c*) 3 (*d*) 3
(*e*) -9 (*f*) -9 (*g*) -3 (*h*) -3

26 (*a*) (*i*) 3 and -3; (*ii*) $3i$ and $-3i$; (*iii*) $\sqrt{3}$ and $-\sqrt{3}$; (*iv*) $i\sqrt{3}$ and $-i\sqrt{3}$ (*b*) Yes

Chapter 2

1 *b*, *d*, *f*
(*a*) The word "warm" is a controversial word.
(*c*) The word "big" is a controversial word.
(*f*) The word "between" is sometimes taken to include the numbers at each end and sometimes not to include them.

2 *b*, *c*, *d*, and *f*
(*a*) There could be amoebae elsewhere than on the earth.
(*e*) There are living things other than animals in the United States.

3 *b* and *c*
(*a*) The numbers 1 and 2 are elements of the set.
(*d*) We cannot tell whether or not the set is empty.
(*e*) Astronauts and various animals are elements of the set.
(*f*) The set may or may not be empty at the time you read its description.

4 (*a*) 16
(*b*) $\{a, b, c, d\}$, $\{a, b, c\}$, $\{a, b, d\}$, $\{a, c, d\}$, $\{b, c, d\}$, $\{a, b\}$, $\{a, c\}$, $\{a, d\}$, $\{b, c\}$, $\{b, d\}$, $\{c, d\}$, $\{a\}$, $\{b\}$, $\{c\}$, $\{d\}$, $\emptyset$
(*c*) 32
(*d*) $\{a, b, c, d, e\}$, $\{a, b, c, d\}$, $\{a, b, c, e\}$, $\{a, b, d, e\}$, $\{a, c, d, e\}$, $\{b, c, d, e\}$, $\{a, b, c\}$, $\{a, b, d\}$, $\{a, b, e\}$, $\{a, c, d\}$, $\{a, c, e\}$, $\{a, d, e\}$, $\{b, c, d\}$, $\{b, c, e\}$, $\{b, d, e\}$, $\{c, d, e\}$, $\{a, b\}$, $\{a, c\}$, $\{a, d\}$, $\{a, e\}$, $\{b, c\}$, $\{b, d\}$, $\{b, e\}$, $\{c, d\}$, $\{c, e\}$, $\{d, e\}$, $\{a\}$, $\{b\}$, $\{c\}$, $\{d\}$, $\{e\}$, $\emptyset$

5 (*a*) *a* *b* *c* *d*

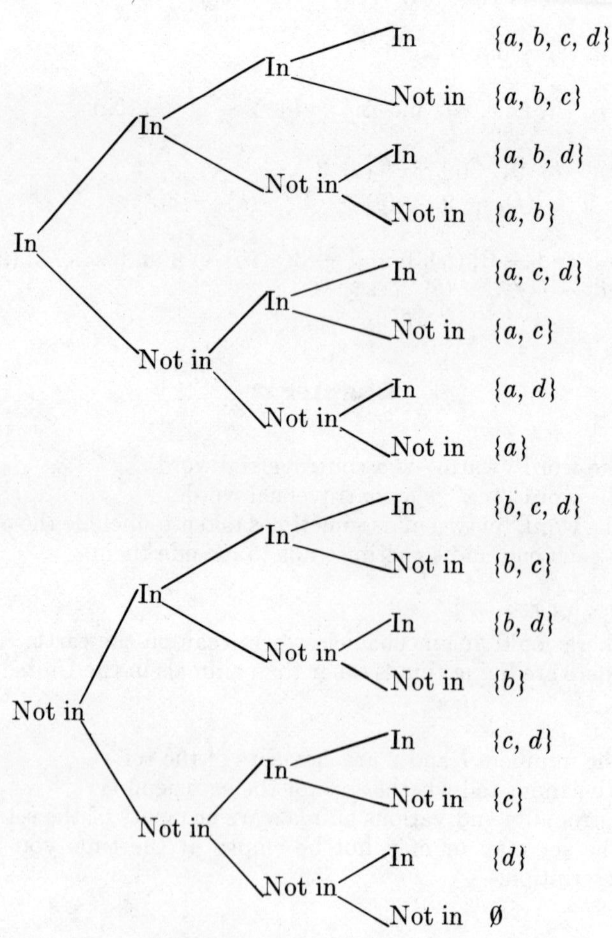

6 (*a*) 2^{11} (*b*) 2,048 (*c*) 2^{20}
(*d*) 1,000,000 (*e*) 2^{30} (*f*) 1,000,000,000

7 (*a*) 10^{60}
(*b*) 1,000,000,000,000,000,000,000,000,000,000,–
000,000,000,000,000,000,000,000,000,000
(This number is called a novemdecillion.)

8 (*a*) ///
/// The sets are equivalent.
(*b*) ///
~~////~~ / The sets are not equivalent.
(*c*) ~~////~~ /
~~////~~ // The sets are not equivalent.
(*d*) ~~////~~ //
~~////~~ /// The sets are not equivalent.
(*e*) ////
//// The sets are equivalent.
(*f*) ~~////~~ /
~~////~~ / The sets are equivalent.

9 (*a*) 1 2 3 4 5 6 7 8 9 10
2 4 6 8 10
The sets are not equivalent.
(*b*) 2 4 6 8 10
1 3 5 7 9 The sets are equivalent.
(*c*) 2 4 6 8 10
1 2 3 4 5 The sets are equivalent.
(*d*) 1 2 3 . . . 1,000,000
2 4 6 . . . 1,000,000 The sets are not equivalent.
(*e*) 2 4 6 . . . 1,000,000
1 3 5 . . . 999,999 The sets are equivalent.
(*f*) 2 4 6 . . . 1,000,000
1 2 3 . . . 500,000 The sets are equivalent.

12 (*a*) 3: red white blue
↕ ↕ ↕
1 2 3
(*b*) 4: f o u r
↕ ↕ ↕ ↕
1 2 3 4

(*c*) 5: 2 4 6 8 10
↕ ↕ ↕ ↕ ↕
1 2 3 4 5

(*d*) 500,000: 2 4 6 . . . 1,000,000
↕ ↕ ↕ ↕
1 2 3 . . . 500,000

13 (*a*) Both 3; rule I (*b*) 3 and 5; rule II (*c*) Both 6; rule I
(*d*) 6 and 50; rule II (*e*) Both 5; rules I and II

15 (*a*) None (*b*) None (*c*) The evens and the odds
(*d*) The evens, odds, and counting numbers

16 (*a*) 1 3 5 7 ⋯ $2n-1$ ⋯
↕ ↕ ↕ ↕ ↕
1 2 3 4 ⋯ n ⋯

(*b*) Each counting number is paired with the odd that is one less than the even which is twice it; each odd number is paired with the counting number that is one-half of the even which follows it.

17 (*a*) 1 2 3 4 ⋯ n ⋯
↕ ↕ ↕ ↕ ↕
0 1 2 3 ⋯ $n-1$ ⋯

(*b*) Each counting number is paired with the whole number that is one less than it; each whole number is paired with the counting number that is one more than it.

19 (*a*) {4, 8, 12, 16, . . . }

(*b*) 4 8 12 16 ⋯ $4n$ ⋯
↕ ↕ ↕ ↕ ↕
2 4 6 8 ⋯ $2n$ ⋯

(*c*) The set of evens is infinite because it has a proper subset to which it is equivalent.

20 (*a*)

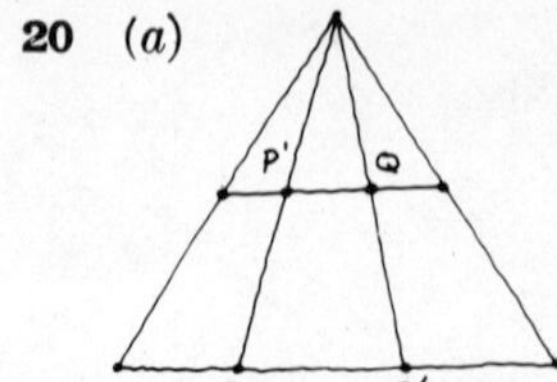

(*b*)

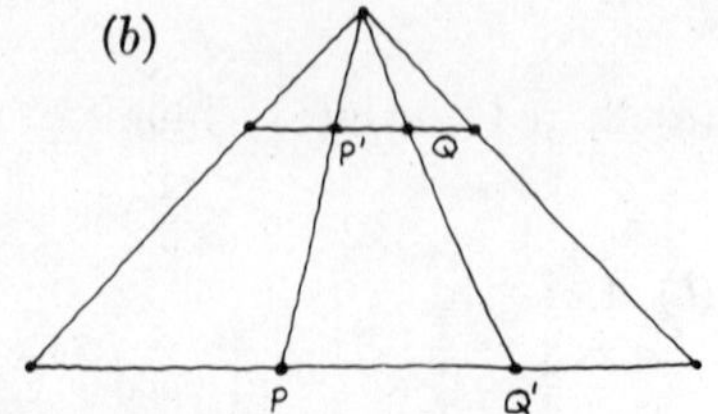

22 (*a*)
$$\begin{array}{cccccc} \frac{1}{1} & \frac{2}{1} & \frac{3}{1} & \frac{4}{1} & \cdots & a/1 & \cdots \\ \updownarrow & \updownarrow & \updownarrow & \updownarrow & & \updownarrow & \\ 1 & 2 & 3 & 4 & \cdots & a & \cdots \end{array}$$

(*b*)
$$\begin{array}{cccccc} \frac{1}{1} & \frac{1}{2} & \frac{1}{3} & \frac{1}{4} & \cdots & 1/a & \cdots \\ \updownarrow & \updownarrow & \updownarrow & \updownarrow & & \updownarrow & \\ 1 & 2 & 3 & 4 & \cdots & a & \cdots \end{array}$$

23 $\frac{0}{1}, \frac{1}{1}, -\frac{1}{1}, \frac{1}{2}, -\frac{1}{2}, \frac{2}{1}, -\frac{2}{1}, \frac{1}{3}, -\frac{1}{3}, \frac{3}{1}, -\frac{3}{1}, \frac{1}{4}, -\frac{1}{4}, \frac{2}{3}, -\frac{2}{3}, \frac{3}{2}, -\frac{3}{2}, \frac{4}{1}, -\frac{4}{1}, \frac{1}{5}, -\frac{1}{5}, \ldots$

24 (*a*) Each decimal from 0.000... through 0.999... is paired with 1 plus that decimal.
(*b*) Uncountable.
(*c*) Uncountable.
(*d*) The set of all reals is uncountable.

25 (*a*) It is equivalent to the set of reals from 0.000... through 0.999....
(*b*) It is equivalent to the set of reals from 0.000... through 1.999.... (Also, it is equivalent to the set of points in a one-inch line segment.)
(*c*) It is equivalent to the set of all reals. (Also, it is equivalent to the set of points in any line segment.)

26 (*a*) $\aleph_0 + 2 = (\aleph_0 + 1) + 1 = \aleph_0 + 1 = \aleph_0$
(*b*) $\aleph_0 + 3 = (\aleph_0 + 2) + 1 = \aleph_0 + 1 = \aleph_0$
(*c*) $\aleph_0 + n = \aleph_0$ for any counting number n.

28 (*a*) $c + 1 = c$ (*b*) $c + \aleph_0 = c$ (*c*) $c + c = c$
(*d*) $c \cdot c = c$
(*e*) The set of reals from 0.000... through 0.999... along with the set of reals from 1.000... through 1.999... is the set of reals from 0.000... through 1.999..., where each of the sets has cardinal number c.
(*f*) 2^c is some transfinite number larger than c.

30 (*a*) If the bibliography lists itself, it is not a bibliography which does not list itself.
(*b*) If the bibliography does not list itself, it does not list all bibliographies which do not list themselves.
(*c*) If the class N is an element of itself, it contains a set which is an element of itself.
(*d*) If the class N is not an element of itself, it does not contain all sets which are not elements of themselves.

31 (*a*) The set of whole numbers {0, 1, 2, 3, . . .}.
(*b*) The set of whole numbers is not finite.
(*c*) Set B contains at least all the transfinite numbers, and the set of transfinite numbers is not finite.
(*d*) Set B contains numbers other than transfinite numbers, such as the cardinal number of the set of transfinite numbers.
(*e*) "Not finite" must include sets such as the set of all transfinite numbers, which cannot be called "infinite" (in the sense that its cardinal number is not a transfinite number).

Chapter 3

1 *a*, *d*, and *e*
(*b*) A question is not true or false.
(*c*) A command is not true or false.

2 The cell under this microscope is an amoeba. Is an element of. The cell is a protozoan. Is an element of. Protozoa are primitive animal organisms. Is a subset of.

3 (*a*) Is identical to.
(*b*) Is an element of.

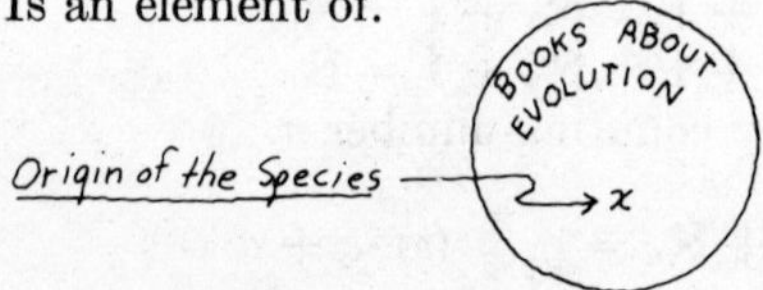

(*c*) Is identical to.
(*d*) Is a subset of.

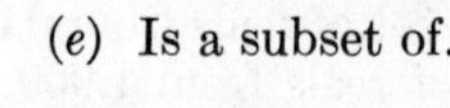

(*e*) Is a subset of.

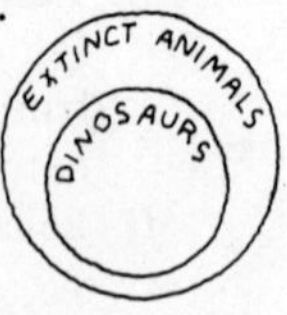

4 (*a*) Audubon is an element of the set of people who painted pictures of birds.

(*b*) The set of birds is a subset of the set of things that live in trees.

(*c*) The set of acorns is a subset of the set of things that grow on trees.

(*d*) The set of acorns is a subset of the set of things that can grow into trees.

(*e*) Kilmer is identical to the person who wrote the poem called "Trees."

5

p	q	$p \veebar q$
T	T	F
T	F	T
F	T	T
F	F	F

6 (*a*) $\sim p$ (*b*) $p \vee q$ (*c*) $p \veebar q$ (*d*) $p \wedge q$

7 (*a*) $\sim (p \wedge q)$ (*b*) $(\sim p) \vee (\sim q)$

(*c*)

p	q	$\sim$	$(p \wedge q)$
T	T	F	T
T	F	T	F
F	T	T	F
F	F	T	F

p	q	$(\sim p)$	$\vee$	$(\sim q)$
T	T	F	F	F
T	F	F	T	T
F	T	T	T	F
F	F	T	T	T

(*d*) The statements have the same truth values.

8 (*a*) George Boole was not a philosopher and not a mathematician.
(*b*) Aristotle was not a philosopher or not a logician.

9 (*a*)

p	q	$(\sim p)$	$\vee$	q
T	T	F	T	T
T	F	F	F	F
F	T	T	T	T
F	F	T	T	F

p	q	p	$\vee$	$(\sim q)$
T	T	T	T	F
T	F	T	T	T
F	T	F	F	F
F	F	F	T	T

The statements are not logically equivalent.

(*b*)

p	q	$(\sim p)$	$\wedge$	q
T	T	F	F	T
T	F	F	F	F
F	T	T	T	T
F	F	T	F	F

p	q	p	$\wedge$	$(\sim q)$
T	T	T	F	F
T	F	T	T	T
F	T	F	F	F
F	F	F	F	T

The statements are not logically equivalent.

(*c*)

p	q	$(\sim p)$	$\vee$	q
T	T	F	T	T
T	F	F	F	F
F	T	T	T	T
F	F	T	T	F

p	q	$\sim$	$[p$	$\wedge$	$(\sim q)]$
T	T	T	T	F	F
T	F	F	T	T	T
F	T	T	F	F	F
F	F	T	F	F	T

The statements are logically equivalent.

(*d*)

p	q	$(\sim p)$	$\wedge$	q
T	T	F	F	T
T	F	F	F	F
F	T	T	T	T
F	F	T	F	F

p	q	$\sim$	$[p$	$\vee$	$(\sim q)]$
T	T	F	T	T	F
T	F	F	T	T	T
F	T	T	F	F	F
F	F	F	F	T	T

The statements are logically equivalent.

12 (*a*)

p	q	$[(\sim p)$	$\vee$	$q]$	$\leftrightarrow$	$[p$	$\vee$	$(\sim q)]$
T	T	F	T	T	T	T	T	F
T	F	F	F	F	F	T	T	T
F	T	T	T	T	F	F	F	F
F	F	T	T	F	T	F	T	T

The statement is not a tautology.

(*b*)

p	q	$[(\sim p)$	$\wedge$	$q]$	$\leftrightarrow$	$[p$	$\wedge$	$(\sim q)]$
T	T	F	F	T	T	T	F	F
T	F	F	F	F	F	T	T	T
F	T	T	T	T	F	F	F	F
F	F	T	F	F	T	F	F	T

The statement is not a tautology.

(*c*)

p	q	$[(\sim p)$	$\vee$	$q]$	$\leftrightarrow$	$\sim$	$[p$	$\wedge$	$(\sim q)]$
T	T	F	T	T	T	T	T	F	F
T	F	F	F	F	T	F	T	T	T
F	T	T	T	T	T	T	F	F	F
F	F	T	T	F	T	T	F	F	T

The statement is a tautology.

(*d*)

p	q	$[(\sim p)$	$\wedge$	$q]$	$\leftrightarrow$	$\sim$	$[p$	$\vee$	$(\sim q)]$
T	T	F	F	T	T	F	T	T	F
T	F	F	F	F	T	F	T	T	T
F	T	T	T	T	T	T	F	F	F
F	F	T	F	F	T	F	F	T	T

The statement is a tautology.

14 (*a*) If I make less than a million dollars, then I will give you half.
(*b*) I make less than a million dollars but I do not give you half.
(*c*) I make a million dollars or more.

15 (*a*) If today is Monday, then you go to your math class. If you go to your math class, then today is Monday. If today is not Monday, then you do not go to your math class. If you do not go to your math class, then today is not Monday.

(*c*) The conditional and the contrapositive are logically equivalent. The converse and the obverse are logically equivalent.

16 (*a*) If you pass this course, then you are good at math. If you are not good at math, then you will not pass this course. If you do not pass this course, then you are not good at math.

(*b*) If you like this course, then you like English. If you do not like English, then you do not like this course. If you do not like this course, then you do not like English.

(*c*) If you fail this course, then you are not good at English. If you are good at English, then you will not fail this course. If you do not fail this course, then you are good at English.

17 (*a*) If today is Monday, then tomorrow is Tuesday. If tomorrow is Tuesday, then today is Monday. If today is not Monday, then tomorrow is not Tuesday. If tomorrow is not Tuesday, then today is not Monday. Today is Monday if and only if tomorrow is Tuesday.

(*b*) All are true.

(*c*) Yes.

19 (*a*)

p	q	$q \to p$
T	T	T
T	F	T
F	T	F
F	F	T

p	q	p	$\vee$	$(\sim q)$
T	T	T	T	F
T	F	T	T	T
F	T	F	F	F
F	F	F	T	T

(*b*)

p	q	$p \wedge q$
T	T	T
T	F	F
F	T	F
F	F	F

p	q	$\sim$	$[(\sim p)$	$\vee$	$(\sim q)]$
T	T	T	F	F	F
T	F	F	F	T	T
F	T	F	T	T	F
F	F	F	T	T	T

(*c*)

p	q	$p \leftrightarrow q$
T	T	T
T	F	F
F	T	F
F	F	T

p	q	$[(\sim p)$	$\vee$	$q]$	$\wedge$	$[p$	$\vee$	$(\sim q)]$
T	T	F	T	T	T	T	T	F
T	F	F	F	F	F	T	T	T
F	T	T	T	T	F	F	F	F
F	F	T	T	F	T	F	T	T

(d)

p	q	$\sim$	$\{\sim$	$[(\sim p)$	$\vee$	$q]$	$\vee$	$\sim$	$[p$	$\vee$	$(\sim q)]\}$
T	T	T	F	F	T	T	F	F	T	T	F
T	F	F	T	F	F	F	T	F	T	T	T
F	T	F	F	T	T	T	T	T	F	F	F
F	F	T	F	T	T	F	F	F	F	T	T

20 (a) If today is Monday, then you go to your math class.
Today is Monday.
———
You go to your math class.

(b) $p \rightarrow q$
p
———
q

(c) $[(p \rightarrow q) \wedge p] \rightarrow q$

21 (a) If today is Monday, then you go to your math class.
You go to your math class.
———
Today is Monday.
The argument is not valid.

(b) If today is Monday, then you go to your math class.
Today is not Monday.
———
You do not go to your math class.
The argument is not valid.

(c) If today is Monday, then you go to your math class.
You do not go to your math class.
———
Today is not Monday.
The argument is valid.

22 (a) $p \rightarrow q$
p
———
q
The argument is valid.

(b) $p \rightarrow q$
$\sim p$
———
$\sim q$
The argument is not valid.

(c) $p \rightarrow q$
q
———
p
The argument is not valid.

(d) $p \rightarrow q$
$\sim q$
———
$\sim p$
The argument is valid.

(e) $\sim p \rightarrow q$
$\sim q$
———
p

or

$p \rightarrow q$
$\sim q$
———
$\sim p$

The argument is valid.

23 (*a*)

p	q	$[(p \to q)$	$\wedge$	$(\sim p)]$	$\to$	$(\sim q)$
T	T	T	F	F	T	F
T	F	F	F	F	T	T
F	T	T	T	T	F	F
F	F	T	T	T	T	T

(*b*)

p	q	$[(p \to q)$	$\wedge$	$(\sim q)]$	$\to$	$(\sim p)$
T	T	T	F	F	T	F
T	F	F	F	T	T	F
F	T	T	F	F	T	T
F	F	T	T	T	T	T

24 (*a*) If you are an intelligent, informed person, then you are the type who subscribes to X magazine.
You subscribe to X magazine.

You are an intelligent, informed person.

(*b*) $p \to q$
q
---
p

(*c*) The argument is not valid.

25 (*a*) If you use brand Y toothpaste, then you will get fewer cavities.
You do not use brand Y toothpaste.

You will not get fewer cavities.

(*b*) $p \to q$
$\sim p$

$\sim q$

(*c*) The argument is not valid.

26 (*a*) If you are somebody, then you like brand Z.
You do not like brand Z.

You are not somebody.

(*b*) $p \to q$
$\sim q$

$\sim p$

(*c*) The argument is valid (but the first statement is false).

27 (*a*) (*b*)

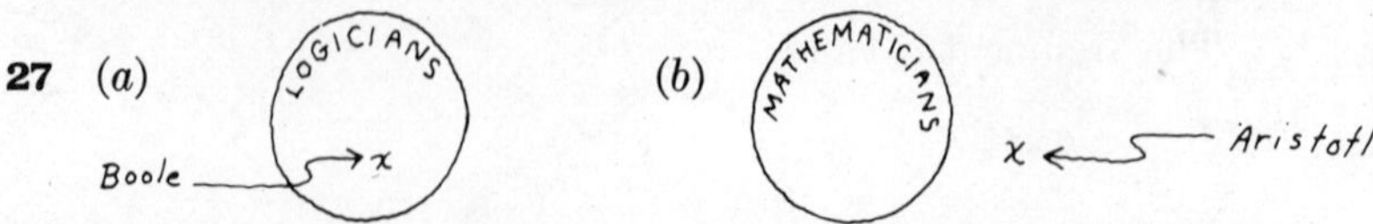

28 (*a*) $\forall x{:}p$

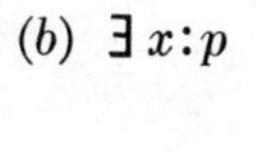

(*b*) $\exists x{:}p$

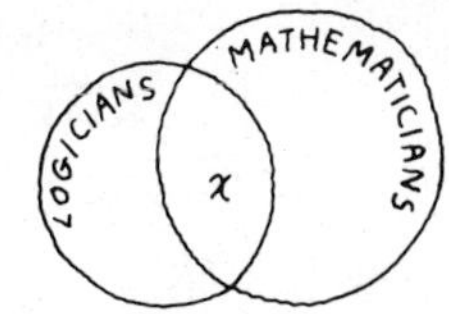

(*c*) $\forall x{:} \sim p$

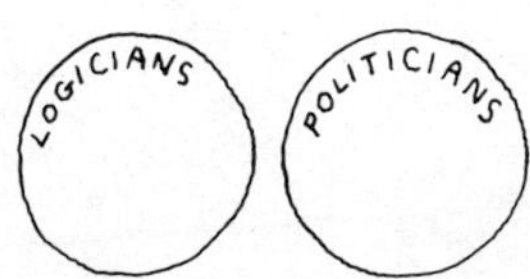

(*d*) $\exists x{:} \sim p$

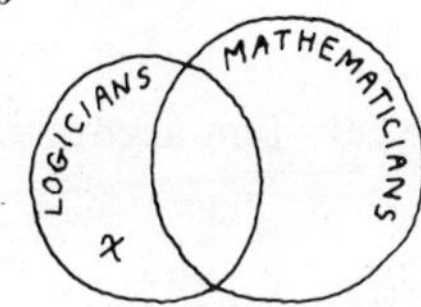

(*e*) Same as part *c*.

30 (*a*) George Boole was not a mathematician.
(*b*) Aristotle was a mathematician.
(*c*) $\forall x{:}p$
$\exists x{:} \sim p$
Some logicians are not philosophers.
(*d*) $\exists x{:}p$
$\forall x{:} \sim p$
All logicians are not mathematicians, or
No logicians are mathematicians.
(*e*) $\forall x{:} \sim p$
$\exists x{:}p$
Some logicians are politicians.
(*f*) $\exists x{:} \sim p$
$\forall x{:}p$
All logicians are mathematicians.
(*g*) Same as part *e*.

31 (*a*) It is false that some banks do insure all deposits in full.
(*b*) No.
(*c*) It is false that all banks do insure all deposits in full.
Some banks do not insure all deposits in full.

32 (*a*) All politicians are humans.
All humans make mistakes.
____
All politicians make mistakes.

(*b*)

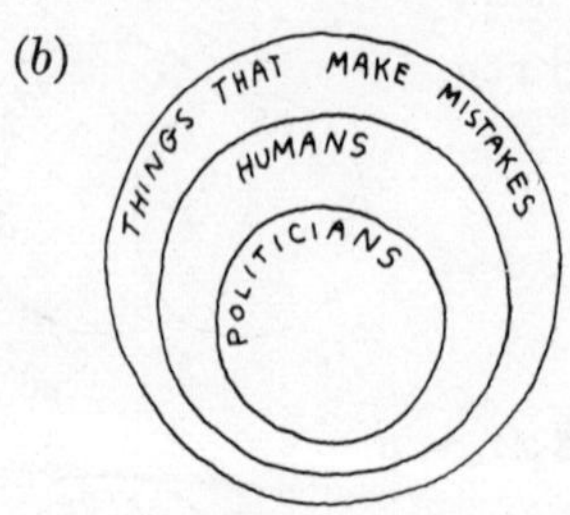

(*c*) The argument is valid.

33 (*a*) All politicians make mistakes.
All humans make mistakes.
All politicians are human.

(*b*)

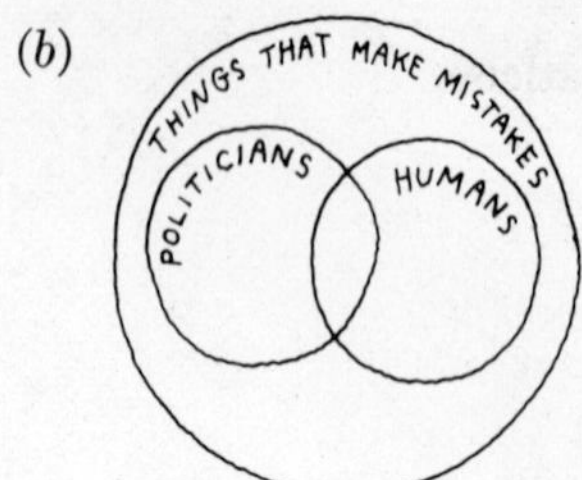

(*c*) The argument is not valid.

34 (*a*) Some politicians accept kickbacks for contracts on urban renewal projects.
Politician P is planning an urban renewal project.
Politician P accepts kickbacks.

(*b*)

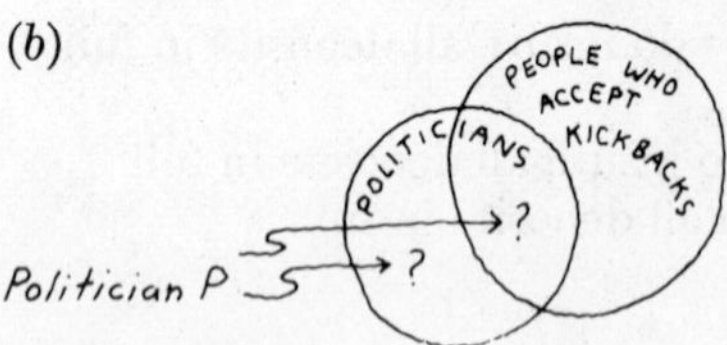

(*c*) The argument is not valid.

35

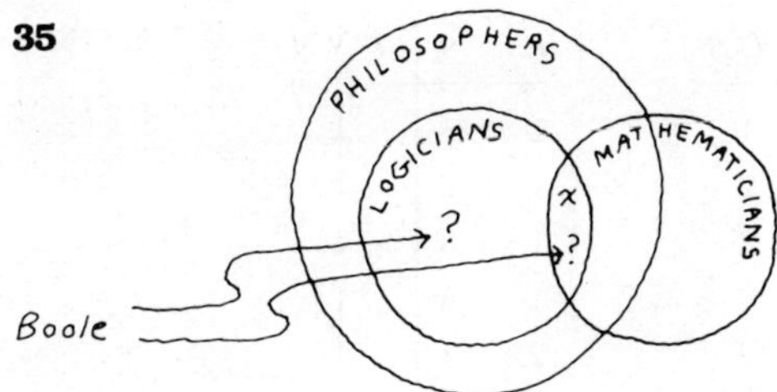

Conclusions *a*, *d*, and *e* are valid.

36

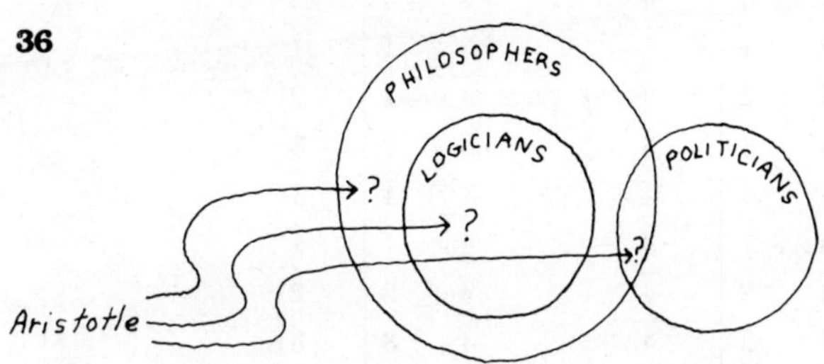

Conclusion *d* is valid.

37 (*a*) 256

(*b*)

p	q	$p \vee q$
T	T	T
T	?	T
T	F	T
?	T	T
?	?	?
?	F	?
F	T	T
F	?	?
F	F	F

p	q	$p \vee q$
T	T	T
T	?	?
T	F	T
?	T	?
?	?	?
?	F	?
F	T	T
F	?	?
F	F	F

(*c*) In the first table, $p \vee q$ is the "higher" of the truth values of p and of q. In the second table, $p \vee q$ is undecidable whenever p is undecidable or q is undecidable, and the same as in two-valued logic otherwise.

38

p	$\sim p$
1	4
2	3
3	2
4	1

p	q	$p \wedge q$
1	1	1
1	2	2
1	3	3
1	4	4
2	1	2
2	2	2
2	3	3
2	4	4
3	1	3
3	2	3
3	3	3
3	4	4
4	1	4
4	2	4
4	3	4
4	4	4

p	q	$p \vee q$
1	1	1
1	2	1
1	3	1
1	4	1
2	1	1
2	2	2
2	3	2
2	4	2
3	1	1
3	2	2
3	3	3
3	4	3
4	1	1
4	2	2
4	3	3
4	4	4

Chapter 4

1

·	0	1	2	3	4	5	6	7	8	9	10	11
0	0	0	0	0	0	0	0	0	0	0	0	0
1	0	1	2	3	4	5	6	7	8	9	10	11
2	0	2	4	6	8	10	0	2	4	6	8	10
3	0	3	6	9	0	3	6	9	0	3	6	9
4	0	4	8	0	4	8	0	4	8	0	4	8
5	0	5	10	3	8	1	6	11	4	9	2	7
6	0	6	0	6	0	6	0	6	0	6	0	6
7	0	7	2	9	4	11	6	1	8	3	10	5
8	0	8	4	0	8	4	0	8	4	0	8	4
9	0	9	6	3	0	9	6	3	0	9	6	3
10	0	10	8	6	4	2	0	10	8	6	4	2
11	0	11	10	9	8	7	6	5	4	3	2	1

2 (*a*) 6, 5, 4, 3, 2, 1
(*b*) {0, 1, 2, 3, 4, 5}, {0, 1, 2, 3, 4}, {0, 1, 2, 3}, {0, 1, 2}, {0, 1}, {0}
(*c*) There is no arithmetic I_0 because I_0 is the empty set.

3 (*a*) 0 (*b*) 0 (*c*) 1 (*d*) 2 (*e*) 0 (*f*) 2 (*g*) 1

4 (*a*)

$(I_6, +)$	0	1	2	3	4	5
0	0	1	2	3	4	5
1	1	2	3	4	5	0
2	2	3	4	5	0	1
3	3	4	5	0	1	2
4	4	5	0	1	2	3
5	5	0	1	2	3	4

$(I_6, \cdot)$	0	1	2	3	4	5
0	0	0	0	0	0	0
1	0	1	2	3	4	5
2	0	2	4	0	2	4
3	0	3	0	3	0	3
4	0	4	2	0	4	2
5	0	5	4	3	2	1

(*b*)

$(I_3, +)$	0	1	2
0	0	1	2
1	1	2	0
2	2	0	1

$(I_3, \cdot)$	0	1	2
0	0	0	0
1	0	1	2
2	0	2	1

(*c*)

$(I_1, +)$	0
0	0

$(I_1, \cdot)$	0
0	0

6 (*a*)

$(I_6 - 0, \cdot)$	1	2	3	4	5
1	1	2	3	4	5
2	2	4	0	2	4
3	3	0	3	0	3
4	4	2	0	4	2
5	5	4	3	2	1

$(I_3 - 0, \cdot)$	1	2
1	1	2
2	2	1

$(I_2 - 0, \cdot)$	1
1	1

(*b*) $(I_3 - 0, \cdot)$ and $(I_2 - 0, \cdot)$

(*c*) $(I_6 - 0, \cdot)$ is not closed under multiplication because 0 is among the entries in the table but not an element of $(I_6 - 0, \cdot)$.

(*d*) There is no arithmetic $(I_1 - 0, \cdot)$ because $I_1 - 0$ is the empty set.

8 (*a*) $(1 + 2) + 3 = 3 + 3 \equiv 1 \pmod 5$
$1 + (2 + 3) \equiv 1 + 0 = 1 \pmod 5$

(*e*) $(1 \cdot 2) \cdot 3 = 2 \cdot 3 \equiv 1 \pmod 5$
$1 \cdot (2 \cdot 3) \equiv 1 \cdot 1 = 1 \pmod 5$

9 (*a*)

a	$a + 3$	0	1	2	3	4
0	3	3	4	0	1	2
1	4	4	0	1	2	3
2	0	0	1	2	3	4
3	1	1	2	3	4	0
4	2	2	3	4	0	1

c	0	1	2	3	4
$3 + c$	3	4	0	1	2
0	3	4	0	1	2
1	4	0	1	2	3
2	0	1	2	3	4
3	1	2	3	4	0
4	2	3	4	0	1

(*b*)

a	$a + 4$	0	1	2	3	4
0	4	4	0	1	2	3
1	0	0	1	2	3	4
2	1	1	2	3	4	0
3	2	2	3	4	0	1
4	3	3	4	0	1	2

c	0	1	2	3	4
$4 + c$	4	0	1	2	3
0	4	0	1	2	3
1	0	1	2	3	4
2	1	2	3	4	0
3	2	3	4	0	1
4	3	4	0	1	2

10 $(a \cdot 1) \cdot c = a \cdot c$
$a \cdot (1 \cdot c) = a \cdot c$

11 (*a*)

a	$a \cdot 3$	0	1	2	3	4
0	0	0	0	0	0	0
1	3	0	3	1	4	2
2	1	0	1	2	3	4
3	4	0	4	3	2	1
4	2	0	2	4	1	3

c	0	1	2	3	4
$3 \cdot c$	0	3	1	4	2
0	0	0	0	0	0
1	0	3	1	4	2
2	0	1	2	3	4
3	0	4	3	2	1
4	0	2	4	1	3

(*b*)

a	$a \cdot 4$	0	1	2	3	4
0	0	0	0	0	0	0
1	4	0	4	3	2	1
2	3	0	3	1	4	2
3	2	0	2	4	1	3
4	1	0	1	2	3	4

c	0	1	2	3	4
$4 \cdot c$	0	4	3	2	1
0	0	0	0	0	0
1	0	4	3	2	1
2	0	3	1	4	2
3	0	2	4	1	3
4	0	1	2	3	4

14 (*a*) Addition tables for $b = 1$.

(*b*)

a	$a + 1$	0	1
0	1	1	0
1	0	0	1

c	0	1
$1 + c$	1	0
0	1	0
1	0	1

(*c*) The only value of b in I_1 is $b = 0$.

15 (*a*)

$0 + 0 = 0 \qquad 0 + 0 = 0$
$0 + 1 = 1 \qquad 1 + 0 = 1$
$0 + 2 = 2 \qquad 2 + 0 = 2$
$0 + 3 = 3 \qquad 3 + 0 = 3$

(*b*)

$1 \cdot 0 = 0 \qquad 0 \cdot 1 = 0$
$1 \cdot 1 = 1 \qquad 1 \cdot 1 = 1$
$1 \cdot 2 = 2 \qquad 2 \cdot 1 = 2$
$1 \cdot 3 = 3 \qquad 3 \cdot 1 = 3$

17 (*a*) $0 + 0 = 0; 1 + 3 = 0; 2 + 2 = 0; 3 + 1 = 0.$

(*b*) 0 is its own additive inverse. 1 and 3 are additive inverses of one another. 2 is its own additive inverse.

(*c*) $1 \cdot 1 = 1; 3 \cdot 3 = 1.$

(*d*) 1 is its own multiplicative inverse. 3 is its own multiplicative inverse.

18 (*b*) $(I_2, +)$: 0 is its own additive inverse. 1 is its own additive inverse.
$(I_3, +)$: 0 is its own additive inverse. 1 and 2 are additive inverses of one another.
$(I_6, +)$: 0 is its own additive inverse. 1 and 5 are additive inverses of one another. 2 and 4 are additive inverses of one another. 3 is its own additive inverse.

19 (*b*) $(I_2, \cdot)$: 1 is its own multiplicative inverse. $(I_3, \cdot)$: 1 is its own multiplicative inverse. 2 is its own multiplicative inverse.
$(I_6, \cdot)$: 1 is its own multiplicative inverse. 5 is its own multiplicative inverse.

(*c*) $(I_2, \cdot)$: 0; $(I_3, \cdot)$: 0; $(I_6, \cdot)$: 0, 2, 3, 4.

(*d*) $(I_2, \cdot)$ and $(I_3, \cdot)$

20 (*a*) $(I_2-0,\cdot)$: 1 is its own multiplicative inverse.
$(I_3-0,\cdot)$: 1 is its own multiplicative inverse. 2 is its own multiplicative inverse.
$(I_4-0,\cdot)$: 1 is its own multiplicative inverse. 3 is its own multiplicative inverse.
$(I_5-0,\cdot)$: 1 is its own multiplicative inverse. 2 and 3 are multiplicative inverses of one another. 4 is its own multiplicative inverse.
$(I_6-0,\cdot)$: 1 is its own multiplicative inverse. 5 is its own multiplicative inverse.
(*b*) $(I_4-0,\cdot)$: 2; $(I_6-0,\cdot)$: 2, 3, 4.
(*c*) $(I_2-0,\cdot)$, $(I_3-0,\cdot)$, $(I_5-0,\cdot)$.

21 (*a*) There are no divisors of zero.
(*b*) Yes

22 (*b*) 2, 3, 4
(*c*) No

23 (*a*)

$(I_7,\cdot)$	0	1	2	3	4	5	6
0	0	0	0	0	0	0	0
1	0	1	2	3	4	5	6
2	0	2	4	6	1	3	5
3	0	3	6	2	5	1	4
4	0	4	1	5	2	6	3
5	0	5	3	1	6	4	2
6	0	6	5	4	3	2	1

$(I_8,\cdot)$	0	1	2	3	4	5	6	7
0	0	0	0	0	0	0	0	0
1	0	1	2	3	4	5	6	7
2	0	2	4	6	0	2	4	6
3	0	3	6	1	4	7	2	5
4	0	4	0	4	0	4	0	4
5	0	5	2	7	4	1	6	3
6	0	6	4	2	0	6	4	2
7	0	7	6	5	4	3	2	1

$(I_9,\cdot)$	0	1	2	3	4	5	6	7	8
0	0	0	0	0	0	0	0	0	0
1	0	1	2	3	4	5	6	7	8
2	0	2	4	6	8	1	3	5	7
3	0	3	6	0	3	6	0	3	6
4	0	4	8	3	7	2	6	1	5
5	0	5	1	6	2	7	3	8	4
6	0	6	3	0	6	3	0	6	3
7	0	7	5	3	1	8	6	4	2
8	0	8	7	6	5	4	3	2	1

(*b*) $(I_7,\cdot)$: There are no divisors of zero.
$(I_8,\cdot)$: 2, 4, 6
$(I_9,\cdot)$: 3 and 6

25 (*a*) $2+0=2$ $\quad 0+2=2$
$2+1=3$ $\quad 1+2=3$
$2+2=4$ $\quad 2+2=4$
$2+3=0$ $\quad 3+2=0$
$2+4=1$ $\quad 4+2=1$

(*b*) $3+0=3$ $\quad 0+3=3$
$3+1=4$ $\quad 1+3=4$
$3+2=0$ $\quad 2+3=0$
$3+3=1$ $\quad 3+3=1$
$3+4=2$ $\quad 4+3=2$

(*c*) $4+0=4$ $\quad 0+4=4$
$4+1=0$ $\quad 1+4=0$
$4+2=1$ $\quad 2+4=1$
$4+3=2$ $\quad 3+4=2$
$4+4=3$ $\quad 4+4=3$

(*d*) $2\cdot0=0$ $\quad 0\cdot2=0$
$2\cdot1=2$ $\quad 1\cdot2=2$
$2\cdot2=4$ $\quad 2\cdot2=4$
$2\cdot3=1$ $\quad 3\cdot2=1$
$2\cdot4=3$ $\quad 4\cdot2=3$

(*e*) $4\cdot0=0$ $\quad 0\cdot4=0$
$4\cdot1=4$ $\quad 1\cdot4=4$
$4\cdot2=3$ $\quad 2\cdot4=3$
$4\cdot3=2$ $\quad 3\cdot4=2$
$4\cdot4=1$ $\quad 4\cdot4=1$

27 (*a*) $(I_2-0,\cdot)$: closure, associative, identity, inverse, commutative
$(I_3-0,\cdot)$: closure, associative, identity, inverse, commutative
$(I_6-0,\cdot)$: associative, identity, commutative
$(I_7-0,\cdot)$: closure, associative, identity, inverse, commutative
$(I_8-0,\cdot)$: associative, identity, commutative
$(I_9-0,\cdot)$: associative, identity, commutative
(*b*) $(I_2-0,\cdot)$, $(I_3-0,\cdot)$, $(I_7-0,\cdot)$

28 (*a*) $(I_2-0,\cdot)$, $(I_3-0,\cdot)$, $(I_7-0,\cdot)$

29 (*a*) $(I_1,+)$, $(I_2,+)$, $(I_3,+)$, $(I_4,+)$, $(I_5,+)$, $(I_6,+)$
$(I_1,\cdot)$, $(I_2,\cdot)$, $(I_3,\cdot)$, $(I_4,\cdot)$, $(I_5,\cdot)$, $(I_6,\cdot)$, $(I_7,\cdot)$, $(I_8,\cdot)$, $(I_9,\cdot)$
$(I_2-0,\cdot)$, $(I_3-0,\cdot)$, $(I_5-0,\cdot)$, $(I_7-0,\cdot)$

(*b*) $(I_1,+)$, $(I_2,+)$, $(I_3,+)$, $(I_4,+)$, $(I_5,+)$, $(I_6,+)$
$(I_2-0,\cdot)$, $(I_3-0,\cdot)$, $(I_5-0,\cdot)$, $(I_7-0,\cdot)$

(*c*) $(I_1,+)$, $(I_2,+)$, $(I_3,+)$, $(I_4,+)$, $(I_5,+)$, $(I_6,+)$
$(I_2-0,\cdot)$, $(I_3-0,\cdot)$, $(I_5-0,\cdot)$, $(I_7-0,\cdot)$

(*d*) $(I_4-0,\cdot)$, $(I_6-0,\cdot)$, $(I_8-0,\cdot)$, $(I_9-0,\cdot)$

30 $(I_{11}-0,\cdot)$ and $(I_{13}-0,\cdot)$

31 (*a*) $1 \cdot (2+3) \equiv 1 \cdot 0 = 0 \pmod 5$
$(1 \cdot 2) + (1 \cdot 3) = 2 + 3 \equiv 0 \pmod 5$

(*j*) No

32 (*a*) $1 \cdot (2+3) \equiv 1 \cdot 1 = 1 \pmod 4$
$(1 \cdot 2) + (1 \cdot 3) = 2 + 3 \equiv 1 \pmod 4$

(*g*) No

33 (*a*) $0 \cdot (0+0) = 0 \cdot 0 = 0 \pmod 1$
$(0 \cdot 0) + (0 \cdot 0) = 0 + 0 = 0 \pmod 1$

(*b*) 1

34 (*a*) All

(*b*) All except $(I_1,+,\cdot)$

(*c*) $(I_2,+,\cdot)$, $(I_3,+,\cdot)$, $(I_5,+,\cdot)$, $(I_7,+,\cdot)$, $(I_{11},+,\cdot)$, $(I_{13},+,\cdot)$

(*d*) If n is prime, $(I_n,+,\cdot)$ is a field.

35 (*a*) $(C,\cdot)$ has the closure property because every entry in the table is an element of the arithmetic.

(*b*)

a	**a · i**	1	−1	i	$-i$
1	i	i	$-i$	−1	1
−1	$-i$	$-i$	i	1	−1
i	−1	−1	1	$-i$	i
$-i$	1	1	−1	i	$-i$

c	1	−1	i	$-i$
i · c	i	$-i$	−1	1
1	i	$-i$	−1	1
−1	$-i$	i	1	−1
i	−1	1	$-i$	i
$-i$	1	−1	i	$-i$

(*c*) 1 is the identity for $(C,\cdot)$.

(*d*) 1 is its own multiplicative inverse. −1 is its own multiplicative inverse. i and $-i$ are multiplicative inverses of one another.

(*e*) $(C,\cdot)$ has the commutative property because each pair of corresponding rows and columns is identical.
(*f*) $(C,\cdot)$ is a commutative group because it has the closure, associative, identity, inverse, and commutative properties.

36 (*a*) $(i \cdot j) \cdot j = j \cdot j = -1$ (*f*) No
$i \cdot (j \cdot j) = i \cdot -1 = -i$

37 (*a*) $(Q_a,\cdot)$ has the closure property because every entry in the table is an element of the arithmetic.
(*b*) 1 is the identity for $(Q_a,\cdot)$.
(*c*) 1 is its own multiplicative inverse. -1 is its own multiplicative inverse. i and $-i$ are multiplicative inverses of one another. j and $-j$ are multiplicative inverses of one another. k and $-k$ are multiplicative inverses of one another.
(*d*) $(Q_a,\cdot)$ has the commutative property because each pair of corresponding rows and columns is identical.

38 (*a*)

a	**a · j**	1	−1	i	$-i$	j	$-j$	k	$-k$
1	j	j	$-j$	k	$-k$	−1	1	i	$-i$
−1	$-j$	$-j$	j	$-k$	k	1	−1	$-i$	i
i	k	k	$-k$	j	$-j$	i	$-i$	−1	1
$-i$	$-k$	$-k$	k	$-j$	j	$-i$	i	1	−1
j	−1	−1	1	$-i$	i	$-j$	j	$-k$	k
$-j$	1	1	−1	i	$-i$	j	$-j$	k	$-k$
k	i	i	$-i$	−1	1	k	$-k$	j	$-j$
$-k$	$-i$	$-i$	i	1	−1	$-k$	k	$-j$	j

c	1	−1	i	$-i$	j	$-j$	k	$-k$
j · c	j	$-j$	k	$-k$	−1	1	i	$-i$
1	j	$-j$	k	$-k$	−1	1	i	$-i$
−1	$-j$	j	$-k$	k	1	−1	$-i$	i
i	k	$-k$	j	$-j$	$-i$	i	−1	1
$-i$	$-k$	k	$-j$	j	i	$-i$	1	−1
j	−1	1	i	$-i$	$-j$	j	k	$-k$
$-j$	1	−1	$-i$	i	j	$-j$	$-k$	k
k	i	$-i$	−1	1	$-k$	k	j	$-j$
$-k$	$-i$	i	1	−1	k	$-k$	$-j$	j

(b)

$\mathbf{a}$	$\mathbf{a \cdot k}$	1	-1	i	$-i$	j	$-j$	k	$-k$
1	k	k	$-k$	j	$-j$	i	$-i$	-1	1
-1	$-k$	$-k$	k	$-j$	j	$-i$	i	1	-1
i	j	j	$-j$	k	$-k$	-1	1	i	$-i$
$-i$	$-j$	$-j$	j	$-k$	k	1	-1	$-i$	i
j	i	i	$-i$	-1	1	k	$-k$	j	$-j$
$-j$	$-i$	$-i$	i	1	-1	$-k$	k	$-j$	j
k	-1	-1	1	$-i$	i	$-j$	j	$-k$	k
$-k$	1	1	-1	i	$-i$	j	$-j$	k	$-k$

$\mathbf{c}$	1	-1	i	$-i$	j	$-j$	k	$-k$
$\mathbf{k \cdot c}$	k	$-k$	j	$-j$	i	$-i$	-1	1
1	k	$-k$	j	$-j$	i	$-i$	-1	1
-1	$-k$	k	$-j$	j	$-i$	i	1	-1
i	j	$-j$	k	$-k$	-1	1	$-i$	i
$-i$	$-j$	j	$-k$	k	1	-1	i	$-i$
j	i	$-i$	-1	1	k	$-k$	$-j$	j
$-j$	$-i$	i	1	-1	$-k$	k	j	$-j$
k	-1	1	i	$-i$	j	$-j$	$-k$	k
$-k$	1	-1	$-i$	i	$-j$	j	k	$-k$

40 (a) $(Q,\cdot)$ has the closure property because every entry in the table is an element of the arithmetic.

(c) 1 is the identity for $(Q,\cdot)$.

(d) 1 is its own multiplicative inverse. -1 is its own multiplicative inverse. i and $-i$ are multiplicative inverses of one another. j and $-j$ are multiplicative inverses of one another. k and $-k$ are multiplicative inverses of one another.

(e) $(Q,\cdot)$ does not have the commutative property because the row and column for i, $-i$, j, $-j$, k, and $-k$ are not identical.

(f) $(Q,\cdot)$ is a group because it has the closure, associative, identity, and inverse properties.
$(Q,\cdot)$ is not a commutative group because it does not have the commutative property.

Chapter 5

1 (*a*) $(R_4,*)$ has the closure property because every entry in the table is an element of R_4.
(*c*) I is the identity for $(R_4,*)$.
(*d*) I is its own inverse. R_1 and R_3 are inverses of one another. R_2 is its own inverse.
(*e*) $(R_4,*)$ has the commutative property because each pair of corresponding rows and columns is identical.
(*f*) $(R_4,*)$ is a commutative group because it has the closure, associative, identity, inverse, and commutative properties.

2 (*a*)

$(R_3,*)$	I	R_1	R_2
I	I	R_1	R_2
R_1	R_1	R_2	I
R_2	R_2	I	R_1

(*b*)

$(R_6,*)$	I	R_1	R_2	R_3	R_4	R_5
I	I	R_1	R_2	R_3	R_4	R_5
R_1	R_1	R_2	R_3	R_4	R_5	I
R_2	R_2	R_3	R_4	R_5	I	R_1
R_3	R_3	R_4	R_5	I	R_1	R_2
R_4	R_4	R_5	I	R_1	R_2	R_3
R_5	R_5	I	R_1	R_2	R_3	R_4

3 (*a*) $(S_4,*)$ has the closure property because every entry in the table is an element of S_4.
(*c*) I is the identity for $(S_4,*)$.
(*d*) I is its own inverse. R_1 and R_3 are inverses of one another. R_2, V, H, D_1, and D_2 each is its own inverse.
(*e*) No. The row and column for R_1, R_3, V, H, D_1, and D_2 are not identical.
(*f*) Yes. No. It does not have the commutative property.

4 (*a*)

$(S_3,*)$	I	R_1	R_2	D_1	D_2	D_3
I	I	R_1	R_2	D_1	D_2	D_3
R_1	R_1	R_2	I	D_2	D_3	D_1
R_2	R_2	I	R_1	D_3	D_1	D_2
D_1	D_1	D_3	D_2	I	R_2	R_1
D_2	D_2	D_1	D_3	R_1	I	R_2
D_3	D_3	D_2	D_1	R_2	R_1	I

(*b*) Yes. No. The row and column for R_1, R_2, D_1, D_2, and D_3 are not identical.

5 (*a*)

*	I	R_1	R_3
I	I	R_1	R_3
R_1	R_1	R_2	I
R_3	R_3	I	R_2

(*b*)

*	I	V	D_1
I	I	V	D_1
V	V	I	R_3
D_1	D_1	R_1	I

(*e*) Closure

6 (*a*)

*	V	H	D_1	D_2
V	I	R_2	R_3	R_1
H	R_2	I	R_1	R_3
D_1	R_1	R_3	I	R_2
D_2	R_3	R_1	R_2	I

(*b*) Closure, identity, commutative
(*c*) No

7 (*a*) 1, 2, 3, 6

(*b*)

*	I
I	I

*	I	D_1
I	I	D_1
D_1	D_1	I

*	I	D_2
I	I	D_2
D_2	D_2	I

*	I	D_3
I	I	D_3
D_3	D_3	I

*	I	R_1	R_2
I	I	R_1	R_2
R_1	R_1	R_2	I
R_2	R_2	I	R_1

$(S_3,*)$

8 (*a*) $(I_5,+)$: 1, 5
$(I_6,+)$: 1, 2, 3, 6

(*b*) $(I_5,+)$

+	0
0	0

$(I_6,+)$

+	0
0	0

+	0	3
0	0	3
3	3	0

+	0	2	4
0	0	2	4
2	2	4	0
4	4	0	2

9 (*a*) 4, 6
(*b*) $(I_5-0,\cdot)$: 1, 2, 4
$(I_7-0,\cdot)$: 1, 2, 3, 6

(*c*) $(I_5-0,\cdot)$

·	1
1	1

·	1	4
1	1	4
4	4	1

$(I_7-0,\cdot)$

·	1
1	1

·	1	6
1	1	6
6	6	1

·	1	2	4
1	1	2	4
2	2	4	1
4	4	1	2

10 (*a*) 3, 4, 5, 6, 4, 6, 3, 6, 4, 8
(*c*) $(I_3,+)$, $(I_5,+)$, $(R_3,*)$
(*d*) No. An odd number does not have a factor of 2.

12 $(I_3,+)$, $(I_4,+)$, $(I_5,+)$, $(I_6,+)$, $(R_3,*)$, $(R_4,*)$

13 (*a*)

$\circ$	u	a	b
u	u	a	b
a	a	u	u
b	b	u	u

x	$x \circ a$	u	a	b
u	a	a	u	u
a	u	u	a	b
b	u	u	a	b

z	u	a	b
$a \circ z$	a	u	u
u	a	u	u
a	u	a	a
b	u	b	b

14

*	I	R_2	V	H
I	I	R_2	V	H
R_2	R_2	I	H	V
V	V	H	I	R_2
H	H	V	R_2	I

*	I	R_2	D_1	D_2
I	I	R_2	D_1	D_2
R_2	R_2	I	D_2	D_1
D_1	D_1	D_2	I	R_2
D_2	D_2	D_1	R_2	I

15 (*a*) 1 is its own inverse. 2 and 3 are inverses of one another. 4 is its own inverse.
(*b*) Cyclic
(*c*)

·	1	2	4	3
1	1	2	4	3
2	2	4	3	1
4	4	3	1	2
3	3	1	2	4

(answer is not unique)

(*d*)

·	1	3	2	6	4	5
1	1	3	2	6	4	5
3	3	2	6	4	5	1
2	2	6	4	5	1	3
6	6	4	5	1	3	2
4	4	5	1	3	2	6
5	5	1	3	2	6	4

(answer is not unique)

16 (*a*) 0 and 3 (*b*) I, D_1, D_2, D_3
(*c*) No. $(I_6,+)$ has two elements which are their own inverse but $(S_3,*)$ has four. $(I_6,+)$ is commutative but $(S_3,*)$ is not.
(*d*) $(I_8,+)$ is commutative but $(S_4,*)$ and $(Q,\cdot)$ are not. $(S_4,*)$ has six elements which are their own inverse but $(Q,\cdot)$ has two.

17 (*a*) $R_1 * R_1 = R_2$; $(R_1 * R_1) * R_1 = R_2 * R_1 = R_3$; $[(R_1 * R_1) * R_1] * R_1 = (R_2 * R_1) * R_1 = R_3 * R_1 = I$

(*b*)

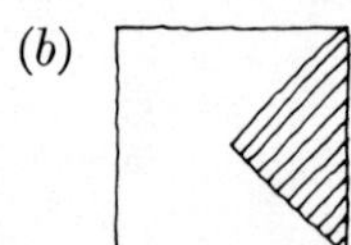

18 (*a*) Fourfold (*b*) Eightfold (*c*) Tenfold (*d*) Square
(*e*) Regular pentagon (*f*) 180-degree rotation, or a reflection

19

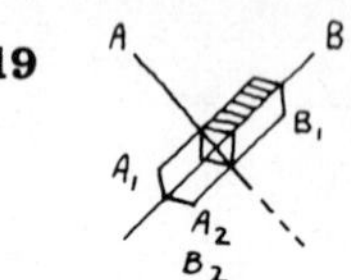

20 (*a*) Twofold (*b*) A — B, A_1, B_1

Chapter 6

1 (*a*) $\{HHH, HHT, HTH, HTT, THH, THT, TTH, TTT\}$
(*b*) (*i*) $\{HHH\}$; (*ii*) $\{HHT, HTH, THH\}$; (*iii*) $\{HHH, HHT, HTH, THH\}$; (*iv*) $\{HTT, THT, TTH\}$; (*v*) $\{HHH, HHT, HTH, THH, HTT, THT, TTH\}$; (*vi*) $\{TTT\}$; (*vii*) $\{TTT\}$.

2 (*a*)

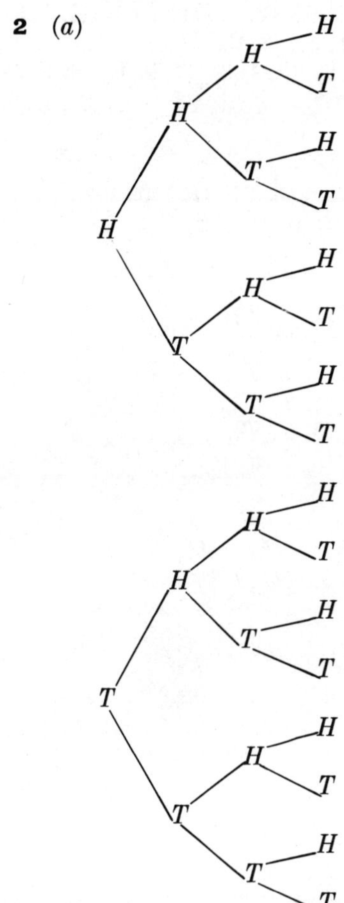

(*b*) {*HHHH*, *HHHT*, *HHTH*, *HHTT*, *HTHH*, *HTHT*, *HTTH*, *HTTT*, *THHH*, *THHT*, *THTH*, *THTT*, *TTHH*, *TTHT*, *TTTH*, *TTTT*}

(*c*) (*i*) {*HHHH*}; (*ii*) {*HHHT*, *HHTH*, *HTHH*, *THHH*}; (*iii*) {*HHHH*, *HHHT*, *HHTH*, *HTHH*, *THHH*}; (*iv*) {*HHTT*, *HTHT*, *HTTH*, *THHT*, *THTH*, *TTHH*}; (*v*) {*HTTT*, *THTT*, *TTHT*, *TTTH*}

3 (*a*) {(1, 6), (6, 1), (2, 5), (5, 2), (3, 4), (4, 3)}
(*b*) {(5, 6), (6, 5)}

4 (*a*) 216
(*b*) (*i*) {(1, 1, 1)}; (*ii*) {(1, 1, 2), (1, 2, 1), (2, 1, 1)}; (*iii*) {(1, 1, 3), (1, 3, 1), (3, 1, 1), (1, 2, 2), (2, 1, 2), (2, 2, 1)}; (*iv*) {(1, 1, 4), (1, 4, 1), (4, 1, 1), (1, 2, 3), (1, 3, 2), (2, 1, 3), (2, 3, 1), (3, 1, 2), (3, 2, 1), (2, 2, 2)}.

5 (*a*) (*i*) {*H*, *T*}; (*ii*) $\emptyset$.
(*b*) (*i*) {1, 2, 3, 4, 5, 6}; (*ii*) {1, 2, 3, 4, 5, 6}; (*iii*) $\emptyset$.
(*c*) (*i*) {(1, 1), (1, 2), . . . , (6, 6)}; (*ii*) $\emptyset$; (*iii*) $\emptyset$.

6 (*a*) 24
(*c*) {(1, *H*, *H*), (1, *H*, *T*), (1, *T*, H), (1, *T*, *T*),
(2, *H*, *H*), (2, *H*, *T*), (2, *T*, *H*), (2, *T*, *T*),
(3, *H*, *H*), (3, *H*, *T*), (3, *T*, *H*), (3, *T*, *T*),
(4, *H*, *H*), (4, *H*, *T*), (4, *T*, *H*), (4, *T*, *T*),
(5, *H*, *H*), (5, *H*, *T*), (5, *T*, *H*), (5, *T*, *T*),
(6, *H*, *H*), (6, *H*, *T*), (6, *T*, *H*), (6, *T*, *T*)}

7 (*a*) 16
(*c*) {*aaaa*, *aaab*, *aaba*, *aabb*, *abaa*, *abab*, *abba*, *abbb*, *baaa*, *baab*, *baba*, *babb*, *bbaa*, *bbab*, *bbba*, *bbbb*}

8 (*a*) $\frac{1}{8}$; (*b*) $\frac{3}{8}$; (*c*) $\frac{1}{2}$; (*d*) $\frac{3}{8}$; (*e*) $\frac{7}{8}$; (*f*) $\frac{1}{8}$; (*g*) $\frac{1}{8}$.

9 (*a*) 16 (*b*) $\frac{1}{16}$ (*c*) $\frac{15}{16}$

10 (*a*) $\frac{1}{6}$ (*b*) 1

11 (*a*) $\frac{1}{6}$ (*b*) $\frac{1}{3}$ (*c*) $\frac{1}{2}$ (*d*) $\frac{1}{2}$
(*e*) $\frac{2}{3}$ (*f*) $\frac{5}{6}$ (*g*) 1 (*h*) 0

12 (*a*) $\frac{1}{6}$ (*b*) $\frac{1}{18}$ (*c*) $\frac{1}{36}$
(*d*) 0 (*e*) $\frac{1}{6}$ (*f*) $\frac{5}{6}$

13 (*a*) $\frac{1}{4}$ (*b*) $\frac{1}{13}$ (*c*) $\frac{1}{52}$ (*d*) $\frac{4}{13}$
(*e*) $\frac{1}{2}$ (*f*) $\frac{3}{13}$ (*g*) $\frac{8}{13}$

14 (*a*) *a*, *b*, *d*, *f*

15 (*a*) $\frac{1}{2}$ (*b*) $\frac{2}{13}$ (*c*) $\frac{4}{13}$ (*d*) $\frac{3}{13}$
(*e*) $\frac{8}{13}$ (*f*) 1 (*g*) $\frac{1}{2}$

16 (*a*) $\frac{11}{16}$ (*b*) $\frac{5}{16}$ (*c*) 1

17 (*a*) $\frac{1}{2}$ (*b*) $\frac{1}{16}$

18 (*a*) $\frac{1}{6}$ (*b*) $\frac{5}{6}$ (*c*) $\frac{125}{216}$

19 (*a*) $\frac{1}{16}$ (*b*) $\frac{1}{16}$ (*c*) $\frac{1}{4}$ (*d*) $\frac{1}{4}$ (*e*) $\frac{1}{169}$
(*f*) $\frac{1}{169}$ (*g*) $\frac{1}{2,197}$ (*h*) $\frac{1}{2,197}$ (*i*) $\frac{1}{2,704}$ (*j*) $\frac{1}{2,704}$

20 (*a*) $\frac{13}{204}$ (*b*) $\frac{1}{17}$ (*c*) $\frac{13}{51}$ (*d*) $\frac{25}{102}$ (*e*) $\frac{4}{663}$
(*f*) $\frac{1}{221}$ (*g*) $\frac{8}{16,575}$ (*h*) $\frac{1}{5,525}$ (*i*) $\frac{1}{2,652}$ (*j*) 0

21 (*a*) $\frac{1}{2}$ (*b*) $\frac{1}{1,024}$ (*c*) $\frac{1}{36}$ (*d*) $\frac{1}{46,656}$

22 (*a*) $\{HH, HT, TH, TT\}$ (*b*) $\{HH, HT, TH\}$
(*c*) $\{TT\}$ (*d*) $\frac{3}{4}$ (*e*) $\frac{1}{4}$

23 (*a*) $\frac{1}{2}$ (*b*) $\frac{1}{2}$ (*c*) $\frac{1}{4}$ (*d*) $\frac{3}{4}$

24 (*a*) $\frac{1}{6}$ (*b*) $\frac{5}{6}$ (*c*) $\frac{625}{1,296}$ (*d*) $\frac{671}{1,296}$

25 (*a*) $\frac{1}{36}$ (*b*) $\frac{35}{36}$ (*c*) $(\frac{35}{36})^{24}$ (*d*) 0.4914

26 (*a*) 6 (*b*) 24 (*c*) 120 (*d*) 720

27 (*a*) 5,040 (*b*) 30 (*c*) 720 (*d*) 665,280

28 (*a*) 56 (*b*) 336

29 (*a*) 40,320 (*b*) 259,459,200

30 (*a*) 210 (*b*) 15 (*c*) 1 (*d*) 924

31 (*a*) 28 (*b*) 56

32 6,435

33 (*a*) 133,784,560 (*b*) 1/33,446,140

34 1/64,974

35 (*a*) 1 (*b*) 4 (*c*) 6 (*d*) 4 (*e*) 1
(*f*) $\{HHHH\}$, $\{HHHT, HHTH, HTHH, THHH\}$, $\{HHTT, HTHT, HTTH, THHT, THTH, TTHH\}$, $\{HTTT, THTT, TTHT, TTTH\}$, $\{TTTT\}$

36 (*a*) 1, 6, 15, 20, 15, 6, 1
(*b*) $\binom{6}{6}, \binom{6}{5}, \binom{6}{4}, \binom{6}{3}, \binom{6}{2}, \binom{6}{1}, \binom{6}{0}$

(*c*)

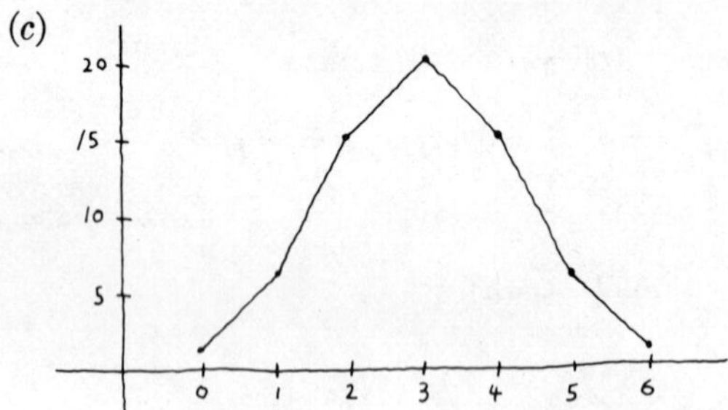

37 (*a*) 1, 10, 45, 120, 210, 252, 210, 120, 45, 10, 1

(*b*)

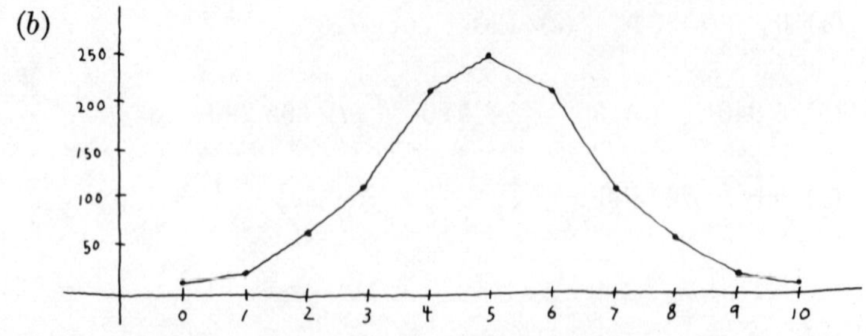

38 (*a*) 83.25 (*b*) 90 (*c*) 91.25 (*d*) 90
(*e*) The mean reflects student B's better overall performance.

39 (*a*) 65.2 (*b*) 75 (*c*) 80 and 40
(*d*) Exactly half the students are above the median, and half below.

Chapter 7

1 *a* and *d*; *b* has a line with a vertex at only one end; *c* has a line with both its ends on the same vertex.

2 *b* and *c*.

3 *b* and *d*.

4 (*a*) 3, 3, 2 (*b*) 4, 4, 2 (*c*) 4, 5, 3 (*d*) 6, 8, 4
(*e*) 2, 3, 3 (*f*) 3, 4, 3 (*g*) 3, 6, 5 (*h*) 5, 9, 6

6 (*a*) Basic network: $2 - 1 + 1 = 2$
(*b*) Draw an arc to a new vertex: $3 - 2 + 1 = 2$
(*c*) Draw an arc to a new vertex: $4 - 3 + 1 = 2$
(*d*) Draw an arc on two existing vertices: $4 - 4 + 2 = 2$
(*e*) Draw an arc on two existing vertices: $4 - 5 + 3 = 2$
(*f*) Put a vertex on an existing arc: $5 - 6 + 3 = 2$
(*g*) Draw an arc on two existing vertices: $5 - 7 + 4 = 2$
(*h*) Draw an arc on two existing vertices: $5 - 8 + 5 = 2$

8 (*a*) 4, 2, 1 (*b*) 4, 0, 1 (*c*) No (*d*) No

9 (*a*) 4, 6, 5 (*b*) 4, 4, 1 (*c*) No (*d*) No
(*g*) Yes (*h*) $4 - 6 + 4 = 2$, $4 - 4 + 2 = 2$

11 There are four odd vertices.

12 (*a*) All the vertices are even.
(*b*) All the vertices are even.

13 (*a*) four even, two odd; traversable
(*b*) seven even, none odd; traversable
(*c*) five even, four odd; not traversable

14 All are traversable.

19 (*a*)

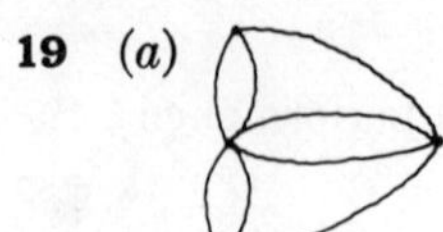

(*b*) Yes; begin on one shore and end on the other.

20 (*a*)

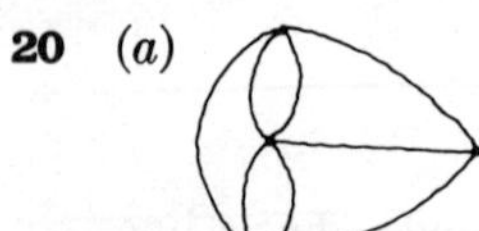

(*b*) Yes; begin on one island and end on the other.

21 (*a*) (*b*)

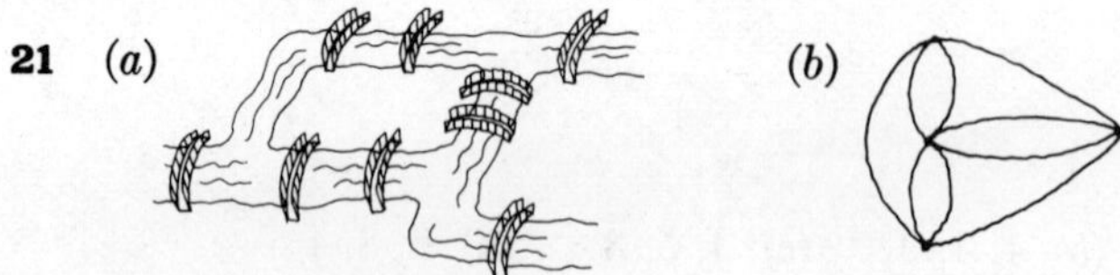

(*c*) Yes; begin at any vertex.

22 (*a*) (*b*)

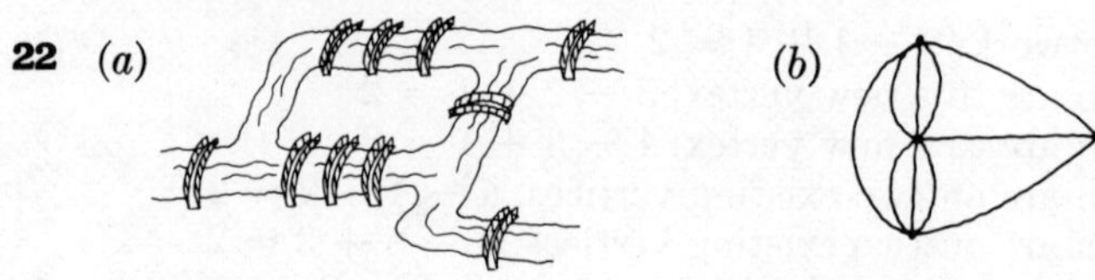

(*c*) No

23 (*a*) There are not exactly two arcs on each vertex. There are not exactly two regions.
(*b*) The arcs are not straight line segments.
(*c*) There are not exactly two arcs on each vertex. There are not exactly two regions.
(*d*) There are not exactly two arcs on each vertex.

24 *a*, *c*, *e*, and *f*

25 (*a*)

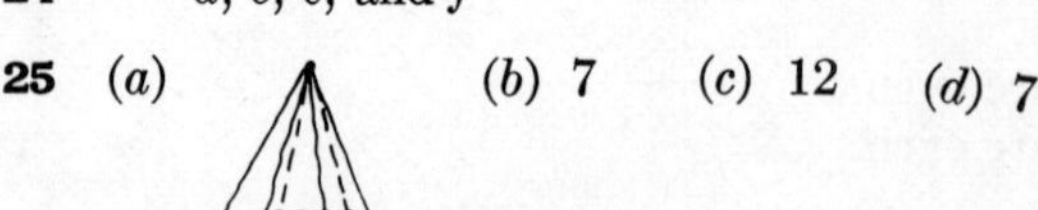

(*b*) 7 (*c*) 12 (*d*) 7

26 (*a*) 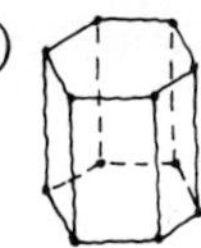 (*b*) 12 (*c*) 18 (*d*) 8

27 (*a*)

		Vertices	*Edges*	*Faces*
	n	$n+1$	$2n$	$n+1$
Triangular	3	4	6	4
Quadrilateral	4	5	8	5
Pentagonal	5	6	10	6
Hexagonal	6	7	12	7

28 (*a*)

		Vertices	*Edges*	*Faces*
	n	$2n$	$3n$	$n+2$
Triangular	3	6	9	5
Quadrilateral	4	8	12	6
Pentagonal	5	10	15	7
Hexagonal	6	12	18	8

30 (*a*) 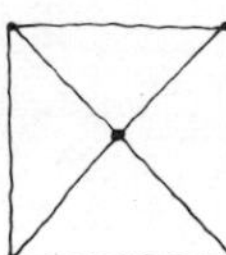 (*b*) 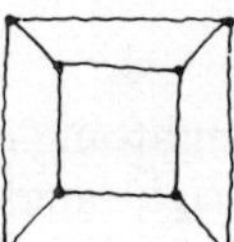 (*c*) 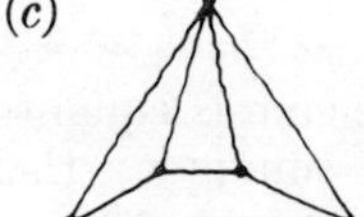

$5 - 8 + 5 = 2$ $8 - 12 + 6 = 2$ $5 - 8 + 5 = 2$

31 (*a*) 8 (*b*) 14 (*c*) 9 (*d*) 3
(*e*) $8 - 14 + 9 - 3 = 0$

32 (*a*) 10 (*b*) 21 (*c*) 16 (*d*) 5
(*e*) $10 - 21 + 16 - 5 = 0$

33

	Vertices	*Edges*	*Faces*
Tetrahedron	4	6	4
Hexahedron	8	12	6
Octahedron	6	12	8
Dodecahedron	20	30	12
Icosahedron	12	30	20

37 (*a*)

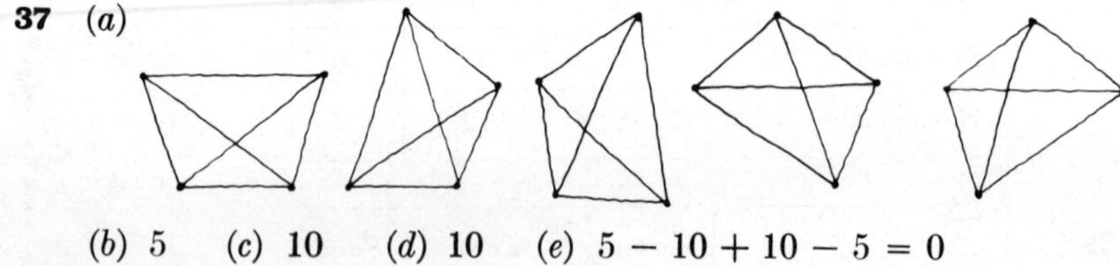

(*b*) 5 (*c*) 10 (*d*) 10 (*e*) $5 - 10 + 10 - 5 = 0$

Chapter 8

1 (*a*) The lines meet on the side toward *N*.
(*b*) The lines meet on the side toward *M*.
(*c*) The lines are parallel.
(*d*) 1.

2 (*a*) The sum is equal to two right angles.
(*b*) The sum is less than two right angles.
(*c*) The lines meet on the side toward *N*.

3 (*a*) The lines are parallel.
(*b*) The lines would appear to meet in a point.
(*c*) *N* is the vanishing point.

4 (*a*) $AB = AC$; $\measuredangle DAB = \measuredangle DAC$; AD is equal to itself.
(*b*) Triangles ABD and ACD are congruent (congruent triangle theorem); therefore, $\measuredangle ABD = \measuredangle ACD$.

5 (*a*) $\measuredangle AEC + \measuredangle AED = \measuredangle AED + \measuredangle DEB$ (common notion 1); therefore, $\measuredangle AEC = \measuredangle DEB$ (common notion 3).

(*b*) $\measuredangle AED + \measuredangle AEC$ is equal to two right angles. $\measuredangle AEC + \measuredangle BEC$ is equal to two right angles. $\measuredangle AED + \measuredangle AEC = \measuredangle AEC + \measuredangle BEC$ (common notion 1); therefore, $\measuredangle AED = \measuredangle BEC$ (common notion 3). (Answer is not unique.)

6 (*a*) Absolute and Euclidean (*b*) Absolute and Euclidean
(*c*) Absolute and Euclidean (*d*) Euclidean only
(*e*) Absolute and Euclidean (*f*) Euclidean only
(*g*) Absolute and Euclidean (*h*) Euclidean only

7 (*a*) $\measuredangle AGH = \measuredangle GHD$ (*b*) $\measuredangle BGH = \measuredangle GHC$

8 (*a*) $\measuredangle AGH + \measuredangle BGH$ is equal to two right angles. $\measuredangle DGH + \measuredangle CHG$ is equal to two right angles. $\measuredangle AGH + \measuredangle BGH = \measuredangle DHG + \measuredangle CHG$ (common notion 1) and $\measuredangle AGH = \measuredangle DHG$; therefore, $\measuredangle BGH = \measuredangle CHG$ (common notion 3).
(*b*) $\measuredangle AGH = \measuredangle EGB$ (vertical angle theorem, Exercise 5) and $\measuredangle AGH = \measuredangle DHG$; therefore, $\measuredangle DHG = \measuredangle EGB$ (common notion 1).
(*c*) $\measuredangle AGH = \measuredangle DHG$; therefore, $\measuredangle AGH + \measuredangle BGH = \measuredangle BGH + \measuredangle DHG$ (common notion 2). $\measuredangle AGH + \measuredangle BGH$ is equal to two right angles; therefore, $\measuredangle BGH + \measuredangle DHG$ is equal to two right angles (common notion 1).

9 (*a*) The Playfair postulate.
(*b*) $\measuredangle DAB = \measuredangle ABC$ and $\measuredangle EAC = \measuredangle BCA$ (parallel theorem). $\measuredangle DAB + \measuredangle EAC + \measuredangle BAC = \measuredangle ABC + \measuredangle BCA + \measuredangle BAC$ (common notion 2). $\measuredangle DAB + \measuredangle EAC + \measuredangle BAC$ is equal to two right angles; therefore, $\measuredangle ABC + \measuredangle BCA + \measuredangle BAC$ is equal to two right angles (common notion 1).

10 (*a*) $AD = BC$ and $AE = EB$. $\measuredangle DAE = \measuredangle CBE$ (since both are right angles, postulate 4); therefore, triangles ADE and BCE are congruent (congruent triangle theorem).
(*b*) $DE = CE$ (congruent triangle theorem); therefore, $\measuredangle CDE = \measuredangle DCE$ (isosceles triangle theorem).
(*c*) $\measuredangle ADE = \measuredangle BCE$ and $\measuredangle CDE = \measuredangle DCE$; therefore, $\measuredangle ADE + \measuredangle CDE = \measuredangle BCE + \measuredangle DCE$ (common notion 2). $\measuredangle ADC = \measuredangle ADE + \measuredangle CDE$ and $\measuredangle BCD = \measuredangle BCE + \measuredangle DCE$; therefore, $\measuredangle ADC = \measuredangle BCD$ (common notion 1).

11 EG could meet CD at a vanishing point.

12 (*a*) FH is not perpendicular to EG.
(*b*) EG and CD would appear to meet at infinity.
(*c*) FH would appear to be perpendicular to both EG and CD at infinity.

13 (*a*) Euclidean (*b*) Lobachevskian (*c*) Lobachevskian
(*d*) Euclidean (*e*) Lobachevskian (*f*) Euclidean
(*g*) Lobachevskian (*h*) Euclidean

14 (*a*) The angle of parallelism is less than a right angle.
(*b*) $\measuredangle BAP$ is greater than the angle of parallelism.
(*c*) AM is parallel to CD, but $\measuredangle BAM$ is not equal to $\measuredangle ABC$.
(*d*) AB is perpendicular to CD but not perpendicular to AM.
(*e*) The lines are ultraparallel.

15 (*a*) $\measuredangle ABM + \measuredangle MBA + \measuredangle AMB$ is less than two right angles and $\measuredangle AMB = 0$; therefore, $\measuredangle ABM + \measuredangle MBA$ is less than two right angles.
(*b*) $\measuredangle NAM + \measuredangle AMN + \measuredangle ANM$ is less than two right angles and $\measuredangle AMN = 0$ and $\measuredangle ANM = 0$; therefore, $\measuredangle NAM$ is less than two right angles.
(*c*) 0.

16 (*a*) Euclidean (*b*) Riemannian (*c*) Euclidean
(*d*) Riemannian (*e*) Euclidean (*f*) Riemannian
(*g*) Euclidean (*h*) Riemannian (*i*) Riemannian
(*j*) Riemannian

17 (*a*) $\measuredangle ABC + \measuredangle CAB + \measuredangle BCA$ is greater than two right angles and $\measuredangle ABC$ is equal to one right angle; therefore, $\measuredangle CAB + \measuredangle BCA$ is greater than one right angle.
(*b*) $\measuredangle ABC + \measuredangle ACB + \measuredangle CAB$ is greater than two right angles and $\measuredangle ABC + \measuredangle ACB$ is equal to two right angles; therefore, $\measuredangle CAB$ is greater than 0.
(*c*) Four right angles.

20 (*a*) Spherical (*b*) Elliptic (*c*) Both (*d*) Both
(*e*) Spherical (*f*) Elliptic (*g*) Spherical (*h*) Elliptic

21 (*a*) Elliptic (*b*) Euclidean and hyperbolic
(*c*) Elliptic (*d*) Euclidean and hyperbolic
(*e*) Euclidean, hyperbolic, and elliptic
(*f*) Euclidean (*g*) Hyperbolic (*h*) Elliptic
(*i*) Euclidean and hyperbolic (*j*) Elliptic

Index

Numbers in italic indicate pages on which photographs appear.